Skywatchers, Shamans & Kings

ALSO BY E. C. KRUPP

Beyond the Blue Horizon: Myths and Legends of the Sun, Moon, Stars, and Planets

Echoes of the Ancient Skies: The Astronomy of Lost Civilizations

Archaeoastronomy and the Roots of Science (edited)

In Search of Ancient Astronomies (edited)

FOR CHILDREN

The Moon and You

The Big Dipper and You

The Comet and You

for

Dr. Robert J. Chambers and Dr. George O. Abell,
who piloted us through the universe
with only the sky for a net

ACKNOWLEDGMENTS

Certainly some power was at work in bringing this book to the shelves of stores and libraries and to the hands of its readers. An honest acknowledgment of the real source of that power must inevitably track it past the author and back to those who influenced and sustained the effort. What power any of us may possess inevitably comes from somewhere else—genetics, friendship, education, and experience. Understandably, then, I acknowledge the support and assistance of family, friends, teachers, and colleagues.

Celestial power is only understood through the symbols that assert its presence and influence, and for that reason I have found it useful—actually, essential—to seek audiences with the sky in the cultural corridors of its power on earth. This means traveling to monuments and museums where relics of celestial power may still be seen. I have, by this time, seen more than 1,300 ancient and prehistoric sites in person. Not all of them involve the sky or reveal our ancestors' interaction with it, but enough of them possess celestial connotations to eradicate any doubt about the influence that zone of our environment has had upon us.

Getting to such places—Inner Mongolia, the indigenous lands of Chile's Mapuche Indians, off-trail ruins in Mesoamerica, or the landmine-encircled temples of Cambodia—isn't always easy, and I have had lots of help. Yvette Cloutier, who owns and operates ETA/Piuma Travel has, for many years now, taken my impossible itineraries seriously and gotten me where I need to go. Likewise, Mary Dell Lucas, the risk-taking tour-organizing proprietor of Far Horizons, has also found the untraveled paths through the jungles and the mountains and the deserts on behalf of my curiosity. Many of these trips have been developed under the auspices of U.C.L.A. Extension, and in particular, Dr. Eve Haberfield and Karen Prinzmetal have encouraged and defended the unlikely destinations these expeditions have included. These field study tours provide opportunities for research. They are subsidized by an audience of adventurous travelers who accept adversity as a reasonable price to pay for uncommon experience and firsthand knowledge. We offered the first of these programs in 1976, and all of those who have enrolled and traveled with me over the years have my gratitude, sympathy, and respect.

Ted Pedas has continued to invite me to lecture for passengers on the unconventional cruise programs he contrives for Sun Line and Orient Lines. Without his willingness to include me in the successful rendezvous with the October 24, 1995, total solar eclipse in the South China Sea, I would not have been able to walk the grounds of Angkor Wat in time for this book, and I would have missed the meaning and importance of the cosmic metaphors of the Khmer royal temples.

I am also indebted to many professionals who, sharing this interest in ancient astronomy, have also helped me hit the road. Dr. Rolf Sinclair, Dr. Ray White, and Dr. George Coyne enhanced my understanding of ancient Rome through their invitation to participate in "The Inspiration of Astronomical Phenomena," a conference hosted by the Vatican Observatory in summer, 1994. A little earlier, Dr. Melvin L. Fowler, with assistance from the Cahokia Mounds Museum Society, got me back to west-central Illinois for the May 10, 1994, annular eclipse of the sun and for an update on recent research through participation in a special symposium, "The Ancient Skies and Sky Watchers of Cahokia: Woodhenges, Eclipses, and Cahokian Cosmology." In spring, 1992, Stanislaw Iwaniszewski, Arnold Lebeuf, and Mariusz Ziolkowski welcomed me to Poland as part of the international symposium, "Time and Astronomy at the Meeting of Two Worlds," and so kept me in touch with developments that might have otherwise eluded me.

Von Del Chamberlain's books and research papers are rich sources of information on American Indian sky lore, and his presentations at conferences are consistently informative and entertaining. The greatest pleasure, however, is hiking and riding horses with Von Del to the hard-to-get-to sites of the American Southwest. I have benefited from his knowledge of and enthusiasm for the land and its sky on three such campaigns and look forward to more.

For almost two decades now, I have been empowered by the advice and company of experts on ancient skywatching who also double as friends. Professor Anthony F. Aveni continues to drive the field forward through his commitment to genuine interdisciplinary study. John B. Carlson's recent insights on Mesoamerican astronomy reflect only a fraction of his varied interests as Director of the Center for Archaeoastronomy. David Dearborn came back, as near as anyone I know, from the dead to inject life and enthusiasm into the research community. He and the late LeRoy Doggett, beyond their own research, invested editorial energy in the quarterly news bulletin, issued at solstices and equinoxes by the Center for Archaeoastronomy. Alexander Marshack, well-known for his work on upper paleolithic symbolic systems, transformed archaeological perceptions of our Ice Age ancestors. He has readily encouraged my attempts to synthesize what is really at work in the cultural expression of celestial themes from the Stone Age to the New Age. With insight and originality,

Arlene Benson and John Rafter keep squeezing the astronomy out of ethnography and rock art in California and adjacent territory.

An unexpected invitation from Eleanor Cross Harrison and Dr. Stephen D. Siemens to contribute a paper to an American Anthropological Association session on "Configurational Approaches to Culture through Analogy" forced me to come to grips with an alien discipline of knowledge. That exercise in turn allowed me to appreciate an entirely different aspect of the sky's imaginative impact.

Dr. Richard E. W. Adams, Dr. Peter H. Keller, Nancy Cattell, and the irrepressible Gary Wirth all generously supplied photographs of places that even I regard as particularly remote ends of the road.

Once again, I am a privileged beneficiary of the graphic resources of Griffith Observatory in Los Angeles. I am beholden to Joseph Bieniasz for illustrations prepared for research papers, for Observatory programming, and for Observatory publications—especially the *Griffith Observer.* Their inclusion in this book enhances and clarifies otherwise arcane material. In addition, many of the photographs processed for the Observatory's extensive picture file were also available to me. The high technical standards of Anthony Cook, Daniel Marlos, and Lisa Auerbach ensure photographic professionalism at Griffith Observatory.

Cultivating an undergraduate's appreciation for the full spectrum of human endeavor, Pomona College, in Claremont, California, applies the power of its professors to humanize the sciences and put rigor in the humanities. They make a permanent investment in each generation of students and ask only for the interest, to be paid in an examined life. The principal on the debt I owe to all of my teachers at Pomona remains outstanding—even after 30 years, but I have tried to honor them with a high interest rate and continuous reinvestment. The late Dr. Robert J. Chambers, my astronomy professor and advisor, generously continued to extend my credit long past Pomona and graduate school and throughout the years of our continuing friendship.

At U.C.L.A., the late Dr. George O. Abell, like Bob Chambers before him, put the power of opportunity at my disposal and helped me negotiate the powerful currents of graduate school in astronomy. He, too, remained a good friend after the period of formal schooling ended and continued to endorse the curious niches my career has occupied.

More than 20 years ago, Dr. Gibson Reaves, Dr. John Russell, and the late Dr. Seymour Chapin exerted an unanticipated power on me through the informal meetings of the Los Angeles Society for the History of Astronomy. By example, they persuaded me to approach the cultural component of astronomy in a more formal way and helped set the stage for the archaeoastronomical inquiries that would follow.

There must have been times when both my agent Jane Jordan Browne, who owns Multimedia Product Development, Inc., in Chicago, and Emily Loose, my editor at John Wiley & Sons, Inc., wistfully embraced the philosophy of power so succinctly expressed by Choup Lorn, our guide at Angkor—"Trust is good, but control is better." Despite their uncertainty about my timetable, the real power that propelled this book into publication resides in Jane's integrity and Emily's patience. Jane always acts as the vigilant guardian of everyone's interest and, playing no favorites, her real client is the finished product. As editor, Emily embraced a variety of unknown risks when she offered a contract, but with professional instinct and steely resolve, she allowed me just enough rope to lasso the beast stampeding through the original proposal and corral it for public view.

Probably the most formative power in each of our lives is exerted by parents. My late father, Edwin F. Krupp, and my mother, Florence A. Krupp, clearly knew how to express their influence, but their goal, of course, was relinquishing power to self-discipline. I prefer to credit them with success and regard this new book as additional evidence. My mother thinks I should give it a rest.

My chronic neglect of family affairs would justify censure by my in-laws, Margaret H. Rector and Dr. Robert W. Rector. Anyone else would be tempted to tap the power of guilt, but they operate with a light touch and have endured without complaint a never ending parade of projects. Commitments that consume evenings, weekends, and national holidays and an inventory of deadlines that brutalizes family priorities still have not jaded them. Hope springs eternal in the generous heart.

My son, Ethan H. Krupp, helped me explore many of the odd corners of the world cornered in this book and always understood my insistence on running on empty to collect one more destination and finish one more book. He has, however, at last had enough sense to evade parental power and get out of town. Establishing his own agendas as he moves through college and toward a career, he now carries the burden of freedom. It's right where it belongs. But I hope we get to travel together again.

The last salute properly belongs to my wife, Robin Rector Krupp, who often demonstrates her power by insisting I have usurped it all. But I am not deceived. She moves and shakes the world in her own time and style. And there must be some deep strength that allows her to reside under the same roof with a husband charmed by his own idiosyncracies. Such mysteries can only be understood through the balance of power. It's a pleasure doing business with her.

Contents

INTRODUCTION
Fingerprints of Cosmic Power 1

CHAPTER ONE
The Center of the World 15

CHAPTER TWO
Plugging In to Cosmic Power 43

CHAPTER THREE
Centers of Creation 67

CHAPTER FOUR
Mother Earth 97

CHAPTER FIVE
Agents of Renewal 127

CHAPTER SIX
Shamans, Chiefs, and Sacred Kings 153

CHAPTER SEVEN
Celestial Empires 183

CHAPTER EIGHT
Enlightened Self-Interest and Ulterior Motives 209

CHAPTER NINE
It Pays to Advertise 245

CHAPTER TEN
Upward Mobility 279

BIBLIOGRAPHY 319
ILLUSTRATION CREDITS 347
INDEX 349

Skywatchers, Shamans & Kings

Fingerprints of Cosmic Power

Arriving on the second terrace of a mesa on the north side of Chaco Wash, we ran into one of those fingerprints the universe prompts human brains to leave on the landscape. We were at the east end of Chaco Canyon, in northwest New Mexico. Fajada Butte, the canyon's most distinctive landmark, looked small but unmistakable some distance to the southwest. Petroglyphs carved near the top of the butte by prehistoric Pueblo Indians centuries ago operate as a solstice marker, certainly the best known site in the United States to be linked with ancient astronomy. We were headed, however, on that cloudless June morning, for a different astronomical target—a ledge in the neighborhood of Wijiji ruin.

Wijiji is a multistory apartment. It and many larger buildings in the canyon were abandoned sometime in the twelfth century. They had been built by the ancestors of the Pueblo Indians, who now live elsewhere in the American Southwest. Archaeologists call the prehistoric Pueblo the Anasazi, which is actually a Navajo name. The Navajo moved into the area long after the Anasazi left and found the half-buried and collapsed remains of their monumental architecture.

The ledge near Wijiji is punctuated with Anasazi and Navajo paintings and carvings. An alignment from the ledge to a natural rock pillar southeast across the rincon coincides with winter solstice sunrise and argues that the place was originally used by Anasazi sunwatchers and later by the Navajo. It was on the way to the Wijiji "observatory," then, that we encountered a more subtle and more recent nod to the sky.

Small stones gathered from the sandy terrace above the canyon had been arranged recently into a circle about three feet across. It was roughly quartered at the cardinal points by four low stacks of stones. In the symbolic traditions of many Indian people of North America, the circle can stand for the horizon. The horizon is the rim of the world, where the earth makes contact with the sky, and the world's key directions are found there. These directions contain power. They provide a template for terrestrial order and are incorporated symbolically into everything from sandpaintings, ceramics, and architecture. Pueblo Bonito, a five-story D-shaped complex of apartments, plazas, and ceremonial chambers at the west end of Chaco Canyon, is accurately oriented with the cardinal directions. So is Casa Rinconada, a large circular subterranean ritual structure one-half mile across the canyon from Pueblo Bonito.

Far from the trails followed by today's tourists to the canyon we encountered an unexpected hint of the power of the sky. Although the private purpose of the modest shrine eludes us, it mirrors the horizon symbolism of Casa Rinconada and marks the land with a sign of the abiding importance of the relationship between earth and sky.

Ancient Aztec farmers stored grain in thatched clay corn bins. Shaped like overweight Coke bottles, they were as high as a house. I spotted one ancient silo after another looming over the low walls of Chalcatzingo as we drove through the village. Chal-

Four low stacks of rocks quarter this small, off-trail ring in Chaco Canyon, New Mexico. The shape and construction suggest this is a relatively recent shrine, and its design seems to allude to the world's four directions. (photograph E. C. Krupp)

catzingo is small and rural. It is in arid eastern Morelos, 79 miles south of Mexico City. The tall cacti there branch into long fingers that reach skyward. With the rest of the dry scrub, they fulfill a travel brochure's promise of Old Mexico. Even the main route through town is a rough dirt road. It is, however, the gateway to an archaeological site known by the same name. Although the site is famous for elaborate low reliefs carved upon the cliffs of Cerro Chalcatzingo and on the large boulders at its foot, relatively few visitors find their way to these ruins. There is no easy way to get here, and you feel lucky to be in the neighborhood.

Chalcatzingo was quarried until 1940, and unfortunately much of the exposed rock at the base of Cerro Chalcatzingo has been dynamited. Some low outcrops survive, however, and are embellished with enigmatic curvilinear petroglyphs and deep mortars. Ceremonial structures were built on the same slopes, and they include massive platforms that once supported pyramidlike mounds, high-status residences, large stone altars, and massive freestanding sculpture. The hillside they occupy was artificially terraced. It is nestled between Cerro Chalcatzingo and its rocky companion, Cerro Delgado, in the protective embrace of a natural amphitheater created by these two peaks. They rise together like the tips of immense icebergs. Their real bulk seems hidden below the flat, open country around them. This is a theatrical landscape, and its primary landmarks are this matched pair of scene-stealing mountains.

Monument 1 at Chalcatzingo enthrones *El Rey* inside stylized jaws that represent the mouth of a cave. The king bellows from the cave, and his breath emerges in air-current curlicues. He is Chalcatzingo's rain lord in the hall of the mountain king. (photograph Ethan H. Krupp)

Scrambling up the rocky trail to Monument I on Cerro Chalcatzingo is not a trivial bout with gravity. The Valley of Morelos is in Mexico's central highlands and is over 4,000 feet above sea level. There is less oxygen, and the path is steep and slippery with pebbles. I make climbs like these much faster than necessary, carrying enough camera gear to put both my next breath and my next step in jeopardy. Time at the ruins is always a nonrenewable resource in short supply. I am propelled by a desire to experience in person the context, scale, and atmosphere of the world's ancient places. Through visits like this, the accounts I read in books of these remote wonders are transformed into personal pilgrimages.

Monument I is a large relief, nearly 11 feet long and about 9 feet high. A figure with an elaborate headdress is seated inside a C-shaped frame that is thought to symbolize a cave. Scrolls emerge from the cave's mouth, and near the top of the vertical rock face are stylized clouds dropping rain. Individual raindrops and several plants appear below the clouds along with several concentric ring designs. The fancy headgear implies high status, and the fellow wearing it is known as *El Rey*, "the King." The style and the content of Chalcatzingo's rock carvings have prompted archaeologists to link them with the Olmec civilization that was centered in Mexico's tropical Gulf Coast.

In the first half of the first millennium B.C., the local elite at Chalcatzingo adopted Olmec style and ideology. Many elements of high Mesoamerican culture were Olmec innovations, including the centralization of power in the hands of an elite and the construction of the monumental ceremonial centers from which they operated.

Although some believe the Olmec pioneered writing in Mesoamerica, there are no texts carved on Chalcatzingo's rocks to clarify who "the King" is and what he is doing. To understand this sculpture, we have to rely upon the scene's imagery and upon the broad symbolic tradition of ancient Mexico. Clouds and rain are certainly part of the story here, and in Mesoamerica, storm clouds were thought to form and simmer inside the mountains. This association was inspired by the fact that clouds do condense over the mountains, as if they emerged from them. The peaks themselves seem to make contact with the sky where the clouds appear, and springs spout from the mountain slopes. Rainwater is shed by the mountains and funneled down their flanks to the cultivated valleys below. The mountain's interior is a hall of the underworld, to which caves provide entry. The king carved on Chalcatzingo's mountainside is dispensing scrolls of wind or cloud from the underworld cave and out its mouth. He holds a "lazy S" glyph in his arms. The same symbol is used for his throne, and evidence from other sources suggests that it has something to do with clouds. This king is a master of the clouds, perhaps a weather shaman or an ancestor spirit capable of brewing rain and dispatching it into the sky. Although rain was thought to originate in the underworld, it is delivered by the sky. Because it is a celestial commodity that guarantees another season of growth and harvest, the king who controls it also commands the life-sustaining power of the sky.

A similar role is played by the Hopi kachinas. These spirits of the dead reside for half of the year in the San Francisco Peaks, near Flagstaff, Arizona. They return to the Hopi mesas in December, close to the time of winter solstice, and are said to bring the seasonal rains on which Hopi corn cultivation depends.

Now rain doesn't just drop from the clouds. Its connection with the sky is more intricate. Rain is seasonal, and that means it is linked with the passage of the year. Its arrival and departure are timed by the annual travels of the sun between its most northern and southern paths and by the seasonal appearances and retreats of the stars. Calling forth the rain, therefore, demonstrates control of the seasonal power of the sky.

Although the water supply at Chalcatzingo was not especially accessible or abundant, the inhabitants controlled the runoff effectively with terracing, embankments, and channels. Chalcatzingo's soil was not rich, but this control strategy enhanced its agricultural success. In time it became a center of regional importance. David Grove, who began a significant program of excavation there in 1972, also pointed out that Chalcatzingo was well located to exploit the resources of the valley's varied terrain, and it eventually controlled the flow of raw materials to other settlements in the area and beyond. The best route for exchange in the territory between Mexico's central highlands and the Gulf Coast ran through Chalcatzingo's neighborhood, and in time Chalcatzingo also managed the region's trade with Olmec rulers on the coast. Imported symbols of power, and perhaps marriage alliances, buttressed the authority of Chalcatzingo's elite. The ideology of this rulership is on display in the portrait of Chalcatzingo's rain king. He coaxes cooperation from the elements with payments in ritual magic. He operates at the pivot of the universe, the place where the sky, the earth, and the underworld all meet. Chalcatzingo's political power is threaded to the sky by this ancestral ruler. He exercises his supernatural prerogative at the mountain cave portal to the underworld and blows water into heaven.

Archaeologist Kent Reilly suggests that Chalcatzingo's topography also possessed symbolic value that served the objectives of its leaders. This symbolic system included a metaphorical vision of the cosmos based upon the great predators of the watery (underworld), terrestrial (surface world), and atmospheric (celestial world) realms of the Mesoamerican universe. With a well-deserved reputation for ferocity, the caiman, a relative of the alligator, was the model for the earth. Like a vast aquatic saurian, the earth was thought to float partially submerged in the primordial waters of the sea. The scaled ridges on the earth-caiman's back were the world's mountains. Its open mouth was the cave entrance to the underworld and the land of the dead, who in burial are consumed by the earth as the caiman devours its prey. The C-shaped cave entrance symbol in Chalcatzingo's Monument I is actually the stylized jaws of this cosmic caiman.

Reilly identifies an animal carved on the rock face just a few feet to the east of El Rey, the King, as a caiman. Crouched on another lazy S (or cloud) glyph, its head is turned toward the sky. A double scroll surges from its long jaws, and there is a rain cloud symbol above that. Raindrops fall from this cloud and apparently invigorate a squash vine carved to look as if it is growing out of the ground at the base of the rock. Reilly believes the scroll symbolizes the reptile's roar.

At first glance, the animal carved in relief on Cerro Chalcatzingo as Monument 14, near *El Rey,* appears to be a fantastic long-necked, goggle-eyed, flat-headed beast surrounded by severed fingers. Careful examination tells us, however, that the creature is a caiman, a large aquatic reptile related to the alligator. The goggle-eyed "face" spouting from the caiman's jaws probably symbolized the animal's roar. Associated with thunder, that roar is canopied by a rain cloud, and the "fingers" are raindrops falling to earth. (photograph Ethan H. Krupp)

Caimans are tropical, of course, and not found in the dry Valley of Morelos. They are featured prominently, however, in Olmec iconography and bellow loudly in the mating season, which coincides with the beginning of the rainy season on the Gulf Coast. In the first days, thunder booms before the clouds build into rain. The caimans even thunder in response to thunder. They make the noise with a closed mouth. The caiman's throat resonates, and its head and tail curve skyward. Its body throbs and quivers, and the water it throws into the air falls like rain. By its seasonal behavior, the caiman was recognized as an agent of the life-renewing rain and became the symbolic ally of Chalcatzingo's storm lord.

According to Reilly, even the form of the land around Chalcatzingo transformed it into sacred territory. He believes Chalcatzingo's twin peaks reminded the ancients of the caiman's brow ridges. In a sense, Chalcatzingo was itself the great caiman and therefore the earth. Its rain king was enthroned at its center, at the main junction of the universe. Water flowed from the top of the mountain through a trench cut in the rock that carried it right past El Rey and let it fall to the hillside below.

There is, perhaps, another portrait of this rain king at the summit of Cerro Chalcatzingo, more than a thousand feet above the valley floor. His face is sculpted in relief on a large boulder. He wears an odd, pointed helmet and large dangling ear ornaments. A hand and wrist are also carved into the rock above him. Up here, the sky makes first contact with the mountain, and the face may symbolize the shaman-king who acted as an intermediary with the forces of heaven. The draw of these heights persists, for now a large metal cross is lodged in the mountain's crown and probably draws an occasional lightning strike. A few yards away, on a isolated column of rock, a shaman of today has contrived his own shrine to seek audiences with the supernatural and draw power from the sky.

You can see the bronze instruments on top of the Old Beijing Observatory profiled against the sky 46 feet above the cloverleaf intersection of the eastern extension of Changan jie Avenue (the main east-west artery of the Chinese capital) and Jianguo men, which runs north-south. Within walking distance of the large Friendship Store, where tourists may buy everything from carpets to White Rabbit candy, the observatory platform is all that remains of the wall near what was the southeast corner of the old Mongol capital of China. That version of Beijing was built in the thirteenth century by Khubilai Khan, the first emperor of the Yuan dynasty. He was Genghis Khan's grandson, and Marco Polo visited his court.

The astronomer Guo Shoujing was commissioned by Khubilai Khan to erect the first observatory on this spot. It was known as the Terrace for Managing Heaven. The tower we see now in Beijing was built later, in 1442 for the sixth Ming dynasty emperor, Zhengtong, who named it the Terrace for Observing the Stars. Its ancillary buildings included the Hall of Celestial Abstrusity and the Sun Shadow Hall.

In the last quarter of the thirteenth century, the best astronomy in the world was carried on at Guo Shoujing's Beijing observatory and at the 26 other field stations he established from Mongolia to the island of Hainan. Guo Shoujing used engraved metal instruments to measure the lengths of shadows and the positions of the stars and planets. He and the Tong tian calendar makers of 1199 A.D. both measured the length of the solar year as 365.2425 days, only 26 seconds longer than it actually is.

This high-powered thirteenth-century research program was sponsored by the state. The Chinese emperor established the calendar through institutionalized astronomy. He enacted the calendar with public announcements and reinforced it with astronomically timed seasonal ceremony. The accuracy of the Guo Shoujing's estimate for the length of the year exceeded, however, the requirements of a practical calendar, and the precision of Chinese instruments also reached beyond the everyday needs of the empire. What motivated this pursuit of high precision? Celestial certification of imperial power.

The engine of astronomy's development in China was its symbolic value. As a truthful mirror of nature, astronomy was official business, a tool in the service of the social and political agenda of the state.

In imperial China, the sky was the visible domain of Shang di, the transcendent personification of the celestial power and cosmic order. Shang di was Heaven, and the emperor was said to be the Son of Heaven. This ruler claimed divine lineage from the

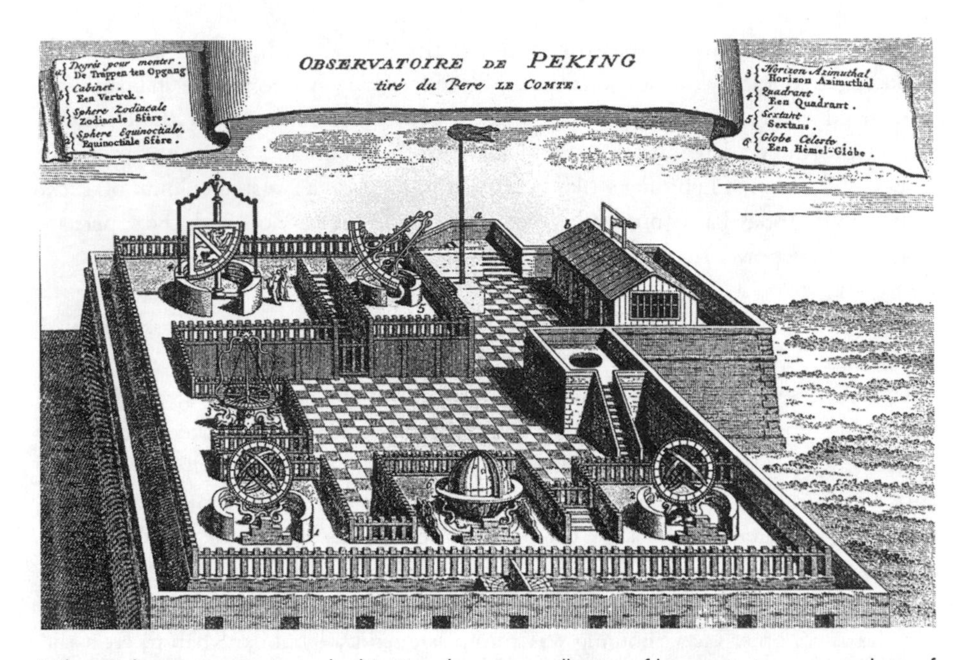

When Father Matteo Ricci reached Beijing, he saw a collection of bronze instruments on the roof of the observatory that were, he said, "well worthy of inspection whether for size or beauty: and we certainly have never seen or read of anything in Europe like them." These old instruments were, in time, replaced by new ones designed and fabricated under the direction of Ferdinand Verbiest, one of Ricci's successors. This view of the observatory's roof terrace shows the facility after the installation of Verbiest's equipment in 1674. Starting at the center of the back row and circling counterclockwise, we have the sextant, the quadrant, the altazimuth, the ecliptic armilla, the celestial globe, and the equatorial armilla. Decades later, two more instruments—the azimuth theodolite and the new armilla were added. In its original form, this illustration first appeared in 1687 in *Astronomia Europaea* by Ferdinand Verbiest. From that time, it was frequently copied and reengraved, as it is here. (collection Griffith Observatory)

sky and operated with cosmo-magical authority conferred by Heaven. He was the primary influence broker with the great forces of nature and the spirits of cosmic order. Although the emperor was insulated from his people by a huge bureaucracy and a pyramid of power, he was part of an ancient tradition of shaman-chiefs who interacted with the supernatural world on behalf of the entire community. Continuing the tradition of sacred kingship, he engaged in ritual reserved as his prerogative. This activity was intended to harmonize the relations between Heaven and Earth, to stabilize the world, and to protect human endeavor.

The emperor's sovereignty was only as good as the celestial mandate he secured from Shang di. The accuracy and the precision of imperial astronomy measured the emperor's adherence to the standards of Heaven. Astronomy reflected the quality of the emperor's virtue and the validity of his power. Under those circumstances, subsidizing precision and accuracy was an exercise in royal self-interest.

The Chinese prepared official histories of their dynasties, and the *Yuan shi*, which details events and concerns during the Yuan period, contains a list of 17 instruments installed by Guo Shoujing at the Old Beijing Observatory. None survive, but copies of four of them are on the grounds of the Purple Mountain Observatory near Nanjing. It is believed the duplicates were fabricated in 1437 by the astronomer Huangfu Zhonghe. The eight instruments now on top of the Beijing Observatory date to the seventeenth and eighteenth centuries and emerge from a different tradition. It takes a close look at the instruments, however, to see what that tradition was.

To reach the upper terrace of Beijing's old observatory, you climb a high, straight set of brick stairs. The stairway makes a 90-degree turn about halfway up, but the mass of the tower itself still completely blocks your view to the right during the ascent. The brick bannister on the left isolates you further in the stairway. You really can't see anything but dark gray bricks and part of the sky, if you happen to glance up. At the top, however, the environment opens up for astronomy. Matteo Ricci, an Italian Jesuit, wrote in 1610 about visiting a similar platform in Nanjing, and reported some of the things the Chinese astronomers did up there:

Beijing's Ancient Observatory is preserved as an historical museum, and the old bronze instruments are still on the roof. Standing at the top of the stairway providing access to the terrace, we are looking southeast. The six-foot celestial globe is in the foreground, and behind it rise the rings of the ecliptic armilla. To the left we can see the altazimuth superimposed upon the azimuth theodolite. (photograph Robin Rector Krupp)

> Here some of the astronomers take their stand every night to observe whatever may appear in the heavens, whether meteoric fires or comets, and to report them in detail to the emperor.

Elevation provides access to the horizon and the sky, but the view from the deck of the Old Beijing Observatory is not what it used to be. New high-rise buildings are usurping the skyline in this quarter of the city. The rings of the armillas used to frame fields of sky; now they frame steel and windows.

The rings on top of the Old Beijing Observatory are subdivided into 360 degrees, and this betrays European influence, despite the dragons, turtles, and other Chinese ornaments. European geometry had long since adopted a convention that originated in Mesopotamia and split the circle into 360 degrees. The inscribed circles on the measuring rings of Guo Shoujing's instruments, on the other hand, were divided into 365¼ "degrees" in accordance with Chinese tradition. The Chinese saw that the sun travels completely around its circular path through the background stars in one year or about 365¼ days and partitioned the circle accordingly. All of the instruments now found on the roof in Beijing were actually fabricated under the direction of the Jesuits, who outperformed the emperor's own astronomers with the predictive superiority of European techniques for astronomical calculation.

Matteo Ricci calculated correctly the impact of seventeenth-century European astronomy on the Chinese emperor. Although he admired the instruments he saw at Nanjing and Beijing, his assessment of the emperor's astronomers was not flattering:

> They have little talent and less learning, and do nothing beyond the preparation of almanacs on the rules of calculations made by the ancients—and when it chances that events do not agree with their computations they assert that what they have computed is the regular course of things, and that some aberrant conduct of the stars is a prognostic from heaven . . .

Astronomy had its ups and downs in China, as it did in Europe, and by the time the Jesuits got there it had deteriorated from the high-water achievement of Guo Shoujing.

Ricci was really the Vatican's point man to bring Christianity to China. Better astronomy, he realized, would provide access to and influence with the Ming Emperor Wanli. The plan worked. Europe's best mechanical clocks and Jesuit eclipse predictions were particularly persuasive. Ricci mastered Chinese and became a paid official in the emperor's retinue. Another Jesuit, Adam Schall, accurately forecasted an eclipse in 1624, and that earned him an appointment on the Board for Calendar Regulation.

Although Ricci and the other Jesuit missionaries never did Christianize China, by 1669, during the reign of Kangxi (the second ruler of the Qing dynasty), they dominated Chinese astronomy. One of Ricci's successors, the Belgian Ferdinand Verbiest, was given the opportunity to renovate the Beijing Observatory. He completed the job in 1673 and was responsible for six of the instruments now on the platform—the equatorial armilla, the ecliptic armilla, the six-foot celestial globe, the rather florid Louis XIV sextant, the altazimuth, and the quadrant, with its marvelous rampant dragon. An azimuth theodolite was added in 1715 by Bernard Killian Stumpf. Ignatius Kögler installed the last Jesuit device, a more elaborate armilla equipped with a traditional Chinese sighting tube, in 1744. The history of these instruments shows the Jesuits were in the driver's seat of imperial Chinese astronomy for over a century. Their status with the emperor ultimately fell, but not through any fault of their astronomy. A dispute within the Church about religious policy in China—between the Dominicans and the Franciscans on the one hand and the Jesuits on the other—was resolved against the Jesuits. Favoring the Jesuits and critical of Rome, Emperor Kangxi gradually relied less on all of the Catholic missionaries.

For a time, more reliable calculations and innovative instruments made astronomy an effective tool of the Church in China because astronomy was so closely allied with imperial power. Even Marco Polo was sufficiently impressed by the astronomical bureaucracy of Khubilai Khan to devote a commentary to it in his *Travels.*

> There are . . . about five thousand astrologers and soothsayers. . . . They have their astrolabes, upon which are indicated the planetary signs, timetables, and aspects for the whole year. The astrologers of each set annually study their respective tables in order to ascertain the course of heavenly bodies and their relative position in each moon. From the paths of the planets in the different signs they predict the weather and foretell the peculiar phenomena each month will produce. . . . Those whose predictions are found to be most often correct are esteemed the greatest masters of their art and are consequently the most honored.

Although Marco Polo misunderstood the details of Chinese astronomy, his assessment of its importance was accurate. The state needed the best astronomy on the market, and three and a half centuries later, when Europeans could provide it, the ethnic background of the astronomers was secondary. Systematic, quantitative astronomy put celestial authority into the hands of the Chinese emperor and ratified his right to rule. Jesuit mastery of superior methods favored their bid for power with the emperor. Together they used the power of the sky to guide and control events on earth.

Visitors to Los Angeles—like medicine men in Chaco Canyon, Chalcatzingo rain kings, and the Chinese emperor's own astronomers—climb closer to the sky to reach a center of celestial power. Driving up Vermont Canyon, they ascend the lower slopes of Mount Hollywood and reach Griffith Observatory. Visible from all over the Los Angeles basin, Griffith Observatory is regarded as a source of accurate, authoritative knowledge. People call, write, and ask questions in person at the Observatory about just about everything.

From the grounds of the Observatory, visitors find a stunning view of the Los Angeles basin that sweeps to the distant horizon. Earth and sky in southern California are united from the Observatory's promenade roof deck, and the bond is strengthened inside the building, where a public planetarium immerses audiences in the night sky through the ingenuity of the Zeiss Planetarium Projector. This device was invented in 1923 and first operated at the Deutches Museum in Munich, Germany, by the Carl Zeiss optical company. Since then, planetarium techniques and auxiliary projectors have grown more sophisticated and dramatic, but a conventional planetarium theater still stages its presentations under a large dome that simulates the vault of the sky. The projected stars and planets create the illusion of the real, starry heavens and allow complex motions to be demonstrated and explained. This sky, however, is still the sky of the ancients. Despite the advancements in astronomical knowledge that tell us we live in an edgeless, centerless cosmos and do not occupy the center of anything, the planetarium theater deceives the eye and engages the brain by putting the audience in the middle of everything.

That middle place is still an anchor. Although people are eager now to understand what happens when comets are slaughtered in collisions with Jupiter, how gravity can make space and time turn in upon themselves to imprison light, and whether the uni-

Griffith Observatory is a celestial landmark in the southern California landscape, and it beckons visitors to make the hilltop pilgrimage to its ramparts and halls for panoramic vistas of the world below and for closer contact with stars. (photograph E. C. Krupp)

verse will expand forever or collapse back upon itself, they also still want to know the geography of the sky. They want to see the constellations contrived by the ancients millennia ago and hear the stories they told about the characters and creatures they painted in stars.

We are no more than a hundred years from the time when anyone could walk outside on any evening and still see the stars, but now the artificial lights that enliven our cities, highways, and towns whitewash the nights and erase the constellations from the sky. The real night sky is now so remote that most people have forgotten what it looks like.

The emptiness of our celestial savings account was confirmed at 4:31 A.M. on January 17, 1994, when everyone in Los Angeles got out of bed at the same time. That's when the Northridge earthquake hit, and many people were propelled out of their houses and apartments by fear and danger. At Griffith Observatory, we started receiving peculiar telephone inquiries within a day, and they persisted for at least a week. People called to ask about the "odd" sky they saw right after the earthquake, and they wanted to know if it had anything to do with the seismic disaster. The staff was puzzled by these calls at first, but after a number of them, it became apparent what people were talking about—the stars. The earthquake had knocked out the power throughout most of Los Angeles, and that put out the lights. From their streets and yards on January 17, residents of Los Angeles saw the night sky, something they never see, except perhaps under the dome of the planetarium.

On the verge of the twenty-first century and enveloped in high technology, we are still moved by the stars. Whether we are drawn by the magnetism of the planetarium's wonders or stunned by the revelations of the real night sky, we are still pilgrims who respond to celestial power. As medicine men, rain chiefs, and sacred kings have receded from the scene, the sky has lost its momentum in the affairs of religion, economics, and government. Its power on the imagination and in the heart, however, is alive and well. Observatories, planetaria, space probes, and orbiting telescopes are now the shrines and temples and palaces where celestial power presides. Like religious power and political power, celestial power has been dispersed and democratized. Almost anyone who wants it has access to a telescope and can see in person the scars on the moon or the secrets of the stars.

Astronomical knowledge confers power. The calendar must be kept. The omens must be read. The ceremonies must be performed. Our itinerary through the rest of these pages is this multifaceted relationship between power and the sky. This relationship involves the concepts of sacred landscape, mythic origins, and cosmovision. These are symbols we have devised to explain how the world works, and the transformations of these symbols are linked with the cultural evolution of societies. The themes of celestial power are, then, an album of self-portraits. As we explore this gallery, we may not recognize all of the names, but the faces should look familiar.

CHAPTER ONE

The Center of the World

I have been to the center of the world dozens of times, and each time it was a different place. I have gazed toward the four corners of the world, where trees, mountains, or gods held up the sky. I have felt the eight winds and seen the colors of cosmic order assigned to the world's directions. I have studied astronomical alignments and symbols at ancient and prehistoric monuments around the globe and learned why the people who built these sacred centers of cosmic power brought heaven down to earth.

In the past, the world's landscape included heaven, the underworld, the land, and the water. Different peoples each identified a special place in the part of the landscape they inhabited as the center of their universe. They always organized the landscape and gave it dimension, direction, and meaning. The world had a top, a bottom, an edge, and a center.

For the Medieval Church, the center of the world was Golgotha, where Christ was crucified. The ancient Hebrews placed it at Jerusalem's Temple Mount. Moslems locate it in Mecca at the Ka'ba. In Babylon it was the ziggurat. For the Romans it was Rome, where all roads were said to lead.

The roads of the prehistoric Pueblo Indians of the American Southwest converged on Chaco Canyon, where community shrines were designed to mirror the cosmos. Southern California's Chumash Indians identified a place near Mount Pinos, north of Los Angeles, as the world's center.

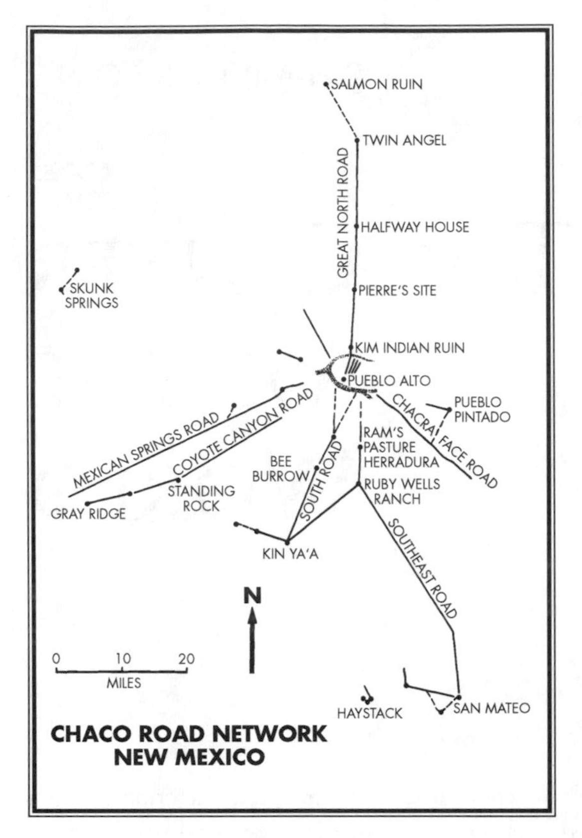

You can sometimes recognize the center of the world in the roads that radiate from it and carry its power to the periphery, the frontier where its influence ebbs. Ancient Pueblo Indians of Chaco Canyon had no wheeled vehicles or beasts of burden, but they built a network of highways that connected it with outlying settlements and shrines. The true function of this road network remains problematic, but ritual use was probably part of the picture. The Great North Road seems to symbolize a dead-straight cosmographic Interstate that transported souls to the realm of the dead in the far north. After 31 miles of sagebrush, it ends abruptly at Kutz Canyon, where the plateau drops into badlands. Twin Angel Mound is also near this terminus, and ruins of a shrine remain on its summit. The mythic realm in the north is equated by some historic Pueblos with the *sipapu,* the place of original emergence into this world. This idea is linked with the concept of birth from the womb of Mother Earth. (Griffith Observatory drawing, Joseph Bieniasz)

Ming and Qing dynasty China put the middle of the world in the middle of Beijing's Forbidden City and put the emperor in the middle of the Forbidden City. The Coricancha temple in Cuzco, Peru, occupied the center of the Inca cosmos, and it was a place where the Inca king conducted sacred ritual. The High King of the Irish Celts presided at Tara, which was said to be the symbolic center of Ireland. In Maya territory, each king was himself the world axis and the center of the world.

In the minds of all these ancient people, the central place was not only the center of the cosmos, it was where creation began, where order and life first emerged. It was also where heaven communicated with earth and where the world was renewed through celestial power. Cities were founded there. Temples were built there. Rituals were performed there. Rulers were enthroned there. Mountainlike pyramid tombs and labyrinthine palaces imitated in architecture the attributes of the center of the universe.

The Aztecs, for example, said they built their capital on a dismal island in an unpleasant marsh because they saw there an eagle devouring the cactus fruit. It was a sign from the gods of their celestial destiny, for the eagle symbolized the sun. The sun was a celestial god that sustained the universe. The cactus fruit stood for the sacrificed human hearts that sustained the sun. Regarding themselves as the primary caretakers of the cosmos, the Aztecs judged that their main pyramid should be oriented to the world's four directions and should occupy the world's heart.

Places like this had power because they matched and reflected the structure of the cosmos itself. In them, the entire universe was miniaturized through symbol. The universe and the shrine shared what those who study the nature of structure would call a homologous relationship. Each element of the shrine mimics—through placement, function, and meaning—a corresponding aspect of the universe itself. Invested, then, with cosmic symbolism, such shrines were sacred venues, places where it was believed the magical power of the universe could be encountered and tapped.

Cosmovision, or world view, is the link between the architecture of the universe, the pattern of nature, the fabric of society, and the personal environment. It integrates these notions and forges a system of relationships between the cosmic and the divine with human society and individual destiny. Ideas about the structure of the world, about the rhythms of time, and about the origin of the cosmos are all combined into a ceremonial landscape.

Now, to understand the power of sacred cities and cosmic shrines we have to understand the power of the cosmos. The ancients recognized that there is really only one thing taking place in the universe, one expression of transcendental power, and that is change. Day transforms into night. Each night alters the shape of the moon. Seasons change. Seeds sprout into the light and gradually grow into mature plants that flower and blow to seed. Through metamorphosis, tadpoles become frogs, and caterpillars become moths. Our lives change. Clothes, language, music, and automobiles all go out of style. New cars inevitably age and betray their mileage. The world changes. Landslides recontour the cliffs. Rivers flood and occupy new grooves. The tide leaves and returns.

Everything changes, but for the ancients, change occurred in an ordered and oriented world. It was a world with a center and with an edge. That boundary might be an echo of the circular horizon, or it might have corners that correspond to the world's key directions.

Because direction organized our experience of the world in a fundamental way, we assigned religious value to it. Acquiring power through its meaning, direction became sacred. People used it symbolically when they wanted to designate special places as sacred. Direction defined the structure of sacred space and the mythical geography of the land. Although finite and earthbound, sacred space imitated the cosmos and operated as a stage for ritual that could attract cosmic power and put it to use on earth.

People have contrived a variety of schemes to organize the space they inhabit, and the most common system quarters the world with four directions. The conventional astronomical symbol for the earth—a quartered circle—is a relic of antiquity. Although not everyone singled out the same four directions or thought of direction in exactly the same way, nearly everyone partitioned the land with some system of direction.

The symmetry of our bodies certainly encourages a preference for four directions. Much of our information about where things are located around us is collected by our eyes and ears, which are closely affiliated with our faces. We face toward one direction and away from another. One side adds right, and the other left. In this way, our anatomy invents geometry.

This kind of directional geometry is exactly what was mentioned in Etruscan city foundation ritual. Etruscan priests read omens from the livers of sacrificed sheep and harmonized their cities with the cosmos at large by orienting the urban plan with a bronze replica of a sheep liver. Through the symbolism of the liver, the establishment of the city's main axis was like divining the will of the gods. The gods were assigned directional zones, and these were inscribed upon the bottom surface of the symbolic liver. With the liver properly held, the priest occupied the center of the new town. That was also the place where the two primary axes of the town crossed. Together, the priest, the liver, and the axes quartered the town. He faced the *pars antica,* and the *pars postica* was behind him. The unlucky *pars hostilis* was on his right. To his lucky left lay the *pars familiaris.*

This is, of course, a highly personalized perception of the rest of the world. Our personal sense of self creates the notion of the center, but without a shared system of references that describe the world around us, culture would evaporate. Something much larger than any individual, yet accessible to each individual, is needed, and people have most often found what they needed in the sky.

For example, we orient our maps and sometimes grid cities and suburban subdivisions by the cardinal directions: north, south, east, and west. The Etruscan priests seem to have done the same thing. They faced south and put north behind them. East coincided with the lucky left. West was "hostile" on the right.

Cardinal directions, whether recognized by the Etruscans or by ourselves, actually originate with the daily—and nightly—rotation of the sky. This motion is, of

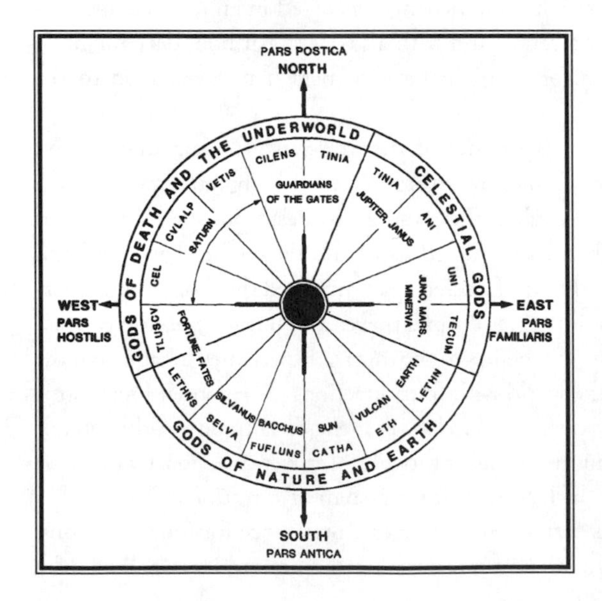

Gods mapped the territory of cardinal cosmic order in the Etruscan world. Directions were linked with deities in a system that oriented the soothsaying, liver-reading priest toward the south and the favorable gods of nature and earth. East was a lucky zone ruled by celestial gods. North and west were hostile realms held by gods of death and the underworld. (Griffith Observatory drawing, Joseph Bieniasz)

course, an illusion, for it is the earth spinning in space that makes it look as if the sky is turning overhead. The center of that daily circular rotation is called the north celestial pole, and it is nothing more than the spot toward which one end of the earth's axis points. That spot is conspicuous in its repose. Although everything else in the sky is moving, it is fixed. And everything seems to be parading in formation around it.

As the hub of the most fundamental movement in the sky, the sky's north pole confers significance to one direction, the direction in which it resides. When we face the north celestial pole, we say we are facing north. Cardinal north is just the corresponding direction on the ideal flat horizon. If we lower our eyes from the north pole of the sky—and from Polaris, a star which by chance is nearby and helps us locate the pole—to the horizon, we are facing, more or less, cardinal north. Cardinal south is just the opposite direction, and cardinal east and cardinal west are halfway between on either side.

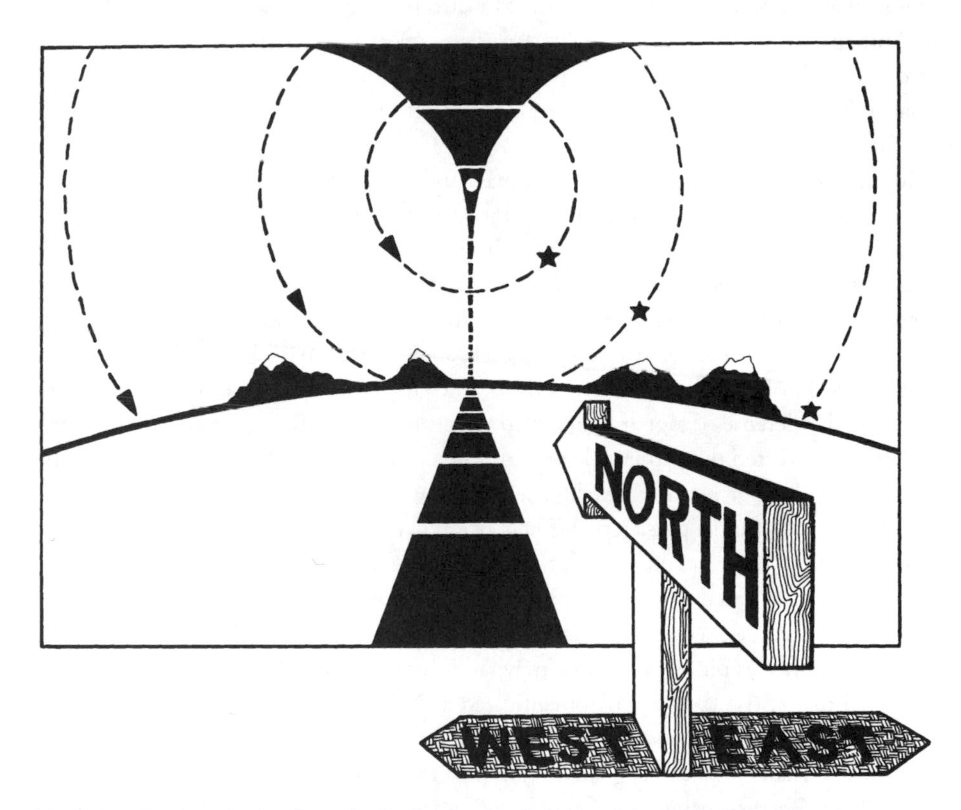

North acquires its meaning from the first fundamental motion of the earth. The earth spins on an axis, and that rotation makes the sky appear to turn around a motionless hub known as the north celestial pole. When you face this center of the counterclockwise circuits of circumpolar stars, you are also looking toward cardinal north, on the horizon. South is directly behind you, but in that direction the paths of the stars look like long, low arcs. Cardinal east is on your right, and cardinal west is on your left. Celestial objects rise out of the eastern half of the horizon, and set behind the western half. (Robin Rector Krupp)

Palaces and temples aligned with the cardinal directions were said to occupy the center of the world. Altars and shrines located outside the symbolically charged center on cardinal axes or in directional zones sometimes stood for the world's four quarters. This does not mean that such places always marked the physical center or the surrounding domains of the world. Rather they marked places of supernatural or spiritual meaning. They mirrored the structure of the cosmos to permit contact with its power. That power of cosmic order energized and organized human affairs. Four sides, cardinal orientation, or directional placement, then, could mobilize any setting into service as a sacred world center or as a world-quarter shrine.

THE FOUR CELESTIAL KINGS

Four Celestial Kings governed the world's directions in China. They were allies of cosmic order, and we find them wherever their supernatural assistance was needed to protect vital institutions. They were posted, for example, in Juyong Pass, on a road that linked a fort on China's northern frontier with the capital.

Six miles beyond Juyong Pass, through the Yanshan Mountains to the base of Mount Badaling, the Great Wall of China stands as a barrier between Mongolia and China. Beijing, the imperial capital of China since the thirteenth century, is about 46 miles southeast of this section of the wall. These days, however, the invading hordes are tourists, not Mongols, and on their way back to Beijing they pass the "tower that bestrides the road." It was known as the Yuntai, or Cloud Terrace, and it is actually a monumental stone platform and ceremonial gateway in the pass. Juyong Pass had been a vital checkpoint on the northeast border since the Han dynasty and a strategic element in the defense of north China, and when the Cloud Terrace gateway was completed, it protected this vulnerable port of entry with images of the cosmic guardians of the world's four quarters. Each Celestial King is sculpted in deep, dramatic relief inside the gate's tunnel passageway. Their eyes bulge. Their jaws clench. Enthroned on the walls, they await intruders with the talismans of their power.

During the Yuan dynasty China was ruled by Mongols. The road once vaulted by the Cloud Terrace used to connect Dadu, the Yuan capital at Beijing, with Khubilai Khan's summer capital, Shangdu, in what is now Inner Mongolia. Shangdu was founded in 1256 A.D. and was romanticized as Xanadu by the English poet Samuel Taylor Coleridge in his poem *Kubla Khan.* The gate was built in 1345, during the reign of the last Yuan emperor, Toghan Temür. The three stupas that once stood on top of the Cloud Terrace gateway reflected the Lamaist brand of Buddhism the Yuan dynasty embraced with enthusiasm.

A *stupa* is a hemispherical Buddhist shrine with a cylindrical base. It is patterned after the ancient burial mounds and usually houses a holy relic, buried within the monument's solid mass. The stupa's shape symbolizes the cosmos. Its dome stands for the sky. From the top, a kind of multilayered umbrella emerges. Its mast is the world

Four Celestial Kings rule the inner walls of the Yuntai, or Cloud Terrace, a ceremonial gate that once controlled traffic on the way to Beijing. This is the south face of the monument, and Garuda, a sacred mythological king of birds borrowed from Hindu belief, presides at the summit of the arch. Small disks that symbolize the sun and moon are lodged above his wings. The wish-granting pearl is suspended from his neck. As the enemy of serpents, he protects the Middle Kingdom from enemies and intruders. (photograph E. C. Krupp)

axis, and the successive parasols refer to higher heavens and homes of celestial gods. If the stupa is enclosed by a wall, the wall is broken at the cardinal points by four gates. Pilgrims circle the stupa clockwise, and their movement is said to reflect the motion of celestial objects.

We find the four Guardian Kings wherever Lamaist tradition found a home. Lamaism is really Buddhism with a Tibetan twist. When Tantric Buddhism from India entered Tibet in the seventh century, it assimilated the indigenous shamanic religion known as Bon and evolved into Lamaist Buddhism. The four kings commonly appear on the walls of entrance halls of Tibetan monasteries. Mural images of them at billboard scale dominate an antechamber on the east side of the Potala, the palace of the Dalai Lama in Lhasa. They are regarded as the guardians of the four cardinal directions. Each holds a symbol of his office, and each has a different colored face. Dhritarashra's face is white. He is the Guardian King of the East and plucks a lute. The red-faced King of the West is Virupaksha. With a snake in one hand, he grips a stupa in the other. Virudhaka brandishes a sword. His blue skin signals he is King of the South. The fourth king, Vaishravana, governs the north. His face is yellow. He lifts the banner of victory and has a pet mongoose. The mongoose has the extraordinary ability to vomit jewels.

Lamaist Buddhism installed the Four Celestial Kings on the entrance wall of the Wuta si, or Temple of Five Pagodas, in Hohhot, the present capital of China's Neimenggu Autonomous Region (Inner Mongolia). From left to right, we encounter (*top*) the Guardian Kings of the South (with sword), East (with lute), (*bottom*) North (victory banner and mongoose), and West (stupa and snake). The first pair flank the left side of the main doorway, and the other two are on the right. This temple was constructed between 1736 and 1795. (photographs E. C. Krupp)

Colorful, painted wood statues of the four kings have their own room in Beijing's Lama Temple. In Chinese, the name of this complex is Palace of Eternal Harmony, and the four guardians of the compass are in the Hall of the Celestial Guardians. They protect the temple from anyone who tries to enter with an evil heart and are joined by Maitreya, the Buddha of the Future, and by Weituo, who is armed with an iron bar. He is the defender of the Buddha and the guardian of the door. In this temple, the King of the North holds an umbrella instead of the banner.

During the Tang dynasty, in 673 A.D., gigantic statues of two of the Four Celestial Kings were carved in hard limestone relief in a grotto at Longmen. Altogether, there are 1,352 grottoes and 97,306 statues here on both banks of the Yi River, about nine miles south of Luoyang in north-central China. Luoyang was the eastern capital of the Tang era. According to tradition, the Tang Empress Wu Zetian contributed 2,000 strings of money from her cosmetics budget to finance the work on the largest Longmen grotto. There, the two kings flank a central image of the Supreme Buddha, which is 56 feet high. Because the Celestial Kings are to the north and south of the Buddha, they are thought to be the King of the North and the King of the South. The north king balances something like a pagoda, or a multitiered parasol, in his upraised palm and holds a demon dwarf in place with his foot.

Virupaksha, the Guardian King of the West, protects his domain in the entrance portico of the Tsokchen, the administrative center and largest building of Sera Monastery, which is located on the northern limits of Lhasa, Tibet. He holds a small bell-like stupa in one hand, and a snake slithers off to the right. (photograph E. C. Krupp)

A giant Celestial King of the North almost makes a Chinese Mount Rushmore out of Ju xian si cave in the Longmen grottoes on the Yi River. The figure to his right is a *Li shi,* or "Defender of the Buddha." (photograph Robin Rector Krupp)

All four of the Heavenly Kings are hardnosed enforcers of cosmic order. Their association with the four cardinal directions tells us they anchor the world with celestial law. Kings who disregarded their ordinances risked the loss of their protection. Kings, then, were obligated to rule in harmony with the greater cosmos, and their power depended on their maintenance of cosmic order in the terrestrial realm.

Although the Four Celestial Kings are part of later Buddhist tradition, they originated in Brahmanic beliefs of India. In the *Upanishads,* sacred Brahmanic commentaries from the first millennium B.C., four directional gods were sometimes mentioned, and later, in Hindu belief, their number increased to eight and encompassed the intercardinal directions. The Buddhists assigned them to the first heaven that surrounded Mount Meru, or Sumeru, the mythic mountain that symbolized the polar axis. Each rules a quarter of the cosmos, governs one of the four continents that borders Mount Meru, and supervises an army of divine guardian soldiers who patrol the slopes of the cosmic mountain. The Celestial Kings were also the masters of the year and the stewards of the seasons. Each one had 90 sons, for a total of 360. This symbolically approximates the length of the solar year (365¼ days), with four seasons 90 days long. Their names, according to the Chinese Buddhists, were Land Governor (east), Far Gazer (west), All Hearer (north), and Growing Grandeur (south). Also known in China as the Diamond Kings of Heaven, they were appropriated by the Taoists, who said they were four genii, the supernatural Mo li brothers.

Mo liqing, the eldest brother, is the western king. The Chinese system of direction correspondences differs from Tibet's, and his face is white. Some of the objects the Diamond Kings hold also differ from those in the hands of the corresponding Tibetan Guardian Kings. Instead of a snake and stupa, Mo liqing possesses a magic lance named Blue Cloud, although sometimes he is shown with a sword. The words on the lance say "earth, water, fire, wind." It prevails with the power of the elemental forces of nature. When Mo liqing swings it in battle, it inspires a black wind to blow. This wind carries thousands of deadly spears, and a gold fire of flaming snakes follows it. Blinding smoke emerges from the ground and traps the enemy army. His spear and serpent storms are thought to symbolize his control of whirlwinds and waterspouts. Also, we sometimes see the Heavenly King of the West with a pearl in one hand and a snake in the other.

The red-faced brother of south is Mo lihong. He has a magic parasol known as the Umbrella of Chaos. It dazzles with jewels, semiprecious stones, and mother-of-pearl, and the inscription on it reads, "shut and open heaven and earth." Any slight movement of the umbrella makes the earth tremor. If Mo lihong lifts it overhead, it hides the sun and the moon. The umbrella seems to have the concession for eclipses, for it cloaks the universe with a cosmic darkness.

A stringed instrument doesn't sound like a powerful weapon, but when Mo lihai plucks his four-string lute, he tunes the winds to his will. One chord blows up a storm. All of his enemies stop to listen, and their camps ignite and burn. His face is blue-green, and he owns the east.

The fourth brother, black-faced Mo lishou, controls the north. He is armed with a couple of whips and an extraordinary creature, which he keeps in a leopard-skin bag. The animal looks like a white rat. Its name means "the sable that can turn into a speckled fox," and when Mo lishou sets it loose, it becomes a white elephant with wings and eats whoever comes its way.

According to the Taoists of China, genii like the Mo li boys are lower-level immortals, guardian spirits with special powers. This Diamond King quartet put a hundred thousand celestial soldiers at the service of the Shang rulers when they were challenged by the Zhou, the dynasty that replaced them. At first, the four great sky generals held the upper hand, but they were defeated by a genie working for the Zhou. He was a master of disguise, and he managed to take their weapons. Then he attacked them with his magical dagger, a blade so bright it blinded the enemy. All four of the Mo li boys were killed, but through their deaths they were elevated to the status of Celestial Kings who control the elements, regulate the rain, administer the winds, supervise the seasons, and watch over the world's directions. Their story is told in *The Art of Deification* (sometimes also known as the *Tale of the Investiture of the Gods*), an ancient narrative about the founder of the Zhou dynasty. From it we learn that the kings of the cardinal directions serve one kingly line until another replaces it. Then the kings work with the new dynasty. The message is clear. The king is allied with the cosmic order revealed in the world's directions.

HOW DEAD KINGS LIE

The Maya ruler Six Sky died at Río Azul on March 6, 502 A.D., and entered the underworld with the power of the world's directions. Río Azul was a large regional administrative and ceremonial center in what is now the northeast corner of Guatemala, and most of the recent excavation there has been directed by archaeologist Richard E. W. Adams. In the Classic era (300–900 A.D.), Río Azul's kings were interred in underground chambers carved into the limestone bedrock. The walls were stuccoed and ornamented with hieroglyphic inscriptions. In Tomb 12, the sepulcher of Six Sky, a different direction glyph was painted in bright red on each of the crypt's four walls. The tomb is oriented to the cardinal directions, and each wall has the proper glyph for the direction in which the glyph appears. Other celestial references accompany these direction emblems. The glyph for "sun" is on the east wall. It can also mean "day." "Darkness," or "night," is written on the west. The north wall carries the word for "moon," and a Venus symbol appears in the south. Four more glyphs, drawn in black in the tomb's corners, stand for the intercardinal directions. The tomb is itself, then, a compass of cosmic order, and it is beneath Río Azul's tallest pyramid.

By placing the dead king in the middle of the world's directions, his city symbolically sealed him in at the center of the world. In myth, this is the same center of the world where the primordial ancestor of the Maya first lifted the sky out of the murky unformed cosmos and created the ordered world we now know. In *Maya Cosmos,* David Freidel and Linda Schele, two experts on Maya writing and civilization, call this creator the First Father. He is equated with Hun ("one") Hunahpu, the father of the Hero Twins whose adventures in the underworld are described in the *Popol Vuh,* the sacred book of the Quiché Maya of highland Guatemala. Hun Hunahpu was killed and decapitated in Xibalba, the subterranean realm of the Lords of Death.

The earliest known text of the *Popol Vuh* is no older than the sixteenth century, but First Father's story probably was told long before the Postclassic period (900 –1520 A.D.). He seems to be the same character that shows up as the Maize God in the art of the earlier Classic era. The Maya relied so much on maize, they said they were created from maize. They linked their origin to a grain god who wore stylized corn in his headdress and who reemerged from the earth like a sprouting stalk of corn.

At the time of Creation, according to hieroglyphic inscriptions found at other Maya sites, First Father broke through the earth's surface at the center of the world and entered the sky after he had raised it into place. There he built a house named Raised-up Sky. The structure First Father fabricated out of the celestial vault was partitioned into eight wings. These eight zones are thought to be the cardinal and intercardinal directions. They radiate from the top of the house, which Schele and Freidel believe is the "heart" of the sky, the north celestial pole. If they are right, First Father's apartments in the sky are really a cosmic compass rose.

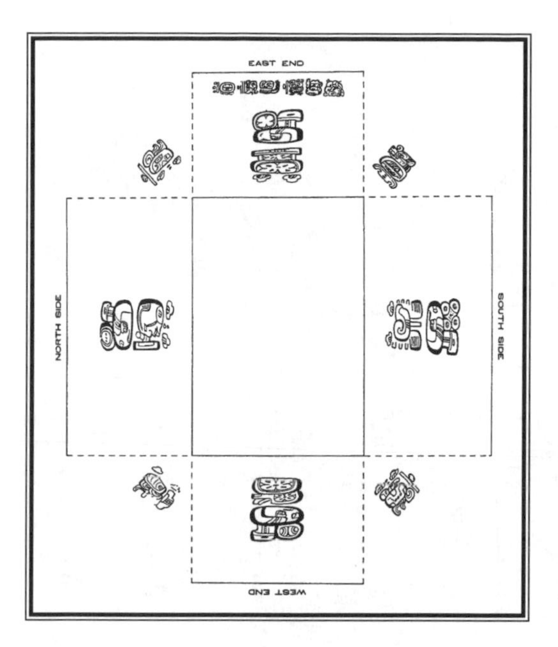

Maya writing charted cosmic order in the realm of the dead at Río Azul, an ancient ceremonial center in Guatemala's tropical forest. These direction glyphs were painted on the walls of Tomb 12, in which a ruler of Río Azul was interred at the beginning of the sixth century A.D. (Richard E. W. Adams, glyph drawings by David Stuart)

The Maya directional glyph for south contains phonetic elements that make it sound like the word for "nadir." This is the artful inscription painted on the south wall of Tomb 12 at Río Azul. The element on the top of the upper glyph, and to the left, looks like a diamond encased in four circles. Usually this symbol refers to Venus. These complexities of Maya writing suggest that the four-square cardinal orientation of the tomb actually may involve a three-dimensional model of the cosmic order. (photograph Richard E. W. Adams)

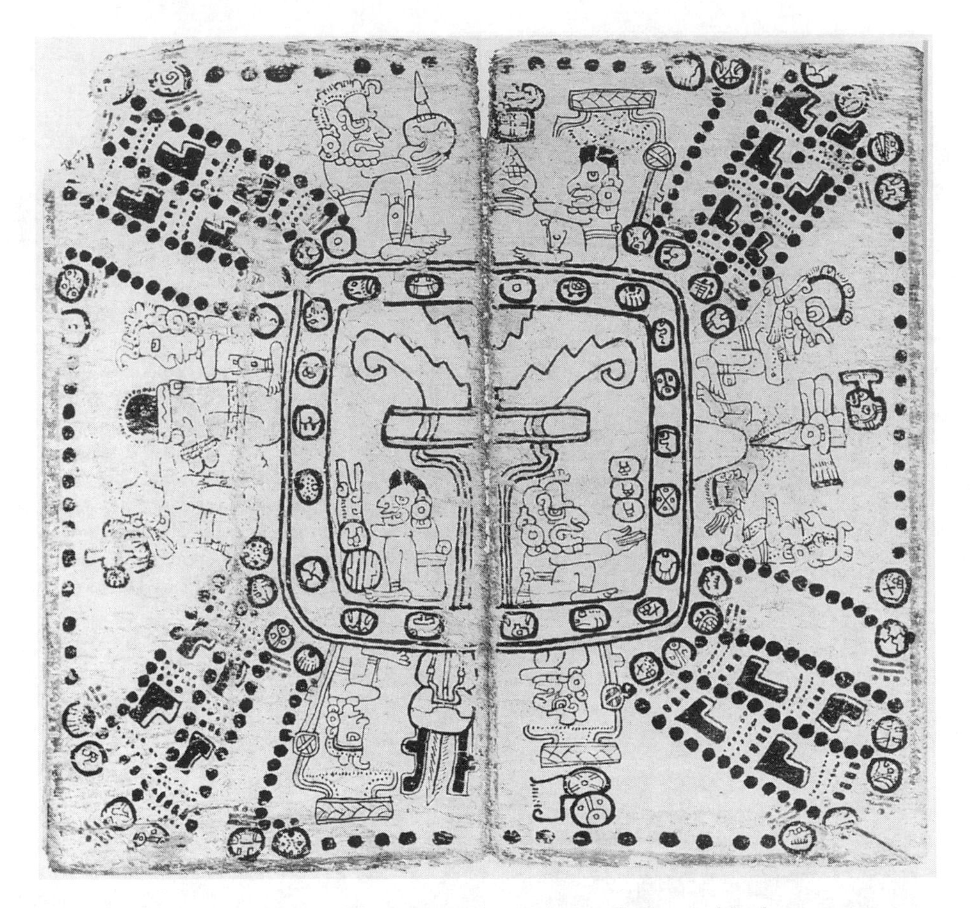

The glyphs and images on two of the pages in the *Codex Madrid,* one of the few surviving Maya "books," are arranged in eight compartments that Mayanists David Freidel and Linda Schele believe symbolize the eight directions of "Raised-up Sky," the heavenly house First Father fabricated at the summit of the world axis. The eight-petaled design is outlined with 260 black dots that stand for the 260 days in the Maya *tzolkin,* or sacred calendar count. This divination almanac is thought to have been composed in northern Yucatán during the fifteenth century. (Akademische Druck-u, Verlagsanstalt, Graz, Austria)

The Maya, however, may not have thought of directions as discrete points in the landscape. Although the earth had four sides and four corners, direction was also defined in terms of the movement of the sun. For example, east—which the Quiché Maya call "sun's coming-out place"—is really the starting point of the sun's daily path. That may be anywhere between the northeast and southeast annual limits of sunrise. It changes daily, and so it is a conceptualization that knits direction to the passage of time.

Also, not everyone agrees that all four of the wall glyphs in Tomb 12 stand for the four cardinal quarters of the world. Linguistic arguments prompt anthropologist Barbara Tedlock to identify the "north" and "south" glyphs as references to the zenith and nadir. If so, the north-south axis of the tomb really means up and down. The vocabulary may actually be interchangeable, depending on context, and with this in mind, Tedlock tracked Venus and the moon back to March 6, 502 A.D. On that night, the moon crossed very close to the zenith. Venus was in the opposite direction, close to the nadir when the moon was overhead. This configuration seems to make more sense out of the north and south walls of Tomb 12 by linking them through the opposition of the moon and Venus to up and down. If this is what the Maya had in mind, they must have imagined different depths of a three-dimensional underworld, an abyss that counterbalanced the dome of the heaven.

The experts may disagree over the details of the meaning of direction glyphs and the organization of space in the Maya universe, but there is no argument about the disposition of the dead king. His status was expressed according to the conventions of Maya elite—through hieroglyphic texts, in a monumental mortuary temple, and in a burial charged with cosmological symbols. Directional references in the tomb of Six Sky put him at the center of the world. There, he mimics First Father. His bloodline makes him the sire of his dynasty. His tomb is the place where the cosmos came to be. He is a target for offerings and a source of supernatural power. Through his descendants, he returns to the world like a new season of corn and maintains world order through the continuity of life.

All of the properties of the sacred center of the world are reflected in the directional magic of Tomb 12. It is a place where power is created, stored, and emitted. It is the place where the world first kicked into gear. There, through the supernatural intercession of the dead ancestor, the world is nurtured, ordered, and cyclically renewed.

ALTARING DIRECTION

The Hopi villages on the mesas of northern Arizona are not governed by kings, but the officers of their religious societies manipulate the symbolic power of direction in ceremonies intended to perpetuate the reciprocal alliance established between them and the spirits. One venue for their offerings and prayers is a "directions altar" sometimes constructed in their kivas, the underground chambers where ceremonies are held.

Alexander M. Stephen, who lived with the Hopi from about 1880 until his death in 1894, kept a detailed journal that documented traditional Hopi life. In it, his description of the Wuwuchim ceremony includes an eyewitness account of the preparation of a directions altar in the Goat Kiva for the use of the Agave Society. Wuwuchim falls in the first month of the Hopi lunar calendar, usually in early November. The Hopi call this month "Initiates' Moon." Adolescent boys are initiated into the adult religious societies during Wuwuchim, although induction is not an annual component of the ceremony. The primary theme of Wuwuchim is preparation for the seasonal renewal that accompanies the return of the kachina spirits to the Hopi villages and the turning of the sun at winter solstice.

Wuwuchim is, however, an appropriate time for initiation of the youth because transformation to adulthood is regarded as a kind of rebirth. Masau'u, the Skeleton Kachina, attends the initiation. He stood at the place of emergence when the First People climbed up from the realm below and into the light of this world, and as the owner of the earth, he welcomed each one. His friendship is inevitable for he is also the spirit of death. Hopi boys emerge from the world of their youth at Wuwuchim and become real people with adult responsibilities and adult knowledge. As youth ends, a new life begins.

On November 10, 1891, Alexander M. Stephen visited the Goat Kiva at Walpi, the Hopi village on First Mesa. At about noon, a member of the Agave Society began to construct the directions altar. He scattered brown sand into a circle centered on the sipapu, a cavity in the kiva floor that symbolizes the place where the First People climbed from the Second World into the Third World—the one just below the surface of our earth.

The sipapu is the place of emergence, and the story of the emergence of the First People is really about the creation of the present world. It echoes familiar processes that deliver new life, including birth, the sprouting of plants from the earth, and the springtime emergence of hibernating animals from their burrows.

As work on the directions altar continued, three lines were drawn across the floor in white cornmeal. They intersected at the sipapu, on which was placed a bowl. Its design incorporated cloud or water symbols. Pollen and spring water were poured into it.

Now water, in Hopi country, is a serious issue. Although the rains return each year, they are marginal and unreliable. Because the Hopi traditionally survived on agriculture in a desert environment, their ceremonial cycle is tied closely to the seasonal

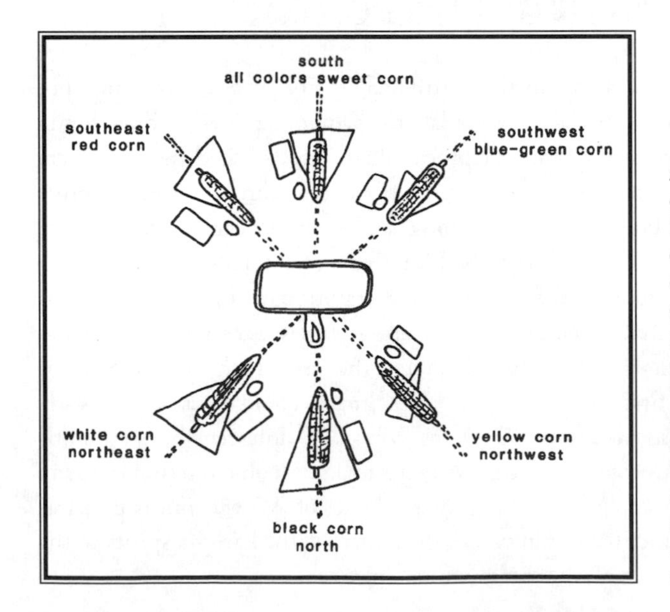

Alexander M. Stephen observed an officer of the Hopi Agave Society build a directions altar at Walpi and documented its cosmographic layout in his journal. A different color distinguished each intercardinal direction. As the zenith-nadir axis, the north-south line incorporated the three-dimensional character of space into the ritual setting. (Griffith Observatory drawing, Joseph Bieniasz, after Alexander M. Stephen)

return of rain. Their religious metaphors are rooted in rain. The kachinas are really spirits of the dead returned to earth, ancestors who materialize as clouds that bring the rain. In this world, the spiritual essence of the kachinas is the rain. When it falls on earth, it creates corn, and corn sustains life. Through the consumption of corn, people metabolize the true nature of the kachinas. For this gift, the Hopi reciprocate with the prayer offerings. They are not worshipping the rain or the clouds or other natural phenomena but instead recognizing them as parts of an integrated cosmos in which a transcendent source of power has made life possible. The spiritual commodity shared between the Hopi and their kachinas is exchanged in the form most suitable to each respective recipient. These exchanges are intended to stabilize the world and Hopi society through the inevitable changes life endures.

Water—the symbolic essence of being—is there, then, on the sipapu to nourish all emergent life. The circular sprinkling of sand around the sipapu represents the world, and the sipapu is said to be the middle of the world. That is where offerings are directed. From it, the gifts and blessings of the spirits are sent.

The lines of cornmeal that radiate from sipapu reach toward the world's directions and so conduct everywhere the spiritual power that emerges from it. These altar directions are not, however, the cardinal directions. For the Hopi, the fundamental pattern of space emerges out of the movement of the sun, which is itself a prime-time player in the Hopi cosmos. Sunrise and sunset at the winter and summer solstice determine the orientation of the two diagonal axes in the directions altar. It is convenient to describe them as northeast, southeast, southwest, and northwest, but they are not true intercardinals. The true intercardinals are defined geometrically and fall halfway between the cardinal directions. Through the year in Hopi country, the sunrise and sunset migrate between northern and southern limits that are less extreme than the intercardinal directions, and they establish the directional frame. The remaining line in the altar symbolizes the two remaining directions—up, toward the zenith, and down, to the nadir.

An ear of corn, feathers, and other talismans are placed on each directional ray, and each direction has its own color of corn and plumage. Yellow is the color the Hopi assigned to the northwest (summer solstice sunset). Blue corn and the skins of a jay and a bluebird belonged to the southwest (winter solstice sunset). The altar's southeast line was accompanied by red corn and the red feathers in the skins of a woodpecker and a robin. For the northeast, white corn, white magpie feathers, and a whippoorwill skin (which also has white feathers) were all deployed. Black, the color of the underworld and the nadir, was brought to the north line of the altar in black corn and crow feathers. The zenith, affiliated with all colors, was represented by the south line. Edible sweet corn and mixed feathers were put there.

Color and direction are actually part of an extended system of correspondences. Through these associations, the Hopi devised principles of symbolic classification, and they threaded the system into all of the other aspects of their lives, including ornamentation, clothing, songs, regalia, and ritual. It is an economical way to express

Chicago's Field Museum of Natural History displays a reconstructed Hopi directions altar. This arrangement departs a little from the setup recorded by Stephen and was used in this way in the women's Oaqöl Society basket dance at the Oraibi harvest ceremony on Third Mesa. We are looking from east to west. (photograph E. C. Krupp)

relationships and is certainly not restricted to the Hopi. We have already encountered color-direction links in the four Celestial Kings of China and Tibet. Other peoples all over the world have fabricated their own templates of correspondences. They extract from them a vocabulary of symbols invested with coherence and meaning. Using them redundantly in all of their enterprises, they advertise the fundamental preoccupations and principles of their lives. The shared understanding enhances social cohesion.

During construction of the altar, the Hopi officers sang, and after the second song a quartz crystal was brought out and carried up the kiva ladder to the ceiling entrance. With it one of the men caught the sunlight and bounced a beam of it into the bowl at the center of the altar. The quartz was then placed in the bowl along with the pollen already sprinkled there. This manipulation of sunlight reinforced the sun's role in the directions altar and injected the sun's power into the most highly charged element of the arrangement—its center. All of this symbolic activity took place inside the kiva, which is itself a model of the world and the underworld. Its four walls correspond to the four directions. Its ceiling is both a ceiling of the realm below and the sky over the land above. Taken together, the directions altar, the rituals performed at it, and the kiva that contains it all, facilitate interaction with the spirits and access to the power of life they possess.

ON EARTH AS IT IS IN HEAVEN

The directional Diamond Kings of China, the cardinal tombs of Río Azul, and the emblems of orientation in the Hopi altar are all symbolic expressions of celestial power. Because they are religious responses, they tell us something about the nature of the sacred. In the palace, in the temple, in the grave, and in ritual, the authority and the agenda of the spirits and the gods were partly proclaimed with emblems of their most substantive concern—the world. In the world view of ancient and traditional peoples, nature itself reveals the world to be an arena of life, a place people can inhabit. And although they, like all living things, endure uncertainty and inevitable death, people recognize in nature a structure of space and time that integrates and unifies the world and makes it a functioning setting for human activity. Although the human beings cannot craft a cosmos, the gods can. They may employ their power to establish the world's order, or they may, for reasons of their own, exercise their will to induce departures from the customary rhythm and organization of the universe.

The sky is not the only place where people see the world's stability on display or compromised, but it is the most obvious. Unlike the earth itself, it cannot be reached on foot. To get there, people had to rely on supernatural power, and that was the business of spiritual specialists. The sky was, then, part of the supernatural terrain. Although it changed, it was eternal. The sun, moon, planets, and stars glowed with an inner power, and the sky itself was personified mythologically. In Vedic India, it was Dyaus. The Greeks called it Zeus, and to the Romans it was Jupiter. These names are related linguistically. All are rooted in an archaic word that refers to the sky and to light.

Similar traditions can be found around the globe. The Kaweskar Indians (or Alacaluf), shellfish foragers and maritime hunters of southern Chile's archipelago, called the world's creator Kólas. His name means "star," and he is all spirit, like the souls of the dead. Although he resides in the sky, far from earth, the world's inhabitants are his concern. He bids souls to inhabit the newborn and so seed them with life. At death, these souls fly back to him. The stars are his eyes, and nothing escapes his view. He knows everything.

Often the personified sky had something to do with the world's creation and the establishment of its order. It was a high, remote god credited with complete awareness and knowledge of the entire cosmos. In the nineteenth century, E. B. Tylor and other pioneering anthropologists equated the notion of an all-powerful god with monotheistic belief and credited the concept only to more advanced civilizations with traditions considered to be evolved. Ideological advancement, in this perspective, marched together with social complexity. It led from the superstition and magic of primitives to the abstract and moral religions of our own time. Although these faiths still harbor a place for the supernatural, those who saw religion in evolutionary terms emphasized the ethical priorities and social forces that govern its development. They minimized the persistence of transcendence in the character of belief. In this century, however, historians of religion undertook comparative studies and incorporated the impact of historical and cultural idiosyncracies. They discovered that many marginal societies like the Kaweskar believed in a supreme but remote god that kept an eye on human affairs.

In fact, Raffaele Pettazzoni, an Italian historian of religion, studied this capacity of the supreme god to know all things and linked this talent with the nature of the sky. In *The All-Knowing God,* he asserted, "The plain fact is that according to the evidence it is mostly sky-gods and astral gods, or gods somehow connected with the heavenly realms of light, to whom omniscience is ascribed."

Knowledge, of course, is power, and cosmic power—the kind of power that drives and governs the natural operation of the world—is understandably affiliated with the celestial realm. From that inaccessible platform with its unparalleled prospect of the territory below, the high god directs the action and puts divine will on display. Sky gods reveal themselves in the order of the world, in the powers of nature, and in the renewal of time. Because the sky conveys a message of directional structure and cyclical time, these concepts are expressed in the institutions of power on the earth below. Our attention may be directed to empires, kingdoms, chiefdoms, or tribes, but in all of them, regardless of their social and political complexity, the celestial canopy informs the centers of terrestrial power with the habits of heaven.

Symbolic centers of cosmic power make sense in the capital cities of empires and in the monuments of kings. These are the institutions of societies that have adapted and thrived by centralizing power. Rulers legitimized their power on earth through demonstrations of celestial approval of their status. Emblems and rituals that advertise supernatural celestial power help, then, to consolidate kingship, but directional order in Hopi ceremonialism tells us that life in village communities is also regulated and explained with celestial references. Even though smaller agricultural communities do not command the political and economic muscle of kingdoms and empires, they, too, are welded to the land. Territory is permanently occupied. Through farming, the powers of nature sustain life. Cyclical celestial events and seasonal renewal reveal the process of creation and change that delivers the groceries. This pattern is centered where people live, and it reaches to the horizon.

Cosmological directionality and the concept of cyclical renewal—especially seasonal renewal—are obviously at home in settled societies, but they are also comfortable in the world of nomadic hunters and herders. The game departs in one season and returns in another. Migrating birds abandon their summer homes when autumn moves back into the neighborhood. Fish return to their spawning waters with the gratitude of the anglers who trap them. Sheep, cattle, and reindeer are seasonally shifted from higher grazing lands to their winter fields.

WHERE SHAMANS FLY

For hunters and herders, access to supernatural power is usually routed through shamans, instead of tribal officers and sacred kings, and the shaman's world view maps the sources of that power. Although the topographical details of the shaman's cosmos may differ from what we have already encountered in China, Mesoamerica, and the American Southwest, the fundamental structure is the same. There is an upper realm that we see as the sky. People reside in the middle world—usually judged to be a flat earth surrounded by water. The lower world is below ground and often entered through caves. All three of these realms host spirits that help, hinder, or ignore the affairs of people. In an ecstatic trance, the shaman's soul travels to these realms. In that state, the shaman talks with the spirits, duels with demons, and bargains for supernatural power on behalf of the people he or she serves. The entire universe is animated by spirits. They occupy heights of heaven and the catacombs of the underworld. Distinctive features of the middle world—mountains, caves, rivers, springs, rocks, trees, and other special places—also house spirits. Their power is exerted in all of the natural phenomena associated with them. Their presence is revealed both by uncanny departures from routine events and in familiar patterns of ordered change.

Tungus shamans performed in fringed robes ornamented with metal symbols of the sun and moon. (photograph E. C. Krupp, Musée de l'Homme, Paris)

Attributing supernatural power to sacred places and populating them with spirits originate in a response to the world we now call animism. E. B. Tylor assigned this belief to the infancy of religious thought and argued that animism was the natural consequence of dealing with two questions: (1) What is the difference between a living being and the dead? and (2) What is the nature of the beings encountered in visions and dreams? The presence of an invisible spirit, or soul, was given the credit for the difference in these states. Natural phenomena—such as the wind, the clouds, thunder, and stars—also were activated by spirits within them. Tylor thought explaining nature was the real function of animist thought. He saw religion as a linear development that began with spirits everywhere and matured into the notion of one god. His simple picture of the evolution of belief was disproved by anthropologists who found abstract high gods in the theologies of ostensibly primitive people. Animism, however, does walk shoulder-to-shoulder with the shamans. Instead of explaining nature, it provides a framework for the acquisition of supernatural power.

Animist beliefs and shamans are found in most parts of the world, but the Tungus peoples of eastern Siberia are the ones who donated the word *shaman.* Tungus peoples include the reindeer-herding Evenks, settled hunters and fishers like the Orochi and Goldi, and others. Together with Manchu, the Tungus languages comprise a branch in the Altaic language family. Russians exploring Tungus territory in the seventeenth century first encountered the term *saman* for someone who moves between the worlds on spiritual errands. First recognized as pagan sorcerers, the Tungus shamans were eventually understood to engineer contacts in the domains of spirits as hunters and gatherers of supernatural power. They are charismatic individuals who slip into something more mystical and then traffic personally with the spirits.

S. L. Shirokogoroff, a Russian anthropologist, lived with the Tungus in the early twentieth century, learned their language, and outlined Tungus cosmological notions in *Psychomental Complex of the Tungus.* The Tungus cosmos is centered, in a sense, on the North Star. Its name among the Evenks means "the middle of the universe." To another Tungus group, it is "the pillar." This pillar is the cosmic axis that connects heaven with the earth and the underworld. According to a Tungus story, this post was erected by Buga, the supreme celestial god, to hold up the sky after he had created it by inflating a bladder. This bladder was intended to hold down the earth as well.

Stability and support are clearly affiliated with the polar axis, and the Tungus regard the sky's rotation around Polaris as an inherent property of the cosmos. Nothing interrupts circumpolar motion. The sky just keeps turning, as if it has always turned. In the minds of the Tungus, the universe has always existed, and the ever-spinning sky may have helped inspire this belief.

The stars turn unceasingly above the earth, which is said to be large, flat, and solid. It is surrounded by the sea and kept from sinking by one or two vast serpents. When the snakes move, the earth quakes. Dead souls and dark spirits live in the underworld.

The sky overhead is really a system of nine heavens, organized in a hierarchy of power. Spirits inhabit the first three zones. The fourth belongs to the sun. Stars and planets occupy the eighth layer, and the moon presides over the ninth. Buga, whose name seems to be rooted in words for sky, transcends even heaven. He is not really a spirit but the power of cosmic order and natural law.

Directional priority is implied by the significance the Tungus attach to the Pole Star and the north pole of the sky, but they do not throw a net of cardinal quarters across the landscape. They recognize celestial order. They see how it migrates to earth. Their geography, however, is more clearly defined by places, landmarks, and routes of travel. Nomadic peoples see the same universe everyone else sees, but they don't necessarily tailor territory according to celestial cues.

A traditional cosmos is embedded in Tungus belief. Some of the details of the supernatural terrain may have been imported from others, particularly Mongol-speaking neighbors, but the core of their world view could certainly represent genuine Tungus tradition, for this kind of cosmos is believed to be a primordial and universal concept. This idea has been especially developed and advocated by Mircea Eliade, the historian of religion who looked for archaic themes in ancient and traditional beliefs. He emphasized the symbolism of world centers, the cyclical renewal of life, and the sacred dimension of space and time. The way people interact with the natural world modifies what is stressed in a world view. Social and political complexity impose requirements that have little meaning in the affairs of hunters, nomads, and village farmers. Just about everyone, however, has—with the help of the sky—transformed the experience of personal identity into a universe structured by its center, directions, boundaries, and time and animated by the power of spirits and gods.

Eliade and others correctly recognized the deep antiquity of these organizing principles of cosmological thought. We might be tempted to attribute them, as Tylor attributed animism, to the childhood of human knowledge, but we should not be

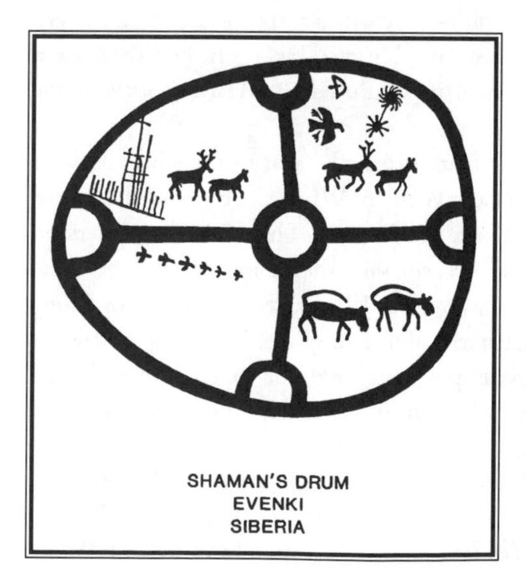

World quartering is not a known component of Tungus cosmovision, but the drums of Tungus shamans are often painted with quartered circle design. This Evenki (also known as Orochen) drum includes what looks like a celestial conjunction in the upper right quarter. The line of birds in the lower left section refers to the seasonal migration of waterfowl and the Milky Way, the celestial path that seems to guide them south at the onset of winter. (Griffith Observatory drawing, Joseph Bieniasz, after a drawing in *Shamanism in Eurasia,* edited by Mihály Hoppal)

seduced by a false understanding of our intellectual development. Old principles of symbolic thought still reside in our brains, despite our sophisticated accounts of inflationary universes and quantum fluctuations. The universe is a lot more abstract than it used to be—and a lot less like everyday reality, but we still move through the world with the anatomy of Ice Age hunters. There is front and back, left and right, up and down, and each of us is still in the center of these directions.

THE WHITE BUFFALO

People, it seems, have always mythologized the landscape, invested life with significance through sacrament and tradition, and attributed symbolic meaning to singular events. They still do. Despite the success of scientific analysis in establishing a physical understanding of what happens in the universe, the news still reports what many judge to be miracles or signs of supernatural intervention. Such events, for many people, mean contact with the sacred and transcendent, and it should not surprise us to find the old, reliable imagery of supernatural power in their company. For example, on September 22, 1994 (coincidentally, the date of the autumnal equinox that year and a key junction of seasonal transition), the *Los Angeles Times* featured a story about the birth of a white female bison calf on a farm in southern Wisconsin. In the tradition of the Sioux, who once were bison hunters of the Northern Plains, a white female bison, or buffalo, calf is a sacred envoy from the spirit world. It symbolizes White Buffalo Cow Woman, the supernatural being who once appeared to the Sioux and gave them the Sacred Pipe, which is smoked as an offering to the Spirit of the World.

The buffalo anchored Sioux life in the open plains. It was a four-legged emblem of a four-direction cosmos. It provided food, hides for clothes and tipis, and other essential materials. In that sense, everything about the buffalo was a gift, and the Sacred Pipe bestowed by White Buffalo Cow Woman solidified the relation between the Sioux and the Creator.

Black Elk, a famous Oglala Sioux holy man, began the story of his life, *Black Elk Speaks,* with the tale of the White Buffalo Cow Woman. When she brought the Sacred Pipe, she stayed four days to teach the Sioux how to live. The 12 eagle feathers on White Buffalo Cow Woman's Sacred Pipe were, she said, the 12 moons, and they were tied with grass that never breaks. According to Black Elk, the four ribbons that hung from his own pipe stood for the four quarters of the universe. Each was a different color. He added, however, that "these four spirits are only one Spirit after all." The eagle feather on his pipe was that "One, which is like a father." The sky is a father. The earth is a mother. And every living thing is one of their children.

In 1947, three years before Black Elk died, Joseph Epes Brown, trained formally in the history of religions, visited him, recorded additional details about White Buffalo Cow Woman, and published them in *The Sacred Pipe.* The Pipe's bowl, taught White

Buffalo Cow Woman, is the earth. All of the plants, all of the four-legged creatures, and all of the birds are symbolized on it. "When you pray with this pipe," she told them, "you pray for and with everything." The north-south axis of the world is the "red road" of purity and life. "Every dawn as it comes is a holy event, and every day is holy. . . ." After she finished instructing the Sioux in the sacred lore of the Pipe, she reminded the Sioux that the world must pass through four ages of time and that she would return at the end. She then left the ceremonial lodge and started to walk around it sunwise. After a short distance she stopped, and the Sioux saw her transformed into a buffalo. She moved a little farther two more times and changed color with each pause. Finally, with her last stop, she nodded to the world's four directions and disappeared.

The original appearance of White Buffalo Cow Woman was a critical moment in the spiritual life of the Sioux, and so the arrival of a white buffalo is greeted with great interest and reverence by them today. Since word about the birth of the calf began to spread, many Indian people have gone to Wisconsin to see it.

New Age America, on its own quest to reconstitute lost spirituality, is enamored with the beliefs of traditional peoples and appropriates them for itself. Lots of people who weren't Indians also made pilgrimages to the white buffalo. Shirley MacLaine, the actress who has authored several books of New Age spirituality, inquired about a visit.

Dave Heider, who owns the farm and the white buffalo calf, has named the baby bison Miracle. Refusing offers to buy the calf, he has been respectful of Indian beliefs and consulted Arvol Looking Horse, a Sioux spiritual leader, about how to handle all the sacred offerings people were leaving on his gate. Arvol Looking Horse is the keeper of the Sacred Pipe, and when he traveled from Green Grass, South Dakota, to see the white buffalo calf, he told Dave Heider to collect the items every four days and burn them. The smoke, he said, "would carry them to the spirits."

We can now easily recognize that this advice is based on the ancient cosmological meaning of the number four. For the Sioux, the number four injects the power of world order into life through directional symbolism. This same idea is shared by many other tribes who are not closely related to the Sioux. It is probably part of the primordial heritage of North America. In our era, however, even proprietary traditions easily migrate across tribal boundaries. The *Los Angeles Times*, for example, quoted a Winnebago version of the Sioux story, and in it, White Buffalo Cow Woman tells one of the two men that first see her that she will return in four days with the Creator's gift. The story really belongs to the Sioux, but the significance of the number four is collective.

Four is not the only number people have selected for the framework of a world order, but it does show up worldwide. It is present in the European tradition and belief, too, and even though it is not part of our scientific approach to the universe, it reappears in popular culture, the haven of the mythology of our age. It is not clear, for example, that L. Frank Baum consulted references on sacred cosmography when he wrote *The Wizard of Oz* (first published in 1900), but the land over the rainbow

Dorothy Gale reached by cyclone, nevertheless, conforms to principles of mythical geography and cosmic order encountered in so many traditional world views. It is an essentially rectangular land, isolated from the world beyond by an encircling desert, the Great Sandy Waste. Oz and its Emerald City occupy the very center, while four autonomous countries each claim one of the surrounding quarters. The Gillikins reside in the north and favor the color purple. To the south we find the Quadlings, whose color is red. East is blue in Munchkin land, and in the west, the Winkies advertise themselves with yellow.

When Walt Disney was working out the topography of "the happiest place on earth," he somehow slipped into traditional cosmology. Main Street, the heart of America, is the umbilical that conveys visitors from the entrance of Disneyland to the navel of the park, where all four of its territories converge—Frontierland, Adventureland, Fantasyland, and Tomorrowland. Together, they quartered the cosmos of the Disney imagination, binding the present to the future through exploration and dream.

Archaic structures of thought are always reemerging, whether we recognize them or not.

HOLLYWOOD AND VINE AND A WORLD-QUARTER SHRINE

According to tradition, if you stand long enough at the corner of Hollywood and Vine in Hollywood, California, you will see everyone you have ever known. Farther to the west, on Sunset Boulevard, there is a vintage piece of novelty architecture, a collection of shops stylized to look like the superstructure of an ocean liner. It is called The Crossroads of the World. Both the world-famous street corner and the nautical storefronts portray Hollywood as a magnet that pulls in the pilgrims following the trail of motion picture glamour. Despite the dispersal of moviemaking from most of the old studios, Hollywood retains its image as the center of the entertainment world. Highly visible relics in Hollywood, such as the heroic Hollywood Sign and the star-studded Walk of Fame, reinforce Hollywood's claim on the imagination. They are places of mythic power. The steep hillside sign advertises the territory of Hollywood over a broad zone of the Los Angeles basin, and the sidewalks of Hollywood Boulevard persuade the receptive tourist that the names inlaid there forge a personal link to the celebrities themselves.

The Walk of Fame runs east-west, like Hollywood Boulevard, and on the west it ends at La Brea. The termination is marked by another Hollywood monument. It incorporates principles of design that would be at home in a world-quarter shrine erected by some ancient civilization or traditional people. At least partially inspired by the Art Deco style of the 1930s, this open pavilion of steel lifts a tapered, four-

The world is still quartered in the name of cosmic order. Although this world-quarter shrine, on the southeast corner of Hollywood Boulevard and La Brea in Hollywood, California, shimmers with a silvery sheen, color-direction symbolism is implied by the multiethnic quartet of glamour queens that support the Marilyn Monroe world axis like caryatids trained at the Acropolis. (photograph E. C. Krupp)

sided spire above the sidewalk and the traffic. Struts and arches argue that this might be the kid brother of the Eiffel Tower, but its modest stature and awkward proportions suggest it is instead the work of an only modestly motivated owner of a beginner's Erector Set. The monument, however, is a tribute to the attractive women who seduced motion picture audiences with beauty, glamour, and sex appeal. Shimmering silver figures of four of these female stars are the corner columns that support the canopy of this sidewalk gazebo. They transform the structure into a symbol of the universe. Its upper tower is the central world axis, and the dome is the sky. A cardinally oriented square on the pavement frames a star, the symbol of celebrity status in the movies. Each corner of the world square hosts one of the women. They are larger than life, and each is attired in a sheer and clinging strapless gown. They invite you to walk under the rotunda and stand on the star in the middle of the Hollywood cosmos, surrounded by images of stunning women. Mae West reigns on the northwest. The southwest belongs to Dorothy Dandridge. Anna Mae Wong is the sovereign of the southeast, and Dolores Del Rio rules the northeast. All the women of the silver screen are deliberately invoked by these four movie queens and their four different ethnic heritages. Because each ethnic group is traditionally associated with a different shade of skin, the Hollywood world-quarter shrine also assigns a color to each of the world's directions. Similar color-direction correspondences belong to traditional cosmologies throughout the world.

The summit of the shrine of course symbolizes the most transcendent zone of the Hollywood universe. Marilyn Monroe presides there in gold—a spellbinding Venus, a high goddess of Hollywood whose milky, knife-pleated dress blows above her legs the way it did in *The Seven-Year Itch.*

The configuration of the universe today is revealed through scientific observation and analysis. We no longer recognize sacred centers and world quarters. They return, however, whenever their old symbolic power adds meaning to the imagery of our time.

CHAPTER TWO

PLUGGING IN TO COSMIC POWER

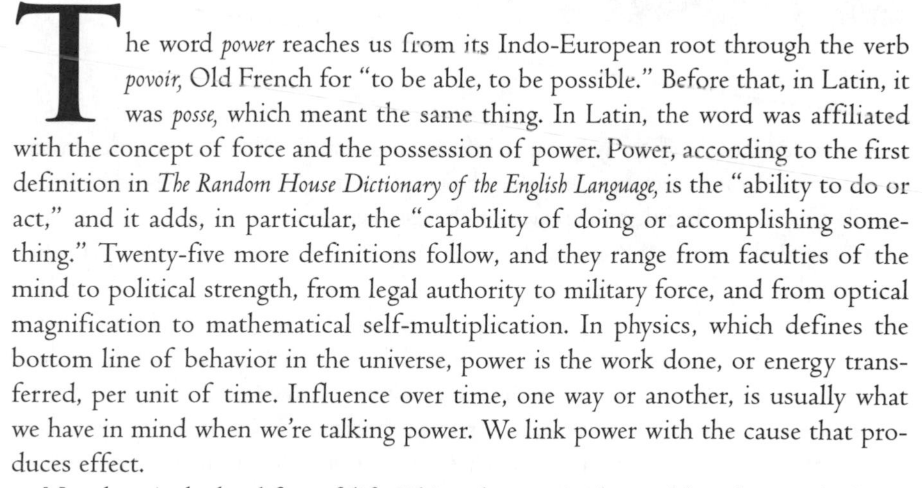

The word *power* reaches us from its Indo-European root through the verb *povoir,* Old French for "to be able, to be possible." Before that, in Latin, it was *posse,* which meant the same thing. In Latin, the word was affiliated with the concept of force and the possession of power. Power, according to the first definition in *The Random House Dictionary of the English Language,* is the "ability to do or act," and it adds, in particular, the "capability of doing or accomplishing something." Twenty-five more definitions follow, and they range from faculties of the mind to political strength, from legal authority to military force, and from optical magnification to mathematical self-multiplication. In physics, which defines the bottom line of behavior in the universe, power is the work done, or energy transferred, per unit of time. Influence over time, one way or another, is usually what we have in mind when we're talking power. We link power with the cause that produces effect.

Now here is the hard fact of life: Things happen in the world, and we can't always do something about them. Events are sometimes beyond our power. The scale can be global—like the impact of a stray asteroid with the earth, national—like a declaration of war or an election return, local—like a crop failure precipitated by bad weather or a county bankruptcy induced by speculative investment, or personal—like an accident, a catastrophic illness, or a broken heart.

Rooted in traditional imagery of sky power, the Comision Federal de Electricidad allows Tlaloc, the fanged and goggle-eyed central Mexican god of storms and rain, to symbolize on its offices in Mérida its effort to bring electricity to the Yucatán countryside with a fistful of lightning bolts. (photograph E. C. Krupp)

A one-day sample of the newspaper informs me that the Russians face overpowering firepower from Chechen rebels, that the power of the Pope assembled millions in Manila to see him, and that the power of the earth convulsively rocked Kobe, Japan.

In our daily confrontations with uncertainty, some people consult what they believe to be signals from powers above. The newspaper horoscope advises Aquarians with an undisguised call for a public display of inner power, "Stand tall, speak your mind, let people know you don't intend to grovel." Nearly halfway around the zodiac, the Geminis are told to permit their "psychic powers to flow freely." I have no confidence in astrology's power to provide pertinent information on anything, but many people take it seriously. Fortunately we have other options for influencing destiny.

We are not completely helpless when it comes to challenges. We contend with reversals and threats with the real powers we command. Nations arm themselves against enemies, and people battle for what negotiation fails to settle. Governments fund, and people deliver, emergency help in the wake of earthquakes, floods, and fires. Insurance helps us respond to the losses from disasters, the cost of repairs, and the expense of medical treatment. Physical strength and the force of personality arm us against personal adversity. Rational, analytic thought makes us problem solvers.

Mobilized through emotion, we call up commitment and resilience on behalf of our goals. Physical power, financial power, intellectual power, willpower, and persuasive power are just some of the power tools at our disposal.

Even the helpless have powers of persuasion. An infant crying from the crib can easily summon attention and assistance from sleep-deprived parents.

Power, however, is not just a response to trouble. Our capacity to imagine the future, to plan and to build, and to carry on normal, day-to-day affairs reflects our ability to alter the outcome. Self-determination—social and individual—is fashionably labeled "empowerment." We spend a lot of time together, advertising our frustrations and our satisfactions, and all of this enterprise reflects our understandable, everyday preoccupation with power.

There are, then, all kinds of power, including political power, economic power, military power, and divine power. We encounter it most often as personal power that emerges from the individual and as social power that organizes and directs the community. We link it with strength, leadership, decision making, and control.

Because power is multifaceted, our relationship with it is complex. We respect it and try to acquire it, but we also mistrust it, especially in the hands of others. "Power," according to the Greek historian Herodotus, "is precarious." We are wary of the power of politicians, and politicians worry about the power of the press. Those who govern walk the corridors of power, and the governed demand power to the people.

We are concerned, of course, not only with the nature of power but with where it leads. In 1887, in a letter to Bishop Mandell Creighton, Lord Acton observed, "Power tends to corrupt; absolute power corrupts absolutely." In *Queen Mab*, the English poet Shelley wrote,

> Power, like a desolating pestilence,
> Pollutes whate'er it touches . . .

and the American historian Henry Brooks Adams added simply, "Power is poison."

On the hardball field of international affairs, Henry Kissinger admitted without shame, "Power is the ultimate aphrodisiac," while Timothy Leary, one of the first flight commanders in the Psychedelic Revolution revealed, "Power is the ultimate turn-on."

Emphasizing social accountability, British statesman Benjamin Disraeli wrote, "all power is a trust." Popular culture concurred a century later when the Amazing Spider-man, a Marvel Comics superhero, realized, "with great power there must also come—great responsibility."

In the 1960s, Flower Power wanted to give peace a chance while those who played power politics advocated maintaining peace through the balance of power. Black Power demanded rights and respect for a minority outside the gates of power, and Norman Vincent Peale sold us on the power of positive thinking. Inspirational advice from Reverend Robert H. Schuller, the minister for the Crystal Cathedral in Orange County, California, comes packaged in print as *Power Thoughts.* He also explains *Living Powerfully One Day at a Time.*

Lao-Tzu, the legendary Chinese philosopher who compiled the principles of Taoism more than 23 centuries ago, thoughtfully warned, "When the people don't respect those in power, then what they greatly fear is about to arrive." Chairman Mao clarified political power for everybody: "It grows," he said, "out of the barrel of a gun." For Mahatma Gandhi, who politically liberated India through passive resistance and nonviolence, power was entwined with "fear of punishment" and "acts of love."

We wear power ties and power suits to power lunches. We brand renegade leaders as power mad. The Department of Water and Power energizes the city, and when the power goes out, we're all in the dark. Our automobiles roll down the highway thanks to horsepower. Powerhouse candy bars deliver an energy assist, and Macintosh PowerBook computers transform deskbound writers into laptop cyberspace warriors.

On television, Superman was "more powerful than a locomotive." On the radio and in the pulps, The Shadow, gangland's doom and a master of disguise, had "the hypnotic power to cloud men's minds." On the silver screen, a carnival balloonist from Kansas, hidden behind a curtain in Emerald City's palace, postured with parlor tricks and a loudspeaker as "the Great and Powerful Wizard of Oz." Saturday mornings and toy stores now bring us Mighty Morphin Power Rangers and monopolize the attention of the kids.

Ralph Waldo Emerson, an American poet and essayist, concluded, "Life is a search after power," but even a short trip to the grocery store confirms it can be found on the supermarket shelves. Carpet Magic Rug Shampoo has "more cleaning power," while Classic Purex All Temperature Detergent offers "cleaning power you can trust." Ultra Purex Concentrated Detergent, on the other hand, is "more powerful." The "double power" in both Surf Concentrated Detergent and Wisk quantifies cleansing vigor for domestic engineers. Those endorsing a hands-on method in household maintenance no doubt will prefer Ultra Era Concentrated Detergent, "the power tool for stains." If, however, you advocate an astringent approach to laundry, call upon Shout stain remover and its "cleaning power of lemon." Also citrically empowered, Lemon Joy Dishwashing Liquid is "so powerful" that a "spoonful cleans a sinkful." The "powerized formula" of Windex Glass Cleaner promises a streak-free shine.

Our preoccupation with power is pervasive. From the wish-fulfillment fantasies of a playing child to the international influence of a world leader, power drives us to accomplishment and frustration.

A USER'S GUIDE TO COSMIC POWER

In a social setting, power is inevitably linked with status, rank, and responsibility, and anthropologist Clifford Geertz argued that for traditional societies, status in this world

The innermost courtyard of Pura Kehen, a Balinese temple compound at Bangli, is accessed through a split gate. Beyond it we can see the temple's tallest and eleven-tiered tower. The gate replicates the two key mountains of Bali, Gunung Agung and Gunung Batur. They are regarded as two halves of the cosmic world-axis peak, Mount Meru, and the pagodalike towers which mimic the multistory mountain are known as *merus*. (photograph E. C. Krupp)

conforms to the pattern of cosmic order. The universe itself is regarded as a hierarchy of power. Society mimics the structure of the world and insinuates human institutions into the natural rhythms of the cosmos. Geertz was analyzing the operation of symbols of royal power in nineteenth-century Bali, but his ideas apply more broadly. Because time and seasons were driven by the sky and because the world's directions were established by the sky, the vocabulary of power incorporated celestial metaphors. The Hindu rulers of Bali were sacred kings, and their palaces were temples where the power of cosmic order was enthroned. Cardinally oriented, the palace included a mound of earth that symbolized the world-axis mountain, and closer to the center of the palace could be found a world-axis shrine. It was called the *Ukiran.* This means "place of the mountain," and it, like the world-axis earthen mound, refers to Mount Meru. As the mountain at the world's center, Meru links the pole of the sky to the heart of the earth. On Bali, the gods are said to reside there. Palace gates are also references to the cosmic mountain. The split in the gate provides entry to the palace and suggests, at the same time, access to the supernatural realm of the gods, a zone the king can enter.

Across the Pacific, in Mesoamerica, the Maya built pyramidal platforms for temples in which their kings, through visions, communicated with the gods as if they were standing in the Mountain of Creation. The levels of these structures were reflections of the layers of heaven or the basements of hell. Their orientation fortified allusions to the ordering power of the daily movement of the sun. Fully costumed in a celestial bird headdress that stood for the Big Dipper, the king impersonated a cosmic axis in the guise of the World Tree. He held a double-headed serpent bar of sky symbols. With the sun god on one end and the Venus god on the other, the regalia of kingship displayed the cyclically driven power of the sky.

The pivot in the ceremonial landscape is the shaman, the chief, or the sacred king. Through calendar, ritual, symbol, and myth, shamans, chiefs, and kings invest the landscape with magical power from the meaning they extract from the sky. Through such transfers of power, they guide and govern the lives of those who reside on earth. The places they occupy and use are charged with magical celestial energy.

Such places may be semiprivate, like the rock shelters where California Indian shamans painted symbols of the sun and the stars. In the remote wilderness beyond Santa Barbara, at Indian Creek, a female figure giving birth to the stars was drawn inside a small cavity in the rocky hillside. The image probably has something to do with seasonal renewal and shamanic transformation.

On the other hand, these places may be public and conspicuous, like the central plaza at Calakmul, in southern Campeche, in Mexico. Between the sixth and ninth centuries A.D., the Maya arranged the platforms and pyramids around that zone according to the cardinal directions to mimic the world's order. A plaza away, they carved a flat, rocky outcrop with figures that appear to refer to the emergence of the

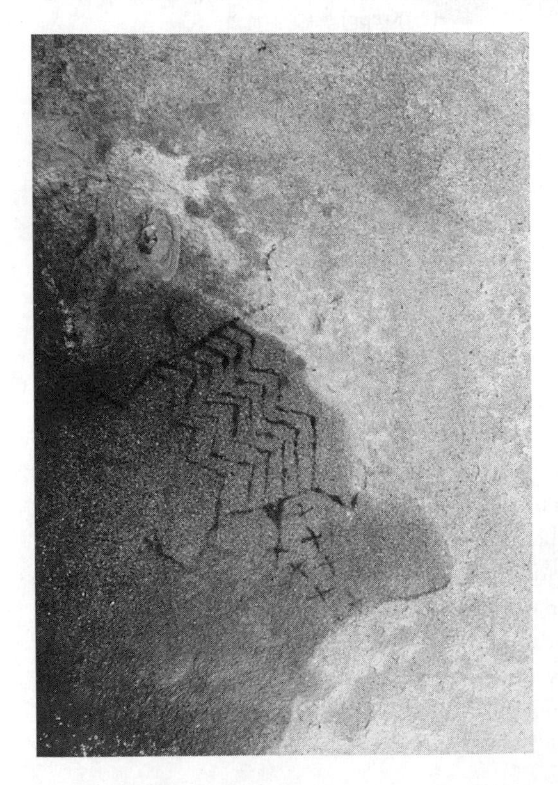

Pictographs at a remote Chumash site in southern California's Los Padres National Forest include a stylized female with a triangle between her legs to symbolize her genital zone. She is giving birth to the stars, a cluster of seven or eight crosses. (photograph E. C. Krupp)

primordial ancestor from the underworld through a crack in the earth-turtle's shell. First Father was reborn as the Maize God, who grew like a corn plant, out of the ground. As near as the Maya could tell, maize was what made life possible in the first place, and they said that the first people were made out of maize. The first appearance of the Maize God, then, is really the story of the origin of the present era in cosmic time.

Places like Indian Creek and Calakmul are junctions between earth and sky. They mirror and model the universe. All over the world power is concentrated in shrines, temples, palaces, and tombs. They are the sacred centers and cosmic modems of the ancient world.

We can understand, then, why cosmovision is an important component of kingship. The sacred places that symbolized the universe helped hold the kingdom together. Even in the less complex societies of village hunters and farmers, the sacred places that defined the world's boundaries, the world's center, and the world's sources of power were the locations of defining moments in creation, narrated by myth.

For shamans, chiefs, and kings, the connection between power and the cosmos was obvious. They were spliced at the sky. That did not mean that all power originated in the sky, but all power had to be rooted in the natural order. That order was most transparent in celestial rhythms and in their intimate resonance with the stability of the world and the seasonal dance of life. Celestial power was cosmic power. It seemed

A low outcrop of limestone in the plaza of Calakmul's Grand Acropolis resembles the earth-turtle's split shell from which First Father emerged as the Maize God in the highland Maya account of Creation. Nude figures carved in relief, including a pregnant female, may be part of the imagery of Creation. That imagery may, in turn, make Calakmul itself the place of Creation and so enfranchise its kings with a primordial charter from the gods. (photograph E. C. Krupp)

to establish the framework of the universe and appeared to guide all of the transformations of nature.

With arrivals and departures that could be anticipated and confirmed, celestial objects installed order into the world. The routes they followed and the itineraries they kept injected direction into the landscape and anchored buoys in the flow of time. Our ancestors figured this structure of space and time is what life required to exist at all, and so celestial power was associated with Creation. Then, with the insistence of cyclical return, celestial events—including the alternation of day and night, the phases of the moon, the seasonal pilgrimages of stars, and the annual migration of the sun—allied the sky with renewal. It was the renewal of time, the renewal of the world, and the renewal of life.

The theme of celestial renewal offered a valuable lesson. Astronomical information had practical value concerning circumstances that change for better and worse and about resources that come and go, but it also told people how the world worked. It was a compelling message for society. By modeling their institutions on the behavior of the cosmos, people could integrate their lives with the natural order. Knowing that they shared the world's dynamic equilibrium enhanced social cohesion. Seasonal ritual and ceremonial reenactments of Creation put people into the cosmic loop. These performances reinforced their understanding of the architecture and conduct of the universe. They were the rebar in the power structure of the community. Society was itself a collaborator on behalf of cosmic order.

Now, when we talk about the ideas of the ancients, or even about our own religious conceptions, we usually distinguish between what we call the natural and the supernatural. And we differentiate between natural power and magical power. Natural phenomena enjoy a physical and scientific relationship between cause and effect. Anything that breaks those rules, however, is supernatural, and scientific analysis banishes the supernatural from rational thought. Spirits and gods, then, are card-carrying members of the Transcendental Lodge of Beings of Supernatural Power and outside of the natural order. The ancients, however, wouldn't see it that way. The cosmos, with all of its spirits and gods, was entirely natural. The miraculous, or magical, was just another element of the environment, an exercise of one channel of power. In fact, everything was miraculous, but some miracles were more common—and some power more reliable—than others.

In an analysis of power in the belief system of the California Indians, anthropologist Lowell John Bean identified the fundamental character of power. Interpreted loosely, these aspects of power can be recognized in many other traditional belief systems. According to Bean, power is what makes things happen, but it is not a mindless, inanimate force. It is conscious. It also inhabits everything, but it is more plentiful or active or accessible in some places than in others. Everything that possesses power competes for more, but nothing can get the upper hand. There is, in a sense, a balance of power in the cosmos as a whole. Finally, the special role of human beings is their interaction with the sources of power. People help preserve the cosmic equilibrium by acquiring and applying knowledge and by engaging in rituals and magic. Symbol and ceremony are the mechanisms of cosmic power. Effective participation requires reciprocation and sacrifice. Offerings were made with the hope that benefits would be received.

Some of the power that seemed to be at work in the world was personified as gods. Storm gods, spirits of the game, and the seasonally-renewed Mother Earth illustrate this deification of nature. More abstract conceptions of divine power were developed as well, but gods were most frequently revealed through encounters with their physical power in the world around us. Their behavior in the sky was evident. The evening reappearance of the Pleiades brought winter rains to the ancient Greeks. In New Zealand, the Maori credited the moon with power for the tides, for they titled Rona, the woman in the moon, the "Tide-controller." Each morning the Egyptian sun god Re was reborn upon the eastern horizon, the winner in an all-night war with the armies of darkness and disorder.

Celestial power was high profile. It was billboarded in heaven, but that visibility didn't make it subject to human control. Although the sky could be watched, it could not be touched. Its power could be tapped but not challenged.

The sun, moon, and stars were powerful, supernatural agents of cyclical change. Their patterned movements made them the stage managers of the earth below. They decided where to locate stage left and stage right. They handled the lighting. They furnished the set, and the actors below followed their cues.

BARGAINING WITH THE SPIRITS

Although we may be inclined to regard the vocation of shamans as a religious response involving unrestricted traffic of the soul, shamans are more concerned with outcome than worship. Producing results requires power, and so, naturally, the acquisition, retention, and management of power keep the shaman alert for opportunity and hazard. Shamanic access to cosmic power is really a matter of commerce across transcendental borders. Recognizing that power is distributed throughout the cosmos, the shaman must be able to travel supernaturally to wherever power resides. Usually this is accomplished by trance. The shaman enters an altered state of consciousness and interacts with the spirits, often assisted by spirit helpers, supernatural allies willing to endorse the shaman's case.

Whether servicing the needs of a wandering soul, seeking a cure for a client captured by disease, negotiating with the spirits for the seasonal return of the game, or on a quest for knowledge of future events, the shaman has to understand the hierarchy of supernatural power. The shaman must know the terrain of this world and of those beyond. Cosmovision, or world view, provides the landmarks for the visionary journeys of the shaman's soul. Departures must coincide with places of unusual power to optimize success.

The best place of all is the center of the cosmos. Primordial creation is usually said to have occurred there. Power is focused there by the topography of space and time. It is a point of contact between the upper realm of celestial objects and sky spirits and the lower realm of subterranean waters and caves and of predatory monsters and spirits of death. In the shaman's universe, the center of everything is usually in our world, the middle realm. The center is rooted and nourishing and safe. It is regarded as navel. It corresponds to home and fire and familiar ground.

When the shaman goes into action, the site of all his mystical endeavor is contrived through symbol to imitate the heart of the world. By making the kingdoms of the spirits part of a sacred cosmic landscape, the shaman is able to move through them with the sophistication of a world traveler, on the lookout for trouble and ready for success. It's always most sensible then to start at the center when embarking into the beyond.

The most familiar and influential description of the mechanisms of shamanic power was formulated by Mircea Eliade in 1951. Eliade was an historian of religion, and in his book *Shamanism, Archaic Techniques of Ecstasy,* he treated shamanism as a religious response. With origins in the primordial paleolithic past, its traditions remained preserved among indigenous peoples who survived on the peripheries of complex societies and modern civilization. To develop his idea, Eliade drew primarily—but not exclusively—from shamanic concepts and practices in northern and central Asia, especially

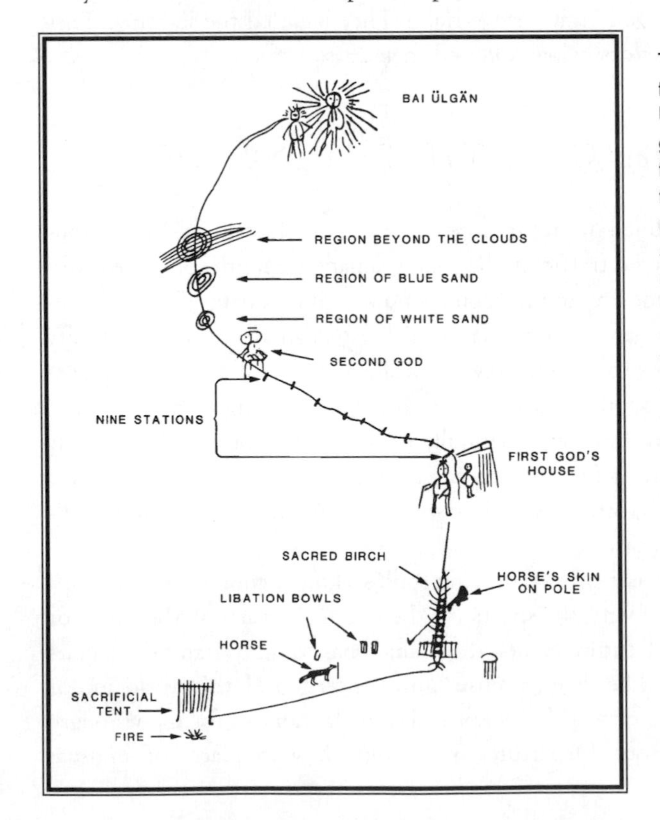

Turkic Altaic shamans in trance rode the road to Bai Ülgän, the supreme celestial god. The point of departure is the tent and hearth, and a little farther up the line (to the right) we encounter a horse sacrifice. Still farther to the right, the route turns straight up, and there the shaman must climb the nine steps of the birch trunk stairway to the sky. Above the top of the transcendental ladder, a deity greets the shaman at the door of his celestial house. Then nine more stations must be crossed. The shaman then negotiates three sandy celestial regions (sets of rings), the third of which is beyond the clouds. Finally, he reaches the radiance of Bai Ülgän and delivers the soul of the horse sacrificed for this audience with the highest god. (Griffith Observatory drawing, Joseph Bieniasz, after Uno Harva)

Siberia, Mongolia, and the Altai Mountains. The shaman's trance, he judged, is what sets the shaman apart from every other spiritual leader, for the trance allows the shaman to deal directly with the spirits on behalf of the living and the dead.

Eliade's consolidation and synthesis of shamanic belief identified its key themes and goals, but others, like anthropologists Nicholas Thomas and Christine Humphrey, have studied shamanism in historical and cultural context and have reached some different conclusions. Shamanism, they advise, is not the fossil fragment of a prehistoric religion but a religious approach that moves into the limelight when time and circumstance favor it. It evolves, in purpose and in practice, to tackle the expectations of the age.

Power, however, is still the issue, and Eliade's examples illustrate how a shaman's ability to cross cosmic boundaries is what permits the shaman to talk with the spirits and influence their policies. According to Eliade, the "shamanic" cosmos of the Turkic Altaic peoples was well suited—with its center and its edge, with its axis and its layers, with its openings and its connections—to accommodate the shaman's ecstatic celestial ascents to Bai Ülgän, the high god, and dangerous subterranean descents to Erlik Khan, the lord of the underworld.

Bai Ülgän's name means "the Great Rich One," and the Altai also called him "White Light." All of the supernatural helpers that give the shaman a boost to the high god are celestial spirits. Because the shaman is engaged in mystic flight to Bai Ülgän's lofty realm, the shaman also mounts an effigy of a goose and imitates its cry. In prayers, he speaks to the Birds of Heaven and affiliates them with the sun and moon. Other Siberians believe the soul, released at death, possesses the form of a bird.

The seven Sons and nine Daughters of Bai Ülgän confirm another theme of shamanic endeavor: the use of mystically significant numbers. Part of their magic must be celestial, for the Altaic sky is said to comprise seven or nine layers.

A birch pole, notched with nine steps and installed at the center of the Altaic shaman's *yurt,* or tent, represents the World Tree. This mythic Tree grows at the mythic Center of the World and is actually the polar axis. It links the world's navel to the hub of the sky. The world axis is the supernatural passageway between worlds and the route of the shaman's soul. Elsewhere in inner Asia it is equated with a pillar, or post, that supports the sky or with the cosmic mountain at the world's center.

The universe itself is sometimes likened to terrestrial architecture. Although explicit evidence for a cosmological metaphor in the Altaic yurt does not exist, the shaman's transformation of his yurt into a miniature cosmos argues for this kind of thinking. The ceiling of the yurt is the dome of the sky. Its floor is the earth. The walls are the horizon, and the pole is the cosmic axis. At the top of the tent, the smoke hole provides a doorway between worlds.

Unlike the priest who officiates in community ritual on behalf of the gods and the sacrificer who serves the gods through offerings, the shaman is a problem solver. He works on contract, spirit by spirit, soul by soul. As crafters of supernatural deals, shamans specialize in magical defense and spiritual cures. They escort the dead and carry messages to the divine. With divination and clairvoyance, they "see" and "find"

The handle of this Turkic Altaic shaman's drum has a face and fringed "sleeves." This figure represents the spirit of the ancestral shaman who is the "master" of the drum and who assists the shaman in his quest for audiences with spirits in heaven and in the underworld. The zone above the iron crossbar is the sky. Symbols painted there stand for the stars and other celestial objects. Bai Ülgän is the high celestial god of this shamanic tradition. His daughters are present in the row of hand-holding stick figures in the upper domain of the drum. (photograph E. C. Krupp, Peter the Great's Museum of Anthropology and Ethnography, St. Petersburg)

what others cannot. When they work for the hunters, they locate the animals and establish reciprocal protocols with the Master of the Game.

Shamans, then, aren't into reverence. They're into results. They are venture capitalists in supernatural power, but they don't option power for its own sake. Productive interaction with sources of power on behalf of the community is the point of their effort. When their people face challenges or encounter threats to survival, the shamans get down to business. Supernatural power, enhanced by magic and obtained through trance, is one of their tools. Ordinary power, acquired through observation of nature and knowledge of the world, is another. Effective handling of either brand of power inevitably confronts the shaman with the structure of the cosmos and the behavior of the sky. We shall look at a few examples to see just how it works.

CLIMBING THROUGH STARS, DROPPING THROUGH ICE

No less an authority than the Smithsonian Institution asserts, in Volume 5 of the *Handbook of North American Indians,* that the most powerful person in each village of the

Yupik Eskimos of southwest Alaska was the shaman. Like all shamans, he moved between this world and the spirit world to cure illness and influence the weather. He persuaded the sea mammals, the fish, and the game birds to return in their proper seasons, and he mobilized the ceremonial life of the community. Yupik communities were small. They relied almost exclusively on hunting, and most of the time each family operated independently. The shaman was their contact with the spirits and the one most familiar with their requirements. To deal with the spirits, he had to go to their neighborhoods, and that meant knowing how the universe was organized.

The Yupik pictured the universe as a series of rings, concentric worlds centered on their village, and the spiritual center of the village was the men's house. From this center, the earth spread out to the horizon, where the cosmos was creased and earth became sky. In the "skyland," normal seasonal associations were reversed. Wrapped for cold weather in summer and dressed in light garments in winter, the spirits of skyland clearly belonged to another world. Its residents included the souls of the game, and shamans would consult with the spirits responsible for them.

Traveling to the sky meant passing through the stars, which the Yupik believed were holes in the sky. In the story of one shaman's visionary journey to the heavens, the sky seemed to fill with falling stars and then descend upon him as he slept on the top of a hill near the village. When the space between the earth and sky left the shaman with nearly no room to move, he climbed through one of the star holes and saw another star-filled sky above him. Those stars, too, were holes lit by light from beyond, and the shaman pulled himself through two more star-perforated heavens

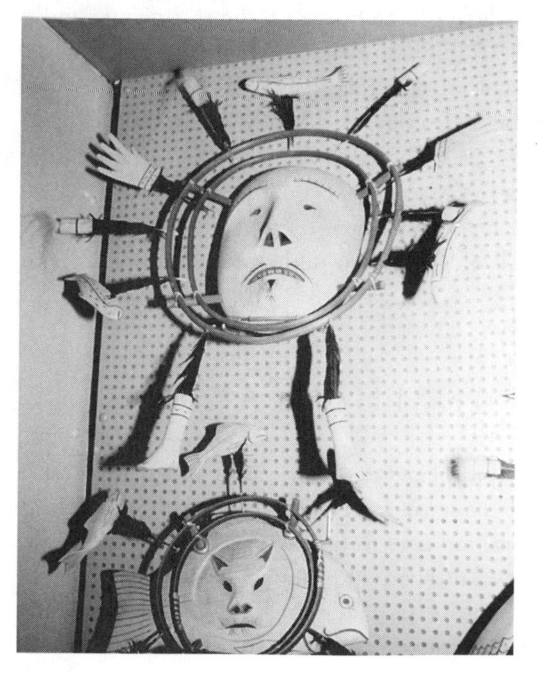

The Eskimo of southwestern Alaska reference the different realms of the universe with concentric rings. The rings put this spirit with a big face and small extremities at the center of the cosmos. The rings stand for the oceanic waters, the earth, and the sky. (photograph E. C. Krupp, University of British Columbia Museum of Anthropology, Vancouver)

and found he had reached the men's house of the spirit village in skyland. Inside were wooden carvings of all of the world's species of fish, birds, and mammals, and each man in the house was in charge of one of them. The men in skyland each controlled the spirits of the game. Great hoops, tasseled with feathers and fastened to rods, were hanging from the ceiling. The hoops stood for the different levels in the sky, and their movements were coordinated with the real skies. Now, the Yupik shaman learned that the sky spirits appreciate the attention of people on earth, and through a dream journey to heaven, he engineered a reciprocal agreement between people on earth and the people in the sky. When the shaman fell back to earth through the hole in the floor of the sky house, he organized a seasonal ceremony for the celestial spirits to ensure the supply of food on earth, and the inside of the men's house was fixed up to look like the house of the sky people.

As a terrestrial counterpart to the men's house of the sky people, the men's house in the Yupik villages is also equipped with portals to other parts of the cosmos. The smoke hole permitted access to the sky and other transcendental realms. When a person died, the body was moved to each corner of the smoke hole and then lowered through it. This process assisted the soul in its transformation from life to death and incorporated symbolic references to the cosmic axis and world directions as a road map for the journey. The spirits of the dead also periodically return to the men's house for important ceremonies, but when they do, they come through the central fire pit in the floor.

During the annual Bladder Festival, which coincided with the winter solstice, the shaman would exit the men's house by the smoke hole to have a chat with the seal people. His magical flight transported him to the ocean and through a hole in the ice, which was the skylight and smoke hole of the house of the seal people under the sea. There the shaman asked the seal people to make themselves available again in the next season. Believing that the souls of the seals resided in their bladders, the Yupik then lifted the bladders from hunted seals out through the smoke hole and carried them, inflated like balloons, to an ice hole. After the air was released, the bladders were slipped back into the sea so the seal souls could return to their underwater home. If good relations between the seals' world and ours were preserved through the shaman's mission, the souls of the seals would return in new bodies in time for the next hunt.

THE COSMOS IN THE CACTUS

Hallucinogenic alkaloids in psychotropic plants, such as the peyote cactus, also launched shamans into the company of the spirits, and Huichol shamans led pilgrims from their communities on real, physical journeys to find the vision-inducing plant. The peyote collectors, or *peyoteros,* traveled through three hundred miles of the Sierra Madre of western Mexico on the hunt. Peyote was said to be Elder Brother, the spirit master of the deer, who allows the Huichol, through visions, "to find their life." Both collecting the cactus and dreaming under the influence of peyote advanced the acquisition of shamanic power.

The Huichol now live in Durango, Zacatecas, and the neighboring states of Jalisco and Nayarít, but their peyote pilgrimages require them to travel to Wirikúta, their traditional homeland in the high desert of San Luis Potosí. It is "the place to the east where the Sun was born." It was also the place of Creation, the place where the First People—the first Huichol—once lived. After hardship in Paradise required the First People to migrate to the west, the First Shaman led these divine ancestors on the first peyote quest back to Wirikúta, the sacred center of the world, to renew their spiritual health. Deprived of the power of Wirikúta, the First People returned to their mythic home to become whole once again with the help of the peyote.

Although the Huichol now are settled cultivators of maize, they recall the nomadic hunting traditions of their ancient ancestors in Wirikúta. They stalk the peyote as if it were a deer. Every shaman-led hunt for peyote in Wirikúta is a return to Paradise and to the primordial beginning. Reentering the Garden with the help of a plant that induces an altered state of consciousness and with a shaman-priest who knows the landscape of space and time, the Huichol are spiritually renewed.

If Huichol campaigns in the sacred realm are really returns to the starting blocks of time, they are also encounters with the inauguration of world order. To get to that ancient era, the border between this world and the beyond must be crossed, and for the peyote seekers it takes place when they enter Wirikúta, the land of peyote. Huichol pilgrims reach this realm by shedding solemn tears at an entrance known as The Vagina and by negotiating blindfolded the Gateway of Crashing Clouds. The uninitiated would not recognize anything special in the landscape there, but the Huichol shaman knows where the frontier to the eternal mythic land is located and how to escort his companions across the threshold.

Barbara G. Meyerhoff and Peter D. Furst, two American anthropologists who took part in the peyote quest, have each written about the ritual and its meaning. Both mention the shaman's gestures to the four directions and the world's center. Power to pass from this world to Wirikúta relies, then, on symbolic references to the underlying structure of the universe.

According to the Huichol, the old fire god they call Tatewari, or Grandfather Fire, long ago planted four great trees at the corners of the world. A fifth, in the middle, holds up the center of the sky. Tatewari, then, ordered the cosmos, and with the help of the Huichol shamans, he keeps the sun in the high heaven where he belongs. The sun is also one of the gods, and the Huichol call him Tayaupá, or "Our Father."

Eating peyote is like having lunch with the gods. It is a life-transforming experience that demands discipline and imposes deprivation to achieve the sense of unity and the divine enjoyed by mystics and shamans. In their visions, the Huichol encounter their gods, and their shamans organize and supervise the process.

Understandably, the Huichol shaman who escorts his entourage to spiritual ecstasy is regarded as an individual of power. This power is inextricably linked to peyote's

A Huichol shaman dons a symbol of power by wearing prayer arrows. The shaman is on the way to the five-pointed star at the end of the road. This white star symbolizes the divine rock-crystal soul of the primordial shaman, who is associated with Creation. To retrieve the power of crystalline spirit, the Huichol shaman must journey to the sky. The four wavy lines wiggling down from the top of the arc he is following are the curtain of sunlight. He must cross the light to reach his destination. (Huichol yarn painting, photograph E. C. Krupp)

ability to inspire perceptions, sensations, and images that persuade the Huichol they have made contact with the gods and are operating in congruence with the cosmos. Returning home with the sacramental plant they have gathered for the rest of their community, the *peyoteros* actually become the First People. They bring their people back into balance by providing the peyote that will be used in the ceremonies designed to perpetuate the continuity of Huichol life.

As the leader of this quest, the Huichol shaman is equated with Tatewari, the First Shaman and Old Fire God. Just as Tatewari organized and stabilized the cosmos with world directions and a place for the sun, the Huichol shaman prepares and protects his fellow travelers. Back in his village, he also steadies and revitalizes his community through his practical knowledge and his spiritual expertise.

Celestial references affiliate a Huichol shaman's knowledge and power with the sky. In Huichol cosmography, the mountains of Wirikúta, the shaman's primary spiritual destination, are said to be the fifth and highest zone of the cosmos and correspond to the celestial territory of the gods. Sun was sung into heaven by the First Shaman, who made the sacred chair on which Father Sun ascended. Huichol shamans have their own special chairs, which derive some of their sacred power from the four upright support rods that mimic the corners of the cosmos. Their ritual drums rest on four legs.

Spirits of deceased shamans are believed to reside in heaven with the sun, but they may return to assist a descendant seeking magical power at the start of his shamanic career. The shaman-to-be must soar through the sun's blazing veil of light and collect

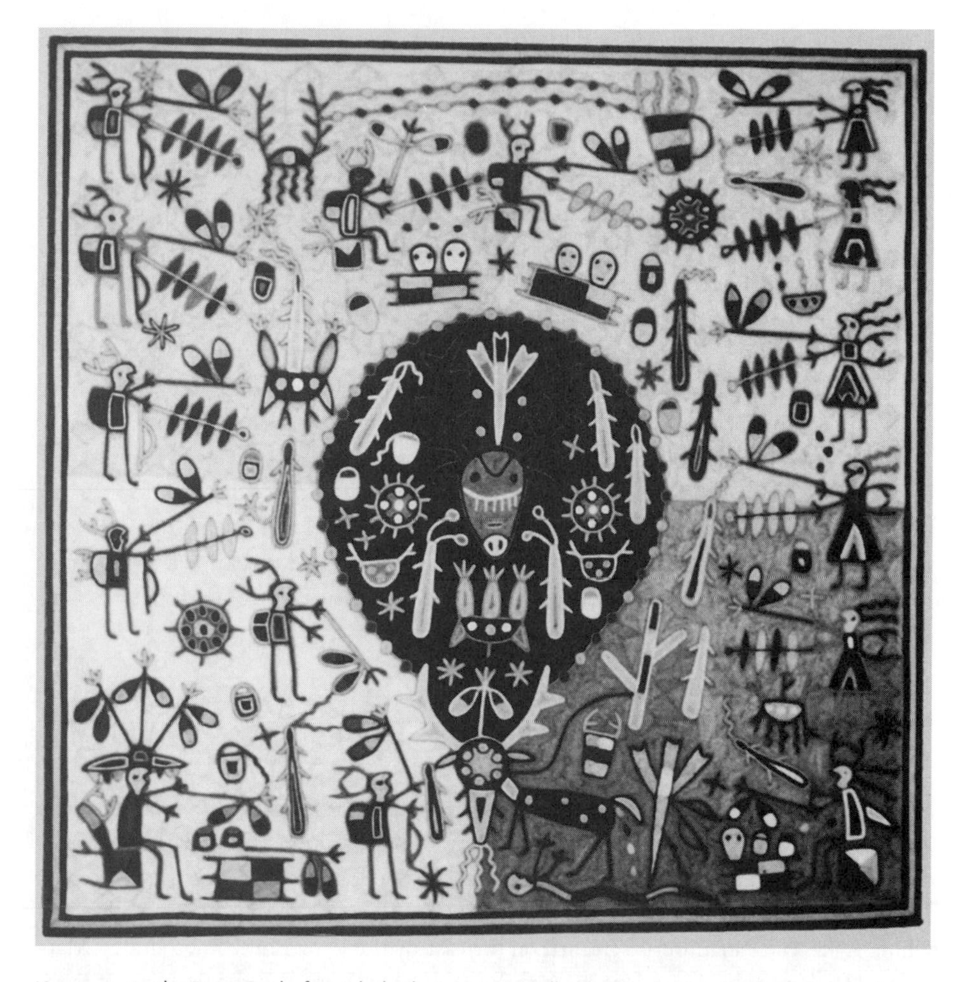

Kauyumarie, the Deer God of Huichol pilgrims, is actually peyote. As the shaman's helper, he resides in Wirikúta, the highest realm of the cosmos and the center of Creation. Four quarters of the world surround the central circle of the Deer God in this Huichol yarn painting. (yarn painting by Cresencio Perez Robles, photograph E. C. Krupp)

a divine crystal from the kingdom of Father Sun. This quartz, which the Huichol say is solidified light, is actually the ancestor shaman.

Celestial and cosmological imagery is also incorporated into the yarn "paintings" for which the Huichol are now so famous. Tayaupá, or Father Sun, appears often in them as a source of shamanic power and as the fertilizing energy of the cosmos. The moon, the earth goddess, and the four cardinal directions are sometimes included as well. Although increasing interest in Huichol art is now satisfied by commercial commissions, the art originated as a vehicle for communicating with gods. Yarn designs began as isolated symbols, but Ramón Medina, the Huichol shaman who worked with Peter Furst and Barbara Meyerhoff, invented a narrative style of yarn painting to illustrate his visions, the gods, and Huichol sacred history. Other Huichol artists have since developed this approach with paintings of extraordinary intricacy. Evolving from the shaman's visionary experience, much of this graphic work documents the shaman's interaction with the spirits and the shaman's relationship to the cosmos. Shamanic ideology drives shamanic ritual art.

Concepts, principles, symbols, and beliefs that define the nature of the world, the character of society, and the place of the individual all reveal ideology. The purpose, in part, of ritual art is restatement of this ideology, for repetition informs and reminds the members of the society what is most important to them. Reinforcement of cultural priorities enhances social cohesion, and the relationships between cultural priorities are understood through the belief system and world view.

What do the mechanisms of ideology have to do with Huichol shamanic power? They let us recognize at least part of the content of esoteric behavior. The power of the Huichol shaman does not operate independently, without cultural references to the perception of cosmic order and to the cycles of time. Huichol traditions allow us to know something about Huichol cosmological thought, and this knowledge permits us to appreciate the shamanic ideology in the myths, in the visions, in the rituals, and in the yarn.

PATHS OF POWER IN THE PANAMINT

Sometimes the journey for power is compressed into a ritual path that represents the kind of long spiritual trail the Huichol followed to Wirikúta. This may be the best way to understand what look like ritual mazes in the Panamint Valley of California. Located in southeastern California's vast Mojave Desert, between Death Valley and China Lake, the Panamint is screened on the east by the Panamint Range and by Maturango Peak and the Argus Range on the west. Few have studied the prehistoric relics of this peripheral region, but Jay von Werlhof, who teaches anthropology at Imperial Valley College in El Centro, California, and conducts archaeological research in association with the Imperial Valley College Museum, has examined, recorded, and analyzed rock alignments and giant complex designs throughout the California desert and beyond. Studying them from the air as well as from the ground, he has determined that the "Panamint Valley contains the largest number of rock alignments uninterrupted by space or imposing landforms known on earth."

California 190 is the only highway in the Panamint, and it cuts straight across the north half of the valley to its only settlement, Panamint Springs. It is dry, rocky, inhospitable territory, but in the late Pleistocene, about ten thousand years ago, its dry lake was filled by runoff from the rain and seasonal melting of the mountain snow. Climatic change started a slow but inevitable contraction of the lake about eight thousand years ago, and for the next six thousand years the foragers of the Panamint arranged their enigmatic rock patterns on the ancient fans of desert pavement above the natural drainage channels that fed the lake. Other signs of the Paleoindian presence in the Panamint have been collected by the archaeologists, but the large, nonfigurative combinations of curving lines and compartments are the most conspicuous evidence of early occupation.

Altogether, there are 32 large rock designs in the Panamint Valley, and whatever motivated the prehistoric Panamint people to set out stones on the desert floor inspired them with a grand sense of scale. The longest stretches 467 feet from one end to the other. Detail and complexity suggest that this earthen art incorporated meaning and function, but direct evidence that might reveal the art's purpose does not exist. Speculation guided by our knowledge of shamanic themes can, however, offer a plausible, if unprovable, interpretation.

The roots of shamanism are believed by most experts to reach back into the paleolithic era. The presence of shamanic techniques throughout the Americas, especially among the hunters, gatherers, and foragers in the most marginal territories, argues that these ideas were transported from Siberia by the primordial ancestors of the American Indians. Von Werlhof sees the Panamint designs as spirit trails. Designed and used by shamans, he believes, they direct passage to a focal point in the ground drawing, sometimes marked by a cairn, a circle, or some other distinctive feature. He equates that spot with an intersection of this realm with the world axis and regards it as a point of contact with the supernatural. Branches, breaks, changes of direction, and enclosures may reflect places in the minijourney that are dangerous or

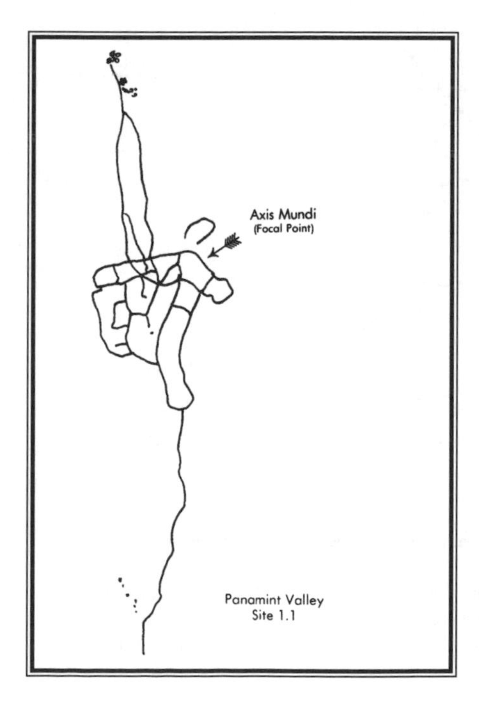

The large rock designs on the desert floor of the Panamint Valley seem deliberately intended to reject customary sensibilities of symmetry and balance. Instead, the convoluted, compartmented ground drawings are more like twisted trails, enclosures, and mazes. The "axis mundi" identified by researcher Jay von Werlhof is not really a geometric cosmic axis that connects earth and sky but a place that operates like a focus of the design and provides contact there with the supernatural realm. (from *Spirits of the Earth, Volume I, The North Desert* by Jay von Werlhof)

require a special approach. They may block some spirits and misdirect others but permit the authorized spiritual agent who knows the route to reach the final destination. Some of these lines of rock seem to have been paths that were walked again and again, perhaps by shamans and initiates. If so, they are examples of ceremonial use. Others may have fulfilled their purpose simply by being built. In that case, the process of construction linked this world with the power beyond.

Ideas about the spirit power of paths and trails are paralleled in ethnography from historic California desert groups, and although there is no direct connection between the recent and the prehistoric past, interaction with spirits in the quest for supernatural power is a reasonable theme in any discussion of the symbolic activity that culminated in these labyrinthine marks on the land.

Viewed as convoluted routes to the supernatural, Panamint ground configurations are like the ritual mazes that were walked in Europe for magical power. John Kraft, a Swedish scholar, has studied the European maze tradition for 20 years and reports that close to 500 ground labyrinths are known in the Nordic countries alone. Most are found near seasonal fishing camps and are probably only a couple of hundred years old. Weather control, enhancing the size of the catch, and securing protection from maritime peril were motives for marching around the maze. Older

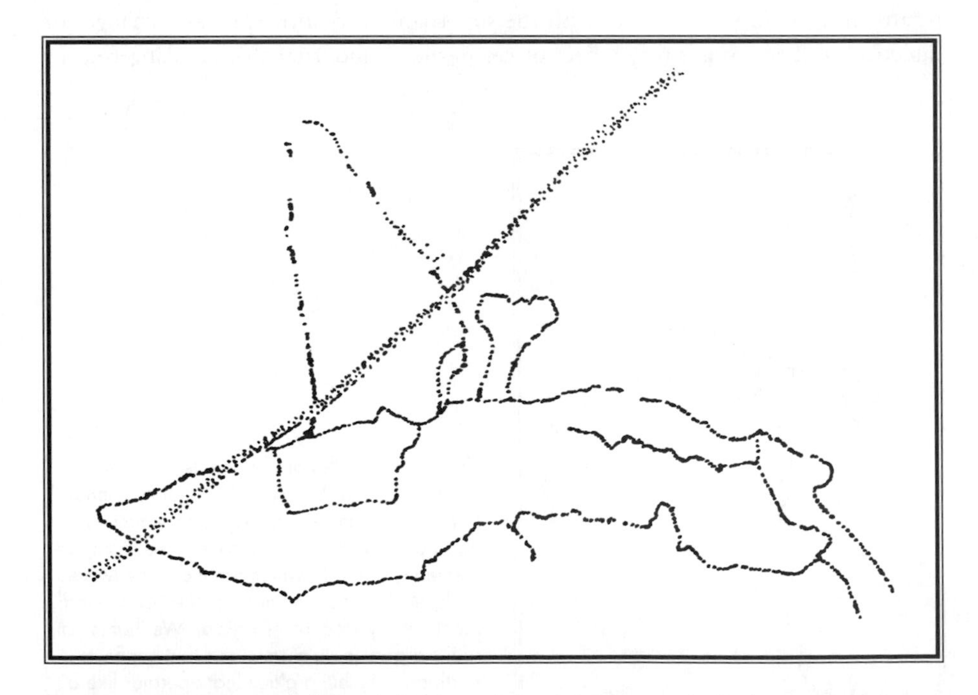

Site 1.15 in the Panamint Valley includes the prehistoric Wildrose trail, which cuts across the rock design. At one of the intersections, white and rose quartz was used to continue the outline across the Wildrose trail, but at others, the path was left open. (from *Spirits of the Earth, Volume I, The North Desert* by Jay von Werlhof)

mazes in the Swedish farm country are associated with prehistoric cemeteries and reflect a spiritual function in pagan cults. Kraft's broader analysis of a widespread European and Asian tradition of the retrieval of a woman or goddess imprisoned in the maze linked the labyrinth with seasonal ritual, the nether world, and seasonal renewal. The maze is an emblem of death and rebirth. Although such specific images and themes cannot be extracted from the stony Panamint paths, they reinforce a fundamental notion about blind alleys, false trails, and the difficult passage through a maze. They all protect a place of special power and limit access to it. That place symbolizes the transitional zone between one world and the next, and the maze is the gateway between them.

It was sometimes possible to pass through that gate by dancing the labyrinth. Ancient authorities tell us that youth on the Greek island of Delos performed a spiral dance around the Horned Altar to honor Apollo. Their tortuous steps mimicked the twisting passages of the original Labyrinth in Crete. Built by Daedalus for King Minos to contain the Minotaur—the monstrous offspring of Queen Pasiphae, the Labyrinth was penetrated by Theseus. This Athenian prince threaded his way to the center, killed the Minotaur, and emerged a hero. The mazelike dance, known as the *geranos*, or crane dance, was, according to tradition, introduced to Apollo's sacred island by Theseus himself on the way back to Athens. Behind this ritual and behind this legend lurks the theme of transformation through an encounter with a source of dangerous power at the center of the maze. Independently, elsewhere around the world, from Indonesia to medieval France, a maze dance symbolizes the journey and metamorphosis of the soul.

In *Sacred Dance*, Maria-Gabriele Wosien asserts that ritual dance simulates the ordered movement of the cosmos and plunges the participants into the time of Creation. Some dances imitate the motion of the sun or the circuits of the stars. Structured movement permits the dancers to transcend this world for an encounter with the divine sources of cosmic order. Rhythmic motion restates the cyclic pulse of sacred time. Dance, then, reestablishes a connection with the power of the gods and magically channels that power to earth.

If the Panamint compartmented rock paths were the work of prehistoric California shamans, they may also have been linked with shamanic dance. Shamans sometimes dance their way into an altered state of consciousness, and when they do, they travel a sacred itinerary. It is the most condensed journey for power. Through rhythmic movement, the shaman taps into what he or she believes is a source of transcendental power.

In Korea, the shamans follow a patterned step into a trance. Most Korean shamans are women, and they dance to ritual music performed on hourglass drums and small gongs by their assistants. Once they reach the ecstatic state, unconscious improvisation takes over. They dance in response to the spirits they encounter. Taking a more formal approach, Taoist adepts in Tang dynasty China actually danced a set of steps intended to let them tread the stars and "pace the void" on their quests for mystic power.

This prehistoric pottery from the seventh century B.C., an Etruscan wine container known as the Tragliatella Oinochoe, is unusually decorated. Two horsemen, a maze, a pair of copulating couples, and a woman in a fishnet dress encircle the jug's waist. A drawing on the other side of the vessel identifies the woman as the goddess Miamnucare. She seems to have had something to do with the moon, with the cyclical renewal of time, and with death. She is standing beside gravestones and holds an object in her hand. The other side of the jug tells us that object is probably a pomegranate, a fruit associated with the land of the dead and with resurrection. The odd, bald figure riding bareback and in tandem on the first horse has an embryonic quality that suggests a developing or newborn child. The X-rated activity of the two couples is an explicit reference to the process that produces new life. The riders' shields have birds on them, and birds were sometimes used to symbolize aspects of seasonal renewal. Everything in the company of the maze then has something to say about cyclical renewal. The maze from which the riders emerge is a classic labyrinth design, and it is labeled *TRUIA* in Etruscan letters. This means "Troy." In antiquity, legendary Troy often symbolized the center of the world, but the Troy maze also represented the journey of life from the center at birth and back to it at death. The center was the point of Creation—the place of birth, death, and rebirth. Taken together, then, the parade on the Tragliatella Oinochoe links the labyrinth and the center of the world with death and rebirth and seasonal renewal. The other images on this container deepen the symbolism, and Charles F. Herberger's analysis leads him to conclude that the labyrinth's themes of world center, creation of life, and seasonal renewal are all references to city foundation rites, ritual regicide, and the renewal of kingship. (Griffith Observatory drawing, Joseph Bieniasz, after John Kraft in *The Goddess in the Labyrinth*)

Of course, in many shamanic traditions, the shaman's own drumming provides supernatural transportation. The Yakuts of Siberia regard the drum as a horse the shaman rides between worlds. Some people decorated their drums with symbols of cosmic power. The drums of Lapp shamans often carry an image of the cosmic axis and the kingdoms it connects. In central Chile, Mapuche shamans quarter their drums with designs that symbolize the cardinal directions and sometimes include emblems of the sun, the moon, and Venus. The Selkup, a Uralic people of Siberia, might see a symbol of the Milky Way on their shamans' drums because the Milky Way was the soul's path to the sky.

This discussion of the shaman's intoxicating drumming and ecstatic dance began with a speculation about odd prehistoric ground designs in an obscure corner of the California desert. Although we can never be certain about the use of the encapsulated paths in the Panamint Valley, they have attributes shamans might exploit in their

Shamans often reach the Other World by dancing themselves into a trance. The dance is propelled by beating a drum, and throughout the world a drum is almost always part of the shaman's spiritual arsenal. This drum belonged to a Mapuche shaman. The Mapuche live in central Chile. Most Mapuche shamans are women, and the drums are always quartered by the world's four cardinal directions. Other symbols, like the four smaller circles here, usually allude to celestial objects such as the sun, moon, or Venus. (photograph E. C. Krupp, drum displayed at the Museo Arqueologico de Santiago)

efforts to acquire supernatural power. They provide the shaman safe conduct in this kingdom and permit him to cross the border into the next. To travel with the spirits like this, the shaman has to know the hierarchy of heaven and the corners of hell. It requires a knowledge of the geography of cosmic order to negotiate transcendental terrain. That knowledge is the shaman's passport to power.

CHAPTER THREE

CENTERS OF CREATION

The stars are great hunters," explained Dabé. He was a South African Bushman, and his ancestors had hunted for ten thousand years on the Kalahari Desert that straddles the Republic of South Africa, Namibia, and Botswana. Long ago, when the world the Bushmen know was created, those starry hunters were among the First People, but now that the Bushmen reside on the earth, the stars chase their quarry in the spirit kingdom overhead. To the Bushmen, the stars are relatives, and that sense of kinship—of belonging to the earth and to the sky—is a reflection of the intimacy with which hunter-gatherers live with the natural environment. They know it well, and that knowledge is one of their tools for survival.

Before feeling the pressure of the more numerous Bantu tribes and the European colonists who followed them, the Bushmen occupied most of southern Africa, from the Cape of Good Hope to the plains of Kenya. Rock paintings attributed to them are found throughout that area. Beginning in the seventeenth century A.D., they were forced to retreat from the Drakensberg range, where some of their most elaborate and elegant pictograph rock shelters are found.

The Bushmen have no calendars, but they recognize the seasons of the stars. Like the Bushmen, the stars are nomads and hunters, and the Bushmen are aware of their comings and goings. Dabé introduced Laurens van der Post to those stalkers in the sky. Van der Post's books on the Bushmen include *The Lost World of the Kalahari* and *The Heart of the Hunter*, and in the second van der Post describes his encounter with the old

Bushman and his small group. They were camped in the thorn and the scrub while on their way toward distant lightning and the promise of a little rainwater. Pointing first to Dubhe, the upper star at the front of the bowl of the Big Dipper, the Bushman said, "That star was a great hunter who hunted in faraway dangerous places in the shape of a lion." The conspicuous radiance of Sirius, he said, was a sign of the hunting ability of the brightest star in the night sky. Sirius sat there in heaven, fat from all the celestial game he caught. Sirius was not, however, the greatest starry hunter. The highest acclaim belonged, as the Bushman explained, to a star that "was not there yet." "It hunted," he said, "in the darkest and most dangerous places of all," so far from them at that moment of the night they could not see it. It would appear, however, "in the early morning when it came nearer on its way home." The greatest hunter was the Dawn's Heart, that is, the planet Venus as the Morning Star. You could spot him just before dawn, when he was "striding over the horizon, his eye bold and shining, and arrow ready in his bow." As he continued forward, "the night whisked around to make way for him." The red you can see advancing before the sunrise is the dust kicked up by his feet. Certainly this is a very anthropomorphic vision of Venus.

Like the stars and the animals and so many other elements of nature, Venus was a living being, a person, not a thing. The Bushmen looked at their neighbors in nature as so many other hunter-gatherers regard them. They are the First People, those who were here at the time of Creation and helped make the world the way it is. The Chumash Indians of southern California also spoke that way—about the sun, the moon, the North Star, the bear, the coyote, the swordfish, the centipede, the eagle, the condor, the datura plant, and many more. In the time of Creation, the animals and all of the other supernatural beings could talk. They all were people then, and they had special power. It is still possible to interact with them and possibly influence their behavior.

Canopus, the second brightest star of the night, is in the southern constellation we call Carina the Keel. The Bushmen call Canopus "Grandmother" and chat with it about the seasons. It is a stellar trademark during the summer nights of the Kalahari, but the star is low or out of sight during the winter. The Bushmen associate it, and Sirius, which more or less accompanies Canopus, with the white larvae of the ants. Regarded as one of the most delectable of foods in the desert, the ant larvae are "carried" by Grandmother Canopus. Offering Grandmother Canopus a little flame from his winter campfire, perhaps as she reappeared in the dawn as the season was beginning to turn, the Bushmen hunter talked to her. He encouraged her to make a case to the sun for more warmth, which both he and Canopus could use to dry the ants' larvae and so bring them to the table. This kind of commerce with the stars, then, is connected with food, and seasons, and foraging. The Bushmen consulted their well-fed relatives in the sky, the great hunters, about the meals to come.

FIRST, CATCH YOUR ELAND

Some spirits, however, may not be reached through campfire conversation. Mantis, for example, was the protector of the antelope, especially the great eland. He was also the first spirit and the creator of the present order of the world. Although he sometimes appeared as the praying mantis, the unintimidated insect the ancient Greeks associated with seers and divination, Mantis also could take the shape of the eland, the largest of the antelope and the animal that loomed largest in Bushmen belief. Mantis created the eland and went out of his way to prevent the Bushmen from killing it. They did occasionally manage to take one down, thanks to the supernatural power to control the game their shamans acquire in a trance. This power supplements the power of Bushmen hunting tricks—imitative sounds and movements intended to attract the eland's interest, the power of their knowledge of the eland's behavior, and the power of the poison on the Bushmen's arrows. When an eland is killed, Mantis, furious with the death of his favorite animal, swoops in on the carcass and disturbs the hunter's spirit with confusion and regret. Other men then join the hunter and, dancing by the dead eland, absorb some of its power. The quarrel with Mantis is diffused further by a shaman, who, in trance, heals everyone present of any supernatural hits they may have taken. It takes, then, special power—the shamanic power of dreams and trances—to negotiate with the spirits, especially the spirits of Creation.

Most accounts of the Bushmen written today use the name San to identify them, although the names of subgroups, particularly !Kung and /Xam, are often employed to distinguish the northern peoples from those in the south. The Dutch gave them the name Bushmen in recognition of their ability to disappear into the bush. The Bantu noticed the same talent and judged it was a product of their supernatural powers, their ability to talk with the spirits. The San believe they access this power in premonitory dreams and in the well-known trance dance. When this ritual is performed, the men circle all night around the women, who are seated around a campfire. The women clap in time and hum, and in time one man, and then another, falls in trance. In part, the reaction is probably due to hyperventilation. The dancers sense a spirit energy rise from deep in the stomach, froth up the spine, and spill into the head. About half of the men are shamans, and when the trance strikes them, they falter and drop to the ground. The others continue to move, from one person to another, as if possessed by a spirit. The dance is really a curing ritual, and diseases, which are mostly judged to be infirmities of the soul, are extracted from the participants by the trembling touch of the dancers.

Meanwhile, on the ground, the unconscious shamans are in a state the San call "half death." In this phase, the spirit enters the earth at a water hole, tunnels through ground, exits the surface, and glides up the threadlike routes traveled by both the gods and the dead as they commute between earth and sky. The celestial destination of the spirit in trance is the place of god. According to one San dancer, you have to make yourself very small to enter that supernatural place, and once there, "You do what you have to do there, and then return." What the San trance dancers experience is an ecstatic shamanic trance, an altered state of consciousness that puts them in supernatural territory and confers on them the power they need to protect and serve their nomadic communities.

As far as power goes, it's nearly an egalitarian world for the San. Their communities are small and mobile. There are no chiefs or leaders. Perhaps the most lofty title bestowed on anyone is "grandfather" or "grandmother," in deference to their age and experience. The oldest members of the camp provide advice, especially about decisions that relate to the timing and direction of movement through their hunting and gathering territory. Consensus rather than coercion resolves communal issues and disputes. There is no one in command. Some individuals have a greater gift for trance and dreams than others and may, therefore, command more supernatural power, but the opportunity for access is open to all of the men. In any case, it is used on behalf of the entire community. The San are too few in number and too dispersed and too open to their personal kinship with the cosmos to exclude or elevate another member of the group from contact with the spirits.

While studying the economics of a remote !Kung band's foraging strategy, anthropologist Richard Lee asked one of the men if they have a headman. With humor he answered, "Of course, we have headmen. In fact, we are all headmen; each one of us is a headman over himself." We talk a lot about rugged individualism in the complex, urbanized, technologized, and commercialized environment of our own lives, but the Kalahari insists on the genuine article.

Normally numbering about 30 persons, an optimum number for social stability and efficient foraging, a !Kung San band gathers and hunts its way from one temporary desert campsite to another throughout the summer. Tempered by experience, the group's decisions are also informed by ongoing observation of the local availability of wild plant resources, which account for nearly 70 percent of the diet, sightings of game, and monitoring the tracks of animals. They understand the land as a domain defined by their migration and bounded by the hunting grounds of other groups.

The land is also defined by its seasons. Southern Africa, of course, is in the southern hemisphere, and so summer extends from October to May. For the San, summer is the wetter season, the time when there is more water to be found. During dry winter, when the temporary water holes give up the ghost, several bands recongregate into a much larger group around the permanent water sources. For the Kalahari Desert nomads, the winter regathering of camps—up to several hundred persons—must seem urban in contrast with the rest of the year. Winter is also the usual season for public activity like the trance dance.

For the San, the world is a territory of moving animals, passing events, water holes, and other landmarks. Places are more important than a center and an edge, and every place is touched by the spirits. It is the cosmos of foraging nomads, who carry the center of the world around with them in their heads.

San culture is software—the dances, the songs, and the stories. Except for their rock paintings, their culture is in their minds. As nomadic hunters, they can't afford the burden of possessions. They transport only the necessities—utensils, weapons, water, and tools. Sometimes their travels bring them to the places they had elected to paint—open-air shrines where images of elands, dancers, and people transformed into animals appear.

Recent detailed studies of the content of San rock art have convinced researchers, particularly J. David Lewis-Williams and Thomas Dowson, that the animals in these paintings are not game the Bushmen intended to hunt but metaphors for the trance in which they tracked the spirits. The eland is a stately, refined creature in which the San see themselves. Unlike many of the other hoofed creatures of the African plain, the eland travels in small groups—like the San hunting camp—not in great herds. The eland's split hoof flattens for better footing in the desert, and the San wear sandals of animal hide to help them negotiate the Kalahari. The San say, in fact, that the eland was created from a discarded sandal. When a !Kung girl has her first menstruation, the women place her in a special shelter and perform the eland dance around her without their usual aprons. They wear eggshell bead tails and mimic the movement of eland cows. Their nudity is a statement of sexual accessibility, which is what the girl's transformation into an adult is about. The women apply eland fat, a substance to which they attribute intense power, to the girl and paint special antelope designs on her face. The one male dancer is a bull eland impersonator. He reveals his genitals to the debutant in a ritual that is part sex education, part coming-of-age, and part endorsement of the continuity of the camp.

The eland in San belief is saturated with supernatural, or shamanic, power. For that reason, elands in the San paintings share many elements with shamanic trance. They are sometimes shown dying, with their hair on end, for the trance is like a death. Blood sometimes sprays from the eland's nose, and the trance dancer's nose bleeds, too. It takes hours for the eland to succumb to the poison in its wound. The men energetically dance for hours, and as overheating and lack of oxygen overtake them, they tremble and stumble as a dying eland. Some dancers appear in the paintings to be half-transformed into elands, and fly whisks and medicine bags, which are shamanic and trance dance paraphernalia, are included in the pictographs. Because the eland has great trance and dream power, the paintings of elands also are sources of power. Enhancing the effectiveness of the shaman's action in the trance may have been one function of the rock art sites. In any case, they document the analogy in San thought. Shamanic power—to travel-out-of-body, to make rain, to cure illness, to coax the game, and to disarm enemies—is like the eland, the great power animal. Portrayals of the eland's death are images of the shaman's trance. The shaman must be a great hunter to kill the eland, and if the stars are great hunters, they, too, must possess enormous supernatural power.

To understand the source of the eland's power, we have to return to the Creation. In the story of the first eland, we see cyclic renewal at work. Cyclic renewal is the pro-

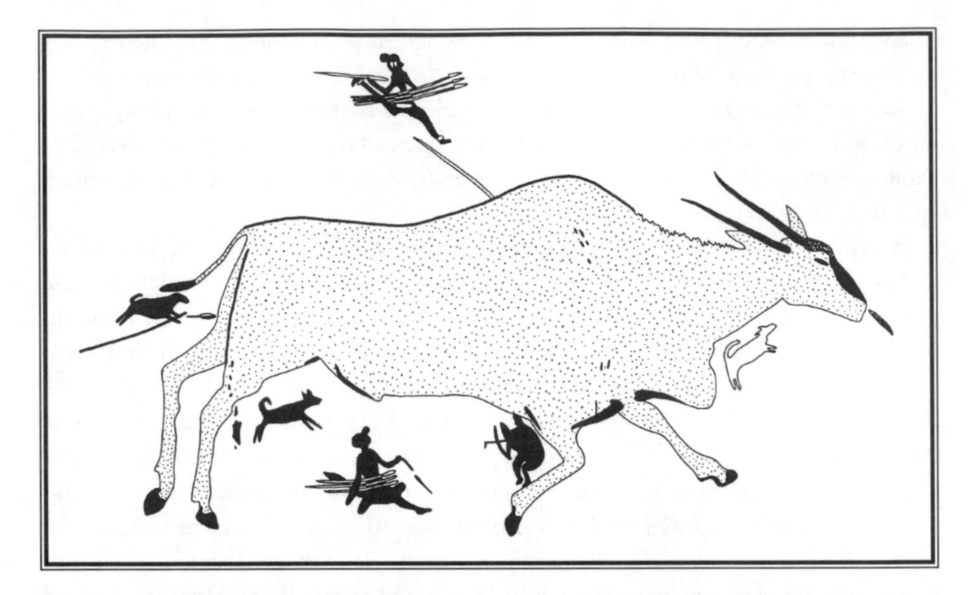

Elands figure prominently in the San rock art of southern Africa. Superficial interpretation of the images would turn them into hunting magic, but the eland hunted here is not a meal. Rather, it is a metaphor for shamanic trance, an altered state of consciousness the San seek to acquire spiritual power. With its head lowered and bleeding from the nose, the wounded eland in this rock painting, from Martinsdale, Barkly East, is dying. Men in trance stagger and sweat like the dying eland, and their nostrils sometimes bleed or foam. (from J. David Lewis-Williams, *The Rock Art of Southern Africa*)

cess of Creation, activated by the supernatural power of the conscious spirit, the Creator who imagines the future and then brings it to pass. This is the power of planning, the power of dreaming, and the power of artful endeavor. It is recognition of connections between things and acting on them. It is a distinctly human capacity, but people get the world started by consigning it first to the Creator. In time, human beings enter the picture and get to share, at least to an extent, this power of spirits and gods. The San acquired their portion from Mantis, the Creator, through the death of his first eland.

Wilhelm Bleek, a German linguist, and his daughter Dorothea Bleek, along with Bleek's sister-in-law Lucy Lloyd, had the vision to perform an indispensable service over several decades on either side of the start of the twentieth century. They became closely acquainted with /Xam Bushmen, collected their traditions and tales, and recorded them for the rest of the world. Several versions of the story that tells how Mantis created the eland exist, but the basic theme is the same. The eland is the symbol of the power of transformation and renewal, the power first orchestrated by Mantis. We find the center of the Bushmen world, not in the geography of southern Africa, but in that story.

Mantis made the eland out of an old shoe. He picked up a sandal discarded by a character named Kwammang-a, whose exact identity is unclear. He is said to be something intangible seen in the rainbow and seems to have something to do with consciousness. He may be the nephew of Mantis.

Mantis dropped the sandal in a pool where reeds grew. There it soaked. Mantis returned to check on it a couple of times and observed its growth. When he discovered eland spoor by the pond, he realized the sandal had emerged from the place of creation and birth to graze, and he knew it had transformed into a living being, a "person." Mantis waited, and when he saw his new creation, he called to it, the way hunters call the game. The eland came over to him, and then Mantis left to find some honey.

The San associate the strong, distinctive, and sweet scent of the eland with honey, and the smell of that aroma is like a transfer of power. When Mantis returns to his eland and rubs it with honey until it shines, he is charging it with supernatural power, the kind the shamans seek in trance. Mantis invests his eland with honey power on two successive nights, each time before the sun rises. The sack in which Mantis carries the honey is reminiscent of the medicine bag the shaman uses, and the nocturnal application of honey parallels the continuation of the trance dance through the night.

After an absence of three nights, Mantis returned to the pool again for another eland encounter. Then, on the next day, he decided to take Ichneumon, who is one of Kwammang-a's sons, to see the eland. The ichneumon is a member of the mongoose family and agile killer of venomous snakes. Mantis, however, is a trickster and doesn't tell Ichneumon exactly what he has in mind. The pair heads for the water hole, where Mantis blankets Ichneumon at midday and directs him to go to sleep. Ichneumon is crafty enough to figure something is up and does not doze. Instead, he catches a glimpse of the eland and calls it to stand. Ichneumon may be smooth, but he doesn't have the Creator's power. The eland ignores him and walks away. Mantis asks Ichneumon what he thinks he has seen, and he replies, "A person is yonder." If this were true, it would be quite a revelation of magical power, but Mantis denies what has occurred. It was not a person and not creative magic but just a piece of the shoe Kwammang-a dropped.

Ichneumon does not believe Mantis and reports what he has seen to his father. Together, he and Kwammang-a go back to have a closer look at the eland. When Kwammang-a saw the antelope, he realized that Mantis was up to his usual tricks. There was new magic afoot in the world, and to acquire some of it Kwammang-a and his son killed the eland on the spot.

They were butchering when Mantis returned. He was very disturbed, not because the eland was dead but because the eland had been killed without him. The eland's death is not exactly the death of an eland. It is the shaman's trance and transfer of power. The eland will return. Death of the older generation is a natural and necessary consequence of the process of creation, but death is not annihilation. The power of cyclic renewal repopulates the world with the next generation of elands. What is really at stake here is the Creator's prerogative and power. Kwammang-a and Ichneumon have appropriated some of that power for themselves and are like the San who, in trances, seek supernatural creative power—the power that cures and that feeds and so sustains life.

It is not exactly a satisfying moment for Mantis. There are now other tricksters in the world, individuals who possess the knowledge and power of life's rhythms. Bitter over what has transpired, Mantis discovers his eland's gallbladder, perhaps dangling on a bush, rejected as the meat was cut after the hunt. Mantis pretends to return home with Ichneumon, Kwammang-a, and the meat, but through a ruse he doubles back to the gallbladder. Despite the warning spoken by the gall, Mantis slits it, and it explodes dark and bitter fluid over everything. Blinded by the gall, Mantis staggers and reaches around in the murk. By chance, he touches an ostrich feather and brushes the bitterness and darkness from his eyes.

Mantis has endured his own transformation and renewal. He tosses the feather that returns his sight into sky and instructs it on its new responsibility:

> You must now lie up in the sky;
>   you must henceforth be the moon.
> You shall shine at night.
> By your shining you shall lighten the darkness for men,
>   until the sun rises to light up all things for men.
> It is the sun under whom men hunt.
> You must just glow for men, while the sun shines for men.
> Under him men walk about;
>   they go hunting; they return home.
> But you are the moon;
>   you give light for men,
>   then you fall away,
>   but you return to life after you have fallen away.
> Thus you give light to all people.

The /Xam who told Bleek this story finished it off with a simple, and explicit, message. "That is what the moon does: the moon falls away and returns to life, and he lights up all the flat places of the world."

The ever-changing moon—growing through half of the monthly circuit and dismembered day by day through the rest, until it reappears again after three or so days of complete absence from the sky—is the foremost celestial emblem of cyclic renewal. It is then an agent of creation and renewal. With poetic redundancy, the !Kung reiterate this fundamental principle of life when they talk about their hunting bow. Of course, it is the invention that extended the reach of the mind to the flesh of the game. Although it is a weapon, an instrument of death, it is also linked with the imaginative spirit of creation. Laurens van der Post asked the !Kung why they see the bow this way, and they told him to look at the new crescent moon. The bow has the shape of the moon and carries within its curve the moon's power of rebirth. Through the contest of the hunt, the San are immersed in the great force that drives the world. The fate of the first eland tells us this force is the process of creation. The origin of the moon at the end of the tale makes sure we get the point. This process is repeated throughout the natural world and in people's lives. Although there is bitterness, darkness, and death, there is rebirth that is eternally activated by the supernatural power of the Creator. That is the power that replaces the elands and all of the other game the San hunt with new animals and that replenishes the larder of the earth with another cache of ant larvae, mongomongo nuts, roots, berries, and all of the other plants in the San diet. It returns water to the earth with seasonal rain. It is no surprise that people covet that power, like Kwammang-a and Ichneumon. They reach for it in all of their interaction with the world and find it in their knowledge of nature and in the visions of the trance. After the Creation, Mantis went away. Although he no longer lives upon the earth, his power is felt, especially when he holds the hunters accountable for his elands. But there is now a little of Mantis in the San. What happened to the eland happens to them. With the eland's power, they, too, are tricksters, transformers, dreamers, and conceivers. The real miracle is human consciousness, for it is the power that allows people to see meaning in their lives. It is ignited by the idea of creation and the promise of cyclic renewal.

It takes power, process, and a place to create the world. The place, like the players, exists in myth and has a counterpart in the landscape, for that, after all, is where the story was staged. As a mobile enterprise, the !Kung hunting camp, perhaps the smallest corporate group in which people operate, has an ambiguous concept of edges and centers, but there are places that are like the center of the world, places of numinous power that kindle a sense of awe in their concrete and theatric revelations of the world's real mystery. This experience, according to historian of religion Mircea Eliade, is the true seed of religious sensibility. Sometimes places like these are marked by people such as the !Kung with paintings and carvings and other signs intended to acknowledge, release, and access the power of the place. In *The Lost World of the Kalahari,* Laurens van der Post detailed his encounter with one of these spots, the Tsodilo (Slippery) Hills in the northwest corner of Botswana.

There are rock paintings in the Tsodilo Hills. The first pictograph van der Post spotted there was an elegant and huge eland bull, but today's Kalahari San take no credit for the paintings. Most of the pictures on the rocks conform to the San pictograph style elsewhere, but there are some Bantu paintings as well. Bantu peoples also attribute sacred power to the site, especially to the water that permanently collects on the summits of the hills. Collectively, the hills are called "copper bracelet of the evening" by the !Kung, who also treat the hills as a family. They name the three largest "Male," "Female," and "Child." The fourth is too young yet to have a name. Each year !Kung groups gather together there for a short stay. Hunters approaching the four distinctive hills were prohibited from killing game. Van der Post's guide to the Tsodilo

Elands and other animals symbolically important to the !Kung Bushmen of the Kalahari are painted on the rock faces of the Tsodilo Hills in Botswana. The eland embodies access to shamanic power. (Griffith Observatory drawing, Joseph Bieniasz)

Hills was not a San, but he knew the place was "the home of every great and ancient spirit" and that the master spirit lives on the central hill. A pair of deep depressions on one section of rock reveals where the Creator knelt on the day he made the world. Near the pond up top, van der Post was shown a tree he was told was "the tree of true knowledge." The hills are inhabited with spirits. The spirits of all the animals occupy separate compartments in the rocks.

Van der Post was on the track of the !Kung and planned to film whatever he could find at the Slippery Hills. Every one of his camera magazines jammed, however, and he never did manage to capture the place in motion pictures. The guide who brought him there was also known as a healer and a prophet, and he elicited information about their difficulties from one of the spirits there. The party had made some serious errors, but van der Post, with the advice of his guide, performed a conciliatory gesture and penned a letter to "The Spirits, The Tsodilo Hills" and apologized for their transgressions. A second interview with the spirits confirmed that the effort was regarded favorably. Later, when they were back at the guide's home, he shared with van der Post his understanding of the flow of power: "The spirits of the hills are not what they were, Master. They are losing their power. Ten years ago they would have killed you all for coming to them in that manner."

The Tsodilo Hills are not the center of the world for the Kalahari Bushmen. They are nomads, always on the move, and they carry the center of the cosmos around with them in their heads. Those solitary hills, however, are one of the places where the world of the eland's power once began.

The eland in the previous illustration is part of a group of animals painted on a large rock panel high on the "Female" hill of Tsodilo in Botswana. It is to the right of the giraffe at the left of the rock face, and there is another eland bull above the giraffe. (photograph Nancy Cattell)

THE WORLD IS YOUR CLAMSHELL

According to the Haida, the Pacific Northwest Coast Indians who live on Canada's Queen Charlotte Islands, it was Raven who actually created the world's first land. Before there was any ground on which Raven could alight, the world was ruled by Power-of-the-Shining-Heavens, the vast, blue god who lived high over the seas, and Raven was his servant. Flying above the broad, unbroken ocean, Raven touched down on "a little thing" that materialized within the swells. Just by speaking, he could make things happen. He said, "Become dust," and that "little thing" turned into sand.

In another version of the story, the Haida say that the ocean's surface was swept out to the clouds by the beating of Raven's wings as he hovered over the waves. The horizon formed where the water met the sky, and the waves pushed aside just beneath Raven's claws were transformed into rocks. As time passed, some of the rocks turned into sand, trees grew, and the Queen Charlotte Islands—the world of the Haida—were created.

The Haida cosmos included a flat, circular earth, surrounded by the oceanic waters of the underworld. Beneath the waters of the ocean, Sacred-One-Standing-and-Moving sat upon a copper box and supernaturally supported the islands upon the ocean's surface. A great central pole was anchored in his breast, and this Pillar of the Heavens held up the dome of the sky above the five layers of heaven.

A great World Flood later forced Raven back into the sky. While airborne once more, the monotony of being alone in the world began to sink in. There might not have ever been any people on earth had Raven not been bored. The waters from the great World Flood had dropped enough to expose some land again, and he caught the Queen Charlotte Islands in his eye. He dropped down to the sand for some beachcombing, but the islands didn't seem to offer much more entertainment than a surf-side stroll.

The absence of any interesting action along the beach annoyed Raven. He turned his beak toward the sky and complained to heaven. Heaven, or one of its emissaries, must have answered, for Raven heard a muted noise that at first he thought originated overhead. As he looked around for the source of Heaven's voice, he suddenly saw a large clamshell, partly wedged in the sand. As he picked it up with his beak and inspected it, he discovered that there were many small wingless and featherless creatures inside. To them, Raven looked like trouble, and they all backed against the far side of the shell. Raven, of course, realized these beings had entertainment value and tried to coax them out. His invitation was at last accepted, and creatures in the clamshell emerged for the first time into the sunlight.

They were, in fact, the Haida. By their own traditions, they were the world's First People and the ancestral inhabitants of the Queen Charlotte Islands and part of the

Raven possesses the creative, transforming power of the shaman in the tradition of the Pacific Northwest. When he paid his call to the unpopulated Queen Charlotte Islands, he discovered a clamshell bulging with human beings and coaxed the First People out into the world. Haida artist Bill Reid carved this event in wood for the University of British Columbia Museum of Anthropology, where it is an anchor for a stunning collection of traditional items from Canada's First Peoples. (photograph E. C. Krupp)

Alexander Archipelago. These islands are just west of the coast of northern British Columbia and southeast Alaska, and settlement on the Queen Charlotte Islands began approximately 9,000 years ago. At the time of first European contact, the Haida population of the Queen Charlotte Islands probably totaled 6,000 persons. About 20 winter villages were established at the time, and their average population was about 300.

Admired for their use of assertive outline in their painting and carving, the Haida turned their winter villages into shoreside groves of totem poles and mortuary poles. The village was the primary organizational unit, and the villages were governed by a system of ranked family authority. The lineage houses of these villages were the real centers of power in Haida life. As the highest-ranking member of the family line, the house chief supervised all of the activity of everyone else associated with the house. Lineages owned resources and properties, and the highest lineage chief was the ultimate arbiter of their use.

Because the Haida emerged from a clamshell, their first appearance in the world is really a birth from the womb of the earth. The shell, after all, was embedded in the ground, and opened like a woman in delivery. The Haida modeled the creation of people on the biological mechanism of human reproduction but made a midwife out of Raven. That was their deference to the supernatural forces that seemed to invest the cosmos with spirits of life. For the Indian people of the Northwest Coast, Raven was a trickster and a transformer, a shaman and a creator, like Mantis for the San. Ravens are actually keen-sighted, fearless, and clever birds that live by their wits. They are noisy and aggressive. They endure—and seem to prefer—wilderness. They eat everything. Intelligence and imitative skill make the raven an ideal stand-in for the trickster who fashions the world in part as the by-product of his own priorities. Raven is the hero who steals the sun, moon, and stars and with them lights the world.

Compared with the nomadic San, the Haida are stay-at-homes, but they do migrate seasonally from their winter villages when the quest for food dispatches them to their summer fishing camps. The San have few possessions and carry everything they own with them. The Haida in turn are wealthy. They redistribute part of their wealth in impressive potlatches, especially to ceremonialize the transfer of chiefly power. The San hunting bands are classless. Haida society is ranked. Neither, however, recognizes a regional leader or a territorial center of power. For the Haida and other peoples of the Northwest Coast, ownership of hunting territories, fishing rights, family houses, and "intellectual" properties—family crests, totem poles, stories, and dances—was established by bloodline. Although each winter village had a "town chief," his power was limited by the house chiefs and lineage chiefs in the same community.

Haida leadership, then, is linked with lineage and is not really spotwelded to the land. Rather, it is stitched through the entire landscape by family connections, and if there is a center of the world, it is the Queen Charlotte Islands themselves. No one owns the clamshell of creation, but the consequences of Raven's creative activity are evident everywhere.

The Haida, like their Pacific Northwest neighbors and like the San of southern Africa, saw the entire land as sacred. It contained places where the spirits were particularly evident and where supernatural power was concentrated and sometimes accessible. Their shamans interacted with the spirits to cure sickness, recover lost souls, and harmonize the community with the rest of the cosmos.

Although the San and the Haida tell different stories about Creation, shamanic power is what first sets up the world for human habitation. This power draws upon nature's endless capacity to transform. Those transformations are evident in the mysteries of birth and the cyclical return of life. Because celestial events also seem to participate in the same kind of transformation and renewal, stories of Creation often incorporate some kind of involvement by the sky along with whatever other powers may be at work. As a bird, Raven belongs to the sky, and with shamanic persuasion he charms the human spirit out of the clamshell of the earth.

DROPPING IN ON THE ROOF OF THE WORLD

Before Buddhism inscribed upon Tibet its own mandala of cosmic order, the Bon tradition of Tibetan shamanism and ritual magic, along with an earlier indigenous animist folk religion, accounted for the origin of the world with enough completely different versions to satisfy every cosmogonic taste. Cosmic eggs, cosmic creatures, cosmic nothing, and cosmic opposition all were given credit for parenting the earth. In one story, a female spirit, or *klu-mo,* of the subterranean waters was spontaneously spawned from the yawning vacancy that was the universe before there was anything in it. The Tibetans saw her body in the earth and sky. The day began when she opened her eyes, and when her lids dropped in sleep, night returned. The crown of her head was the sky, and her eyes were the sun and the moon. Four of her gleaming white teeth were planets that also migrated across the top of her head. Twelve more teeth shined as the stars of the 12 mansions through which the moon moved each month. The storm emerged from her mouth as her tongue darted in the lightning and as her voice spoke in the thunder. She exhaled the clouds in her breath. Tears, of course, slipped off her face as rain, and her nostrils released the winds. Hail originated in the fat of her tongue and fell upon her fleshy body. Her body was the earth itself, and its oceans were her blood. The rivers ran in her veins. Her bones could be seen in every mountain.

The universe has to be something—or at least has to be like something—and the human body is friendly territory. Earth and sky are often personified in ancient creation myths as the body of a living being. Although this supernatural being is not really human, human attributes help to describe it. People usually try to use familiar imagery to model abstract ideas.

Sometimes creation myths like this were localized in the natural environment. Fastening events to specific features of the landscape allows the primordial past to make constant contact with the present. All of the circumstances of creation made the world what it is today, and the ongoing influence of the past is maintained through evidence of its persistent presence. This process is apparent in an alternate description of the Tibetan cosmos. In this case, the world is the body of a man-eating tigress. When captured and killed, she was transformed into a trophy. The land was upholstered with her stripes, and that territory was known from then on as the Spotted Plain of the Tiger. Her tail was turned into the gorges on the rMa River, and her head became Ger-mjo, a sacred mountain. Finally, this myth also equated the people and horses of the clans that lived in the region with more of the anatomy of the cosmic tigress. This creation myth, then, also upheld territorial claims to this area and validated the prevailing precedents.

In the seventh century A.D., Queen Wen cheng discerned the body of a demoness in the Tibetan countryside. She was a Chinese import, a Tang dynasty Chinese princess married to King Songsten Gampo, who by tradition was the first Buddhist ruler of Tibet. Wen cheng's geomantic knowledge revealed the head of the shedemon in the east and her feet in the west. Perceived as an agent of chaos and a challenge to Buddhist order, the demoness was restrained by the king. With Wen cheng's guidance, he built a temple on each shoulder and on each hip to hold her in place. Trandruk Temple, near Tsetang, about 113 miles south and east of Lhasa, nails her left shoulder down in central Tibet. Katsel Temple is northeast of Lhasa, near Medrogungkar, and it also restrains the demoness. Two more "Extremity Subduing" temples are in the west—Buchu Temple in Kongpo and Dram Temple in Tsang. With the four world quarters schematically mapped into the temples and onto the countryside, Buddhism overpowered this supernatural incarnation of the wild spirit of Tibet. The Johkang, Tibet's most sacred temple, marks the earth-filled lake in Lhasa that was the heart of this demoness and the center of the inspirited land.

The Tibetans naturally put Tibet at the center of the world, and their homeland—a mountainous plateau that averages 15 thousand feet above sea level—is a lot closer to the sky than most of the rest of the world. Its Himalayan highlands make Tibet "the Roof of the World." In a passage of a ninth-century Tibetan poem, discovered in a manuscript in the Buddhist grottoes near Dunhuang, China, Tibet's place in the cosmic landscape is reduced to the essentials:

> The center of heaven
> The core of the earth
> This heart of the world,
> Fenced round by snow.
> The headland of all rivers
> Where the mountains are high,
> The Land is pure.

Lhasa's Jokhang Temple is like a stake in the heart of the she-demon spirit of Tibet before Buddhism. In that sense, it is the center of the Tibetan world. Along with 12 other temples, it locks the demoness and the land in place and preserves the new spiritual order. (photograph E. C. Krupp)

Integrity, inaccessibility, the origins of life-sustaining rivers, and contact between the sky and the land are, apparently, the attributes of the center of the world. Defining the structure of the universe in terms of the ordered movement of the sky, as so many other peoples did, the Tibetans modeled the heart of the world on the hub of the sky—the north celestial pole—and they coupled the earth to the sky with Sumeru, the world-axis mountain of cosmological myth. At the center of the universe, Sumeru also performed as the navel of the earth, and in time, Tibetans came to regard Mount Kailas in western Tibet, as the visible, finite, and terrestrial form of Sumeru.

At the summit of Kailas, earth was in direct touch with the sky. Lamaist texts liken Kailas to an umbrella with eight supporting ribs for the eight world directions—cardinal and intercardinal. The umbrella, in turn, is a celestial metaphor for the vault of the sky, and in Buddhist tradition the sky is symbolized both as a parasol and as an eight-petaled lotus. Because Kailas holds up the sky, it is also an eight-petaled lotus in blossom upon the earth. Its cardinal gates are defended by the same talismanic animals that police the four directions in China. A total of 360 gods populate the mountaintop palace, and by their number they symbolize the year. The traditional Tibetan lunar calendar counts the days of the year as 360 and adds seven intercalary months over a 19-year interval to keep the phases of the moon in step with the seasons. With a nod toward the calendar, then, the 360 gods that call Kailas their home link the passage of time with the structure of space.

Mount Sumeru, the cosmic-axis mountain at the center of the world, makes regular appearances in the wall paintings of Buddhist monasteries in Tibet. This mural from Gongkar Monastery adheres to convention by illustrating the base of the mountain as a set of concentric squares circumscribed by the circular horizon. Each zone between the outermost square and the circular border represents one of the four "world continents" and world directions. The concentric squares represent the higher and higher stories of Sumeru, and from its summit a series of levels expands through the sky and to the top of the painting. (photograph E. C. Krupp)

The symmetric profile of this mountain and its theatric setting—an immense tectonic stadium—induce an emotional response. The grandeur is amplified by natural cuts and tiers that sculpt a cross on the mountain's south face. Buddhists see this immense scar as the partial imprint of the sacred swastika stenciled on the ice. Mount Kailas is also said to be a ladder providing passage between heaven and earth, and Hindus judge the great south gash to be a stairway to heaven for the god Siva. Like the transcendental halyard that once kept Tibet's heavenly kings moored to the sky, Kailas is the earth's umbilical cord.

Although the Tibetans traced their own origin to the earth, their kings came from the sky. In fact, until Nyatri Tsenpo, the legendary first king, dropped out of heaven and landed on the summit of Lhabab Ri, "the Hill Where God Descended," the Tibetans were not politically united. The Bon account of his arrival is preserved in the Dunhuang manuscript quoted earlier, and it tells us the spot where he touched down is the center of the world. These verses also emphasize the sky king's part in ensuring the productivity of the land, for they confirm that "he came as the rain that impregnates the earth." The mountain itself repeatedly bowed low to him. Trees closed ranks, and springs rippled in his honor. Even the rocks and boulders "saluted him respectfully." This first ruler of Tibet arrived as the son of celestial gods to serve as the "lord of everything under heaven." Later, the Buddhists rationalized this story by eliminating his divine lineage and diluting the impact of his flight to earth. They portrayed him as an abandoned prince who met some Bon sorcerers as he made his way through the Himalayas to the base of a mountain and a place known as the Royal Plain of Four Portals. This name seems to describe a Bon ritual plaza, which also was oriented to the four directions of the world. In any case, the prince and the Bon chiefs spoke different languages. When they asked him who he was, he answered, "A king." The magicians did not understand him, but they continued their interrogation. In response to their interest in the place from which he had come, he pointed to heaven. Now, as far as the Buddhists were concerned, Bon religion was infatuated with the sky. It made sense, then, in this Buddhist tale, that the Bon ritualists interpreted the gesture of the prince as evidence of celestial descent. Recognizing him as a worthy nominee for king, they lifted him upon a wooden throne and carried him back down the mountain to turn Tibet into a monarchy.

Legend associates this king and the first dynasty of Tibet with the Yarlung Valley. The valley is in the eastern part of central Tibet and is regarded as the birthplace of Tibetan civilization. Tibet's first royal capital was allegedly established there by the heavenly line of Yarlung kings. A fairy-tale ridge-top monastery at Yumbulagang, up the valley and about nine miles south of the town of Tsetang, is the traditional location of the first king's first palace.

Pilgrims en route from the Crystal Cave on Sheldruk Mountain pass by Lhabab Ri, the place where the first Tibetan king arrived from the sky, and regained the Yarlung Valley here at the Gontang Bumpa Chörten. This type of shrine symbolizes the world axis and the layered cosmos. Architectural "rays" in the top of this one made a sun out of its summit. The pilgrims approached this monument from the east and circled it clockwise, or "sunwise." (photograph E. C. Krupp)

Without historic or archaeological evidence, it is impossible to assign a date to the first Tibetan king, but the mythical inauguration of kingship is sometimes extrapolated back as far as 500 B.C. Others place it in 237 B.C. or in 130 B.C. Certainly, the present structure at Yumbulagang is nowhere near that old. In fact, it replicates an earlier building destroyed in 1982 in China's Cultural Revolution. Long before that chaotic period, the same building was believed to have been ground zero for the first Buddhist scriptures to bombard Tibet. They actually fell out of the sky and hit the roof like a rogue meteorite in 466 A.D. The foundation, however, is only as old as the seventh century A.D. and the era of king Songsten Gampo. He was 33rd in the line of Tibetan kings and the first Buddhist ruler of Tibet.

Despite the lack of any physical evidence for the first capital, archaeology has confirmed that Tibetan agriculture began in the Yarlung Valley. By reputation, the Yarlung Valley is the most fertile area in Tibet, and the village near Yumbulagang Monastery preserves a sanctified plot of land known as Zorthang, or "the Sickle." Believed to be the first field ever planted by Tibet's ancient farmers, even today local people and other pilgrims carry home a handful of earth from this parcel of land to sprinkle on their own fields in the interest of fertility. At the base of Gonpo Ri, the mountain right next to Tsetang, there is another piece of land also said to be the first cultivated field. Its claim is enhanced by the mountain itself, for Gongpo Ri is one of the four sacred peaks of central Tibet. The other three are Chakpo Ri in Lhasa, Chuwo Ri near the junction of the Tsangpo and Kyi Chu rivers south of Lhasa, and Hepo Ri just east of Samye monastery. Each refocuses the power of the cosmic center and world axis in a local mountain.

Many more places in the Yarlung Valley and the vicinity of Tsetang commemorate the origins of kingship and culture in Tibet. The mountaintop where Nyatri Tsenpo, the first king, first arrived is on the pilgrimage route from Tsetang to Chongye, south of Sheldruk Cave, itself on nearby Crystal Mountain. Caves in the Buddhist kingdoms of the Himalayas are primarily meditation retreats for dedicated pilgrims. Eight of these sanctuaries were activated in the eighth century A.D. by Guru Rinpoche, the famed Tantric master from India responsible for much of Buddhism's influence in Tibet. Because each cave is likened to the center of a mandala, a Buddhist diagram of cosmic order, the pilgrim in retreat taps the spiritual power of the world's center, the power in the world axis, and the power at the world mountain. The cosmographic power of the cave is extended to the points of the compass by four other pilgrimage caves in Sikkim. These, too, are credited to Guru Rinpoche, and each is said to belong to one of the four cardinal directions.

Mythology even seeds the origin of the Tibetans themselves in a cave overlooking the fertile soil of the Yarlung Valley. Known as Monkey Cave, it is near the top of Gongpo Ri. There, an anchorite monkey possessing magical power mated with an especially attractive and aroused "ogress of the cliffs." At first he rejected her advances, but she argued his refusal would condemn her instead to lifetimes of mothering ogre children, and generation after generation, they would bring more misfortune to the

Tradition locates the first palace of Tibet's first king where the Yumbulagang Monastery now is perched on an evocative ridge. (photograph E. C. Krupp)

world. Reluctantly persuaded by her logic, the monkey consulted with Avalokitesvara, the bodhisattva of compassion and the spiritual patron of Tibet. Bodhisattvas are divine beings. By achieving Buddhist enlightenment they are entitled to nirvanic bliss, but they choose to remain, like saints, among the living, to assist every soul to salvation. The monkey was permitted to break his religious vow, and so in this Buddhist fable, the primordial parents met, married, and multiplied with divine blessing.

Also known as Chenrezi in Tibet, Avalokitesvara is, in a sense, the divine protector of Tibet. In fact, the meditating monkey in this story is actually a supernatural incarnation of Avalokitesvara himself, and his demon rock queen is the goddess Tara, Avalokitesvara's female counterpart and a maternal expression of divine compassion. The consequence of the primate-ogress union was six monkey children who eventually proliferated into the whole population of Tibet, but their reproductive success was not matched with agrarian reform. They were still foragers and, for that reason, uncivilized. Avalokitesvara, who had his eye all along on the future of Tibet, intervened again. He climbed to the top of Sumeru, reached deep into the body of the world-axis mountain, and pulled out the seeds of the five different grains. Scattering them over the Tibetan plateau, he turned the monkey children into farmers. They built homes, settled the land, and in time, under one of their celestial kings, adopted the Buddhist faith, as Avalokitesvara intended all along.

Against the backdrop of Chuwo Ri, one of central Tibet's four sacred mountains, on July 18 villagers performed this ceremony for a good harvest to protect the fields from hailstorms. Summer is the season of thunderstorms, and close to harvest the crops are most vulnerable and the peasant economy most at risk. Five piles of white stones were lined up across the river from the mountain, and juniper was burned in offering. The procession marched clockwise around the smoky fires. (photograph E. C. Krupp)

What we see so far in the legendary history of the Tibetan kings and in the mythical origin of the Tibetan people is the role played by the central mountain of the world. It delivers the king. It delivers civilization. It galvanizes the earth with power from the sky. That power originates in the fundamental movement of the sky and is expressed in the stabilizing influence of the king, in the continuity of culture, and in the productivity of the land.

THE MANDALA AND THE MOUNTAIN IN BUDDHIST TIBET

Real Tibetan history actually begins in Chongye, about 17 miles south of Tsetang. Fragmentary remains of a Yarlung capital are identified in the ramparts and walls on the knife-edge ridge above the town. These are the precarious ruins of Chingwa Takste, a castle attributed to the tenth Yarlung king. There is also physical evidence of Songsten Gampo's ancient presence in this region. His monumental earthen tomb is just south of the modern town and a half mile beyond the bridge across the Chongye River. The burial mound is more or less square, 424 feet on a side and about 44 feet high. Nine other Yarlung tombs are within sight on the valley floor and part way up the flank of nearby Mount Mura. There has been a chapel dedicated to King Songsten Gampo and his Chinese and Nepalese Buddhist wives on the summit of his tumulus since the twelfth century, and today's rebuilt shrine is still an active point of pilgrimage. Devotions converge in the prayer flags, sacred juniper hearths, inscribed stones, and a large rock cairn that turn the top of the tomb into a lively open-air shrine.

Still tracing royal lineage back to the sky, Songsten Gampo mobilized central Tibet into an organized state and an expansionist military power. He and his successors managed to subdue and occupy Nepal and even plundered Xian, the Tang dynasty capital of China. They extended the western frontier beyond Shangshung, the land of Mount Kailas and an independent kingdom where the Bon religion was believed to have started. Songsten Gampo collected tribute from southwest China, controlled traffic on the Silk Road, and created an empire in central Asia.

Songsten Gampo also made Tibetan culture much more sophisticated. He introduced a writing system for the Tibetan language, commissioned Tibetan literature and translations of important scriptures from abroad, and crafted a system of laws. He realized that Bon and the old, folk beliefs were deeply entrenched in Tibetan culture, and so, when he introduced Buddhism into the country, he grafted the new reli-

The Dharma Kings of Tibet introduced Buddhism and defended it against the indigenous religious establishment of the Bon priests. Most of these early kings are buried in monumental earthen tombs in the Chongye Valley. From the top of Mangsong Mangsten's burial mound, we are looking northeast to the tomb of Songsten Gampo, the first Buddhist ruler and an expansionist king who turned Tibet into an empire in the seventh century A.D. The shrine dedicated to him on top of his tomb continues to attract numerous pilgrims to this site. (photograph E. C. Krupp)

gion onto the old faith. He also moved the capital from the Yarlung Valley to Lhasa, to the north of the Tsangpo River. His marriage to a pair of devout Buddhists hardened the armor of Buddhism in its contest with the Bon chieftain-priests. He injected Buddhist cosmographic power into the system of temples he devised to restrain the demoness his Chinese princess spotted in the land. Like the ogress of the cliffs who mated with the monkey, this she-demon was actually the spirit of old animist, Bon-practicing Tibet, and it took the template of Buddhist cosmovision to domesticate her and handcuff her to the new world order.

Tibetan Buddhism offered its own more abstract vision of the creation of the world, and it clearly steered away from the earlier animist notions. This story also begins with a great emptiness, but it is a void with philosophical substance. It is timeless. It is blank. It has no impulses, no relationships. It is not just an empty room. It is no room at all. Nevertheless, currents like subtle breezes spontaneously flowed. If this were quantum physics and relativity, we would be talking about virtual wormholes in space-time foam. Instead of the Big Bang, however, this impromptu presence of something in nothing—the original virtual reality—prompts the materialization of the double lightning-bolt scepter, a symbol of the activating energy of creation at maximum gain. Its electricity creates the clouds, and out of the clouds falls rain, enough rain to fill even a primordial ocean. After the waters became as calm and flat as glass, they started to churn, and land foamed up like butter out of cream. A small speck of earth cream grew into a vast mountain, a rocky road to heaven. Rising high enough to scrape the sky, it became Sumeru, the four-sided cosmic axis mountain at the center of the world. In time, its jeweled flanks hosted rivers, trees, and plants. It is ringed by a lake, and the lake is framed by a range of golden mountains. Seven more circular lakes and six more mountain rings complete the concentric topography. In the outer lake, there are four island continents, one for each cardinal direction and each side of the central mountain. Each land has a different shape. To the east, the territory takes the shape of the "half moon." The realm in the west is a disk and so is round like the sun. In the north, the island is square. Ours is the south continent, which is tapered like a keystone and said to resemble the shoulder blade of a sheep. The inhabitants of each of these domains have faces with the same shape as the continents on which they live. The gods of the central mountain were transformed into human beings when they discovered a tasty creamy coating on the surface of the earth. Pleasure compromised their spiritual meditation, and as their divine radiance faded, the world turned dark. The relentless black night retreated, however, when the sun, moon, and stars emerged, but human behavior trailed the darkness. Although there was a fruiting plant for everyone that was renewed with a free lunch each day, people became greedy. Theft and conflict arrived and became permanent houseguests. The world, in fact, turned into what it is today, but people realized their future survival would depend on their ability to organize themselves. For that reason, they all met and selected a king. It was he who taught everyone how to cultivate the land, how to construct a home, and how to live together as a community.

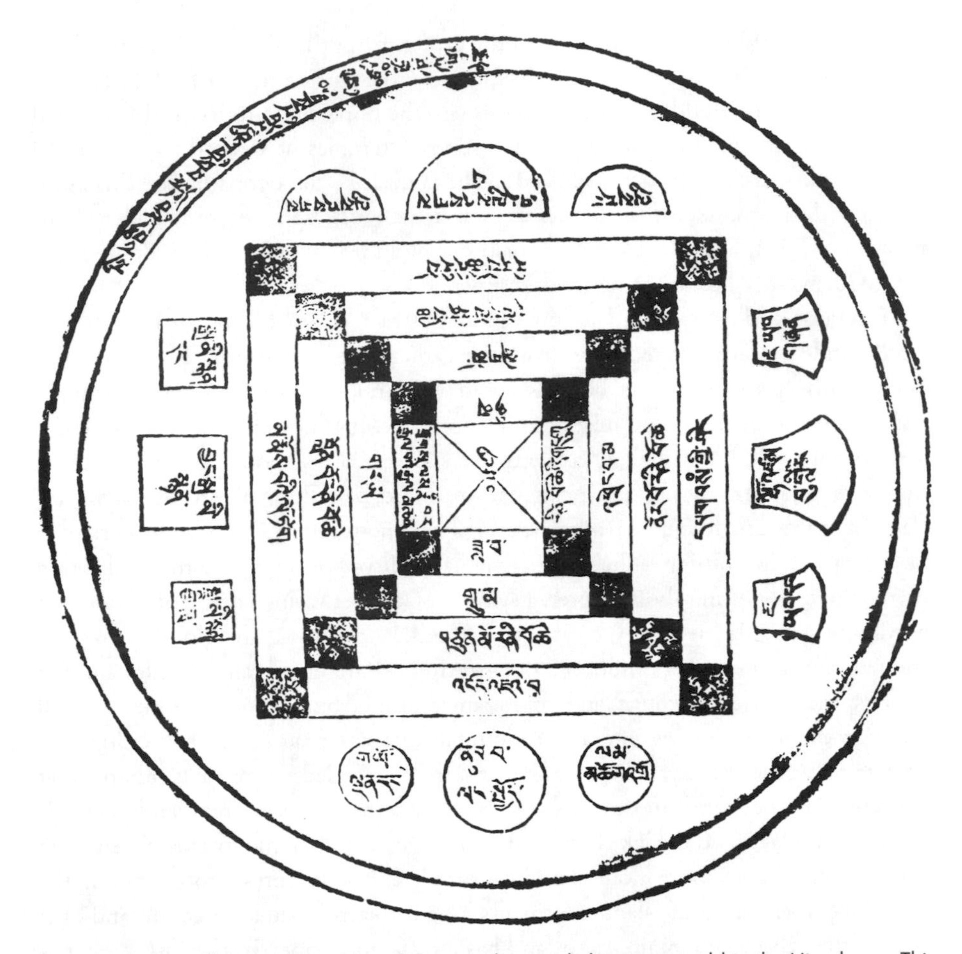

Inexpensive wood-block prints of mandalas, amulets, and charms are sold in the Himalayas. This Nepalese "mandala of the universe" symbolizes the world-axis mountain like the wall painting at Gongkar Monastery, with a set of nested squares. World continents and directions are indicated by the simple geometric shapes on each side of the outer square. East is on top and represented by the half-circles, or "crescents." North, with its squares, is on the left, and south, on the right has tapered "ax blades." Circles for the west are on the bottom. Oceanic waters separate the base of the mountain from the double-circle horizon. (from *Tibetan Tantric Charms & Amulets* by Nik Douglas, Dover Publications, Inc.)

The point of the animist story of Creation told earlier is the physical character of the universe. It describes how we got what we see. In this Buddhist story of Creation, the punch line is the human condition. Yet, like the first, it relies on physical cosmic framework. It is an image contrived from three-dimensional geometry. There is a center and an edge. The perimeter is circular, like the horizon, but the world has four directions and so the inner lands are organized like the four sides of a square. The center is point of creation and the socket for the plug that conducts the divine energy that makes everything work. The source of that energy is the sky—transcendental but ever-revealing, and the earth is hardwired to the sky by the central world-axis polar mountain. When this entire scheme of organized power and being is mapped into two dimensions, it becomes a mandala, a symbolic design of religious devotion and esoteric knowledge. Tibetan monks draw them in colored sands on the temple floor and paint them on the temple ceiling. They contrive them into altar foundations and commit them to instructional scrolls. They print them on paper, fold them up, and wear them as protective amulets. As "psychocosmograms" they diagram the structure of the universe, link the microcosm with the macrocosm, and assist a meditator in the quest of higher states of consciousness. The ultimate destination of the soul is nirvana, the pure transcendent universal mind. Everything that exists originates from this undifferentiated state, and four cardinal points and center of the mandala and the cosmos represent the flow of world-making energy and wisdom from the primordial void. The cosmic mountain is a synapse in the spiritual nervous system of the universe, and the structure of the Buddhist cosmos, then, is the process that makes the cosmos what it is.

These concepts clarify what King Songsten Gampo and Queen Wen cheng had in mind when they pinned the earth demoness of Tibet with a mandala of Buddhist temples. He didn't stop with one quadrilateral enclosure, either. Beyond the Johkang in the heart of his kingdom and the four "Extremity Subduing" temples at crucial demonic joints, he fenced Tibet with two more spiritual enclosures of demon-suppressing monuments. Like the next concentric square in a mandala, an arrangement of four more temples "tamed the border" at the knees and elbows of the reclining demoness. Even more distant from the capital, his four temples intended to "tame the areas beyond the borders" drove stakes through her hands and feet. What looks like an effort to homestead the earth, pacify its demons, and convert the unenlightened was really intended to make the place safe for kingship. Songsten Gampo was not just introducing a new religion, he was disarming the Bon chiefs and priests. His frontier temples conveyed more than a message of religious devotion to Tibet's international neighbors. His civilizing introduction of laws and writing were essential tools of state. Ideology was on the march with power politics. As the primary advocate of the new system of belief, Songsten Gampo could reorganize the established structures of power in Tibet. He centralized power that was once dispersed. He consolidated power in the center of the world with a mandala of ideological monuments that spelled out exactly where the center was and presided there like the world-axis mountain.

The true world-axis mountain, of course, was Mount Kailas far to the west in Shangshung. The language originally spoken in this ancient kingdom was probably part of the Indo-European family and not related to the Sino-Tibetan linguistic branch. Because this area is thought to be homeland of Bon cosmomagical religion, with its tradition of local sacred power mountains and the central world-axis mountain, some have speculated that similar Vedic ideas may derive from the same archaic central Asian roots.

Mount Sumeru, the polar axis of the universe and modeled in bronze, is encountered in the first courtyard past the Hall of the Celestial Guardian Kings in Beijing's Yonghe gong, or Lama Temple. The temple itself is aligned on a north-south axis. Waves of the seven world oceans lap the mountainside. Embedded in its slopes are the four palaces of the Celestial Kings. From the mountaintop, a constellation-studded column continues higher into heaven to the paradise of Sakyamuni, the Buddha of the Present. (photograph E. C. Krupp)

One of the old Tibetan names for Mount Kailas is the Bon Mountain. That clearly links it with Bon belief, and it was also said to be the Soul-Mountain of Shang-shung. Through that status, Kailas preserved the pre-Buddhist belief in the ancestral gods of the local mountain. Each district's peak nurtured the population that lived under its protection, and each embodied the essence of that region's identity. Mount Kailas was also called "the navel of the world" and "the seat of the sky goddess." While the mountain's designations as a celestial umbrella and an eight-petaled lotus were Buddhist images in Tibetan cosmology, this goddess in the center of the sky belongs to the older Bon beliefs. She tells us that Tibetans were talking about a world-axis mountain before the Buddhists brought their own version of it to Tibet.

Throughout Tibet, along pilgrimage routes, loose stones are heaped as offerings into miniature versions of the world-axis mountain. These wayside shrines insert the power and the creative order of the center of the world into the rest of the neighborhood and unify every local territory with the cosmos as a whole. Before the Buddhists began accumulating sacred piles of stones and circuiting them clockwise with the sun, the Bon magicians also stacked the rocks and perambulated in the opposite direction, like the northern stars around the pole and like the moon through the lunar mansions. No matter which route you take, the axis is still cosmic.

Buddhism shared enough cosmography with Bon belief to make a comfortable fit in Tibet. Although the cosmic mountain embodied in the Buddhist mandala's center is not just the organizational axis of earth and sky, but the route of mystical ascent to Enlightenment, it belongs to a world-axis mountain tradition found throughout Asia. It is well developed in the Brahmanic concepts, and Tibet actually acquired its theological detail about Sumeru, from India, where the traditions of Meru, the cosmic center world-axis mountain was part of archaic Hindu belief.

Naturally, the mountain at the center of the world comprises the spirit of the entire world. Its paternal character is balanced by Manasarovar, the sacred "unconquerable" lake below the foot of Kailas. Pilgrims circumambulate its shore clockwise in the same kind of dedicatory journey they make about Kailas, around other holy mountains, and on sacred itineraries through other districts, such as the perimeter of Lhasa. It is the highest body of fresh water on the planet, a maternal pool to which Tibetans trace part of the origin of the Bon doctrine.

The central mountain and central lake are paired in a symbolic polarity that is repeated in the complementary opposition of sky with earth and male with female. Although the mountain appears to be part of the earth, unlike the lake, it is really bonded to the sky. The lake, on the other hand, is a watery pool with depths in the underworld. The Hindu symbolic transformation of these ideas incorporates the mountain-sky-male principle in the pillarlike lingam. The lake-earth-female theme is expressed in the yoni, a basinlike receptacle which the erect lingam sometimes occupies in a temple setting. The insertion of one in the other is a sexual metaphor that mirrors the merging of the two fundamental aspects of the material universe. The lingam is specifically identified with Siva, who is connected with the eternal and absolute generative power in the universe and its capacity for endless transformation. He is the first cause and sits in mystic retreat on the summit of Mount Kailas. The yoni is the vulva, vagina, and womb of Siva's consort. As Devi, she personifies *sakti,* the creative vivifying energy of nature. With Siva, the goddess perpetually renews the cosmos through another cycle of creation. Their conjugal union is the marriage of polarity that creates and completes the cosmos.

In India, Kailas is first recognized as the terrestrial prototype for Mount Meru in the *Vishnu Purana.* The *Puranas* comprise a group of ancient Sanskrit religious texts, and the *Vishnu Purana* was committed to writing at about 200 B.C. More general references to sacred mountains are encountered earlier, in the *Rig Veda.* This compilation of ancient Sanskrit hymns details the oldest Indo-European religious concepts. Probably composed around 1500 B.C., in the era of the Vedic invasion of northwest India, it contains the foundation of Hindu cosmographical ideas, which eventually found their way to Tibet and other parts of central Asia.

Vedic treatises formalized the idea of Meru through the subsequent millennium. The *Taittiriya Aranyaka,* an appendix to these treatises, identifies Mount Meru as the abode of the Adityas, celestial gods who personify the sovereign laws that govern the cosmos and society. Their mother, Aditi, is the primordial immensity of heaven, and she is equated with the night sky. All power originates in the infinite expanse of her darkness.

Male and female symbols are united in the inner sanctuary of a Hindu temple dedicated to the god Siva. Delicately decorated with blossoms and other offerings, the yoni-linga combines the spiritual potential of creation with creative activation of matter. This union is the most profound expression of divinity for Siva devotees, and this yoni and lingam are paired in Mukteshvara Temple, a tenth-century house of the holy in Bhubaneshwar, India. (photograph E. C. Krupp)

By the seventh or sixth century B.C., the *Mahabharata* offered an enhanced vision of Mount Meru. The *Mahabharata* is one of India's two great mythological epics. As the longest narrative of the affairs of the gods in all of world literature, it informs us that Meru has a "blazing appearance." "Abounding with gold and variegated tints," it is "the haunt of the gods." It is "immeasurable," and for persons tainted by sin, it is "unapproachable." "Dreadful beasts of prey inhabit its breast, and it is illuminated with divine herbs of healing virtue." It stands "kissing the heavens by its height and is the first of mountains." Ages ago, the gods met there and were instructed to churn the cosmic ocean into the world we know by Narayana, an avatar of the god Vishnu, who is the pervasive cause behind all things. Elsewhere in the 90,000-verse poem, we learn that Meru is in the Himalayas. It is a "gold mountain" and "the highest of all mountains." Reaching 350,000 miles into space, it "shines like the morning sun and is like a fire without smoke," and it is rooted just as deeply in the earth. The sun and the moon are obligated to circuit it. Conflated with the world's polar axis, Meru makes contact with the sky at the north celestial pole. The Milky Way passes to the side of that stable center of heaven.

According to the geography of the Hindu sky, the Milky Way is a celestial river. Its waters originate beyond the sky, pour through the Pole Star, and flow across heaven, like the Ganges flows across India, until they plunge to earth and splash the summit of Mount Meru. Split into four terrestrial rivers that run to the four quarters of the world, they stream down the mountain's flanks from there.

The Milky Way is the celestial counterpart to the Ganges, and to Hindus, the Ganges River is the purifying mother of salvation. Deified as the goddess Ganga, the daughter of the Himalayas, the entire river was sacred, from source to mouth. Although the source of the Ganges is on the south side of the Himalayas, it is linked by the Karnali River to Lake Manasarovar and the domain of Mount Kailas. The other great rivers of India also originate near Kailas. The Indus flows from a source north of the mountain, and the Sutlej—and in the Vedic period, the Saraswati—flows directly from Rakshas Tal, the holy lake paired with Manasarovar and connected to it by a channel known as Ganga Chu. The Brahmaputra, which dominates the eastern zone of the Indian subcontinent, begins in a glacier just east of Manasarovar.

In Tibet, the Brahmaputra is called the Tsangpo, and it is the highest river in the world. After running southeast for about 300 miles from the far west of Tibet to Shigatse, the Tsangpo bends east through central Tibet. It continues east for another 400 miles, passing south of Lhasa, and divides the Tibetan plateau in two. By the time it reaches Mount Namche Barwa, it is another 300 miles or so east of Lhasa, and there the river starts a great turn. Bending south, it drops 11,000 feet through a long chain of spectacular gorges, waterfalls, and white water into Assam, in eastern India. It eventually merges with the Ganges in Bangladesh and empties through the great delta into the Bay of Bengal.

Coursing down from the supernal heights of the Himalayas, the Tsangpo and the other great rivers of central Asia that drain from the vicinity of Mount Kailas echo the watershed of sky and are regarded as part of the system that integrates heaven and earth. The Tsangpo, Lake Manasarovar, Mount Kailas, and all of the other significant features in the landscape—caves, springs, river sources, local mountains, oracle lakes—possess topographical power. Because each place is linked through analogy to the entire cosmos, the land itself is enshrined. A visit to any of these special places in person is truly an encounter with divine power. Life in this kind of environment is really an exercise in sacred geography.

In this chapter, we have seen how the sacred geography of Creation may be related to social complexity and the distribution of power. Nomadic hunters are always at the center of creation, and power is everywhere. Concentrating it has no adaptive value for migrating foragers. They collect it as they need it, like the meals they gather from the countryside. When people are more settled on the land, property acquires more meaning and greater value. Property—whether tangible or intangible—requires power to maintain and power to appropriate it for use. Family lineage and tribal organization enter the picture here. The story of Creation helps weave people into the lands they occupy and the property they own. Human presence makes the place make sense. Designated caretakers of power introduce rank into the community. Power is not particularly condensed, but neither is it all over the map. With chiefs and kings, however, it's a whole different story. Like the foraging band or the seasonal village, chiefdoms and kingdoms are adaptive responses. Their purpose is the prosperity of the community and the continuity of the culture. The formation of an organized state, however, requires the centralization of power. When that happens, the place of Creation, the structure of the cosmos, and social institutions are more firmly anchored to the center of the world. The landscape of earth and sky is transformed into ideology that explains, and so endorses, the narrow focus of power.

CHAPTER FOUR

Mother Earth

If the sky's capacity for cyclical renewal was the kind of power that affiliated it with Creation, then the same process of cyclical renewal was evident in the earth's seasonal transformations. Our ancestors could see Mother Earth has a gift for restocking the world with life and hosting her houseguests with meals and shelter. Inspired by her hospitality, they recognized a maternal instinct in the land. Seasonality and motherhood also became part of the story of Creation.

The Nature of Mother Earth

Although images of the earth as a mother are not universal, they are common. For the Greeks, she was Gaia, the broad-breasted earth who first gave birth to the starry sky, who in time became her mate. She was Papa in New Zealand, where the Maori said she constantly embraced Rangi the Sky until their children finally pushed them apart and created some breathing room on the planet's bosom.

Mother Earth was not around at the very beginning of the universe for southern California's Luiseño Indians. The cosmos was empty and quiet. The only spirit at large in the universe was Alone-and-Empty, and Alone-and-Empty was probably the universe itself, before form and time and life transformed it. In this primeval darkness, two round beings were created by Pale White, probably a personification of the remote power and mysterious presence of the Milky Way. After three days in a cosmos with no sunrises, the two children of Pale White ignited with life and gender. Experiencing further transformations, they paired and then separated. He became Tukmit, the Sky, and she was Tomaiyowit, the Earthmother, pregnant with all of the world's First People. Her children included not only the first human beings but the animals, the plants, the celestial objects, the Luiseño ceremonial equipment, and everything else in the universe, and she gave birth to them all while reclining on her

Southern California's Luiseño Indians prepared ground displays for ceremonies that initiated the youth into adult life. The design used in the ritual for girls was based on concentric rings, always broken at the north, that stood for fundamental aspects of the life of the spirit. In this replica of a traditional ground drawing, the white outer ring represents the Milky Way. The middle ring is black and stands for the night sky. Blood and the "root" of being are built into the inner red ring. The symbolic figures in the center illustrate "avengers" that keep each person on the straight and narrow. (ground display replica by Paul Apodaca, photograph E. C. Krupp)

back, her legs opened to the north. With a nod to the primordial labors of Mother Earth, the concentric rings of the ground displays drawn for the ceremony that initiated Luiseño girls into womanhood are broken on the north.

Male and female, clasped together in complementary opposition, are the obvious engine of procreation, and they are therefore well suited to operate symbolically as the mechanism that drives the process of cyclical renewal. This is the central notion in the Chinese concept of yin and yang, the dynamic balance of power that oscillates the universe through night and day, winter and summer, wet and dry, hot and cold, darkness and light, aggression and repose, death and birth, earth and sky, and every other pair of equally matched heavyweights that go round after round in the rhythmic circuits of time. Each prevails and then retreats. Every aspect of life is reflected in the male and female partners of creation.

The food that is consumed, the shelter that is occupied, and every endeavor that is undertaken is tempered by the seasonal dimension of the contending partnership, and the character of the season mirrors which half of the partnership is seizing the day. Even water, which everyone realizes is an essential term in the equation of life, plays on both teams. It is dormant in winter, when the ponds freeze. The life of the stream and river subsides, and it's too cold for rain. Water regains its mobility, however, when the world warms up again. Fish reappear. Frogs and turtles retake their turf. The rushes grow.

For most peoples who cultivate the land, the rain that falls from the sky drops like semen from a celestial god to fertilize the earth. The pools and springs from which water emerges and the lakes to which water flows are points of connection with the womb of the earth, the dark, internal cradle of life. Water leaves her body like menstrual blood, the fluid that signals a new cycle of fertility. Water leaves her body like the milk from her breast, the fluid that nurtures new life. As the Chumash see it, water even leaves her body like urine.

When the Navajo medicine man draws Mother Earth in a sand painting, he includes on her abdomen a circle that symbolizes the lake at the center of the world. First Man and First Woman led all of the First People through the body of Mother Earth to the sunlight through those waters. They all once lived in one of the four worlds below our own, and when they finally climbed up to the surface of this world—the Many Colored, or Changeable, Earth—they emerged through the central lake. Four world direction mountains surrounded the lake. Regarded as the place of birth, it is accompanied on the sand painting by the four sacred plants—corn, beans, squash, and tobacco. Each has its roots in the lake, and each grows out toward one of the cardinal directions.

Even in pharaonic Egypt, where the gender of earth and sky were reversed, water still followed a natural course. River water, and not rain, irrigated the fields in most of Egypt. Rain is rare there, and the real source of fertility is the Nile. Accordingly, the Egyptians portrayed the sky as the goddess Nut. Each night she birthed the stars and then swallowed them at dawn. At dawn, the sun emerged from her womb but was devoured by her at dusk, when the stars were again reborn. Staking her hands and feet at the four cardinal directions, Nut spread herself like a tent above her husband Geb. Geb was the earth. He impregnated his celestial wife with the fertilizing power of the river that ran through his loins.

Egypt, however, is the exception, not the rule. Most of the time people mapped a woman's figure onto their image of the earth. Like the Egyptian sky goddess, however, Mother Earth not only creates life, she consumes it. Her body is both a womb and a tomb. New life is created and incubated within the earth and then released into the world outside, but when we bury our dead, they return to the mother that gave them life. The wild green shoots that push their way to the sunlight through her skin of soil blow to seed at the end of their time and fall back again upon the breast that fed them. Seeds are also deliberately planted by farmers in the body of the earth. The furrows cut by spade and plough are her genitals. Hibernating animals—such as bears and snakes—enter caves and burrows when the weather turns cold and the times turn lean. They emerge like new creatures—usually accompanied by a new generation—when the earth thaws in spring, reborn from Mother Earth.

Caves provide access into the very body of the Mother Earth. Springs and pools are points of contact with her fertile, life-giving waters. Natural rock formations and other features of the landscape take the shape of her breasts and her genitals. All of these female features of the landscape were therefore charged with the transforming power of Creation and were used to put that power to work.

The Cave of Creation of Kawaiisu (or Neewoowah) myth is a rock shelter with female architecture. (photograph E. C. Krupp)

In the Tehachapi Mountains, between Bakersfield and Mojave, California, a large rock shelter with pictographs is known to the Kawaiisu Indians as the Cave of Creation. The Kawaiisu language is part of the Numic branch of the great Uto-Aztecan family, and in their own language they call themselves by another name: Neewoowah. In their belief, the First People, who were all animals, gathered at this place in Sand Canyon to create the Kawaiisu people. Grizzly Bear called all of the animals to a conference at which each decided what he would be. These animals—including Squirrel, Coyote, and Owl—may still be seen in the shapes of the large boulders and natural rock formations nearby, and the spirits still reside there. When Andy Greene escorted me to the site, he collected juniper berries along the trail that crosses the site's natural amphitheater and showed me how to offer them to the creatures in the rocks.

With a vertical cleft about 40 feet high, the Cave of Creation resembles a huge vulva, which is apt architecture for the place where people were first brought into the world. In winter, a seasonal waterfall pours down the cleft. The Kawaiisu who told ethnographer Maurice Zigmond the story about the First Animal Congress said the place of creation was marked by a bedrock mortar, and the world just flowed up out of that mortar. There is at least one shallow cavity in the rocky floor of a side chamber that can still be seen. On my visit, I noticed a prayer circle—a quartered ring of small stones—left in reverence below the great female split in the rock.

This world-quarter prayer circle was drawn in small stones just below the cleft of the Cave of Creation. (photograph E. C. Krupp)

THE PRIVATE PLACES OF MOTHER EARTH

Painted Rock, another sacred California Indian site, rises like the largest vulva on the planet out of the Carrizo Plain, near the flank of the Caliente Range. It is an oval, sandstone outcrop about 55 feet high on the south side. The rock encloses a "womb" that is open to the sky. The interior is about 225 feet in the north-south direction and about 120 feet across east-west. Open to the north, its "vaginal" entry is 20 feet wide.

Painted Rock sits in its own valley between San Luis Obispo and Bakersfield, California. Originally, the valley was called the Carrisa Plain, after the high carrisa grass that covered its floor. The celebrated San Andreas Fault runs right along its eastern edge. As a well-known pictograph site in northern Chumash territory, Painted Rock once possessed the largest and most elaborate collection of polychrome Chumash rock paintings known, including one baroque panel 40 feet long, but vandalism and erosion have destroyed most of them. Despite what is lost, there

Painted Rock opens the womb of the world to the top of the sky. This natural formation is on the Carrizo Plain, an inland California valley. Until vandalism and erosion nearly eradicated the pictographs, the great rocky vulva possessed the largest and most complicated collection of paintings in Chumash territory. A Bureau of Land Management officer to the left of the entrance indicates the scale. (photograph E. C. Krupp)

is still plenty to see, including a painting of a lizard superimposed upon a coyote. In the Chumash tale of the origin of human beings, Coyote lobbied energetically for people in the image of Coyote. By imprinting his paw on the rock, he would set the standard for human hands, but at the last moment Lizard slipped his five-fingered foot in first. For that reason, people have hands more like lizards' claws than coyotes' paws.

Probably most of the images were painted sometime in the last thousand years, perhaps only a few centuries ago, and within the era of European contact with the Chumash. Although detailed ethnographic data were collected almost a hundred years ago, particularly by the American anthropologist John Peabody Harrington, there are no known Chumash commentaries about the original purpose and meaning of Painted Rock. In 1910, Myron Angel, an author of several local histories, published a romantic interpretation of the site in his book *La Piedra Pintada—The Painted Rock, a Legend.* His narrative was based upon information passed along to him by one of the Carrizo ranchers. The rancher got the story from a Mohave Indian who worked as his foreman and who heard it from his mother. The Colorado River territory of the Mohave is a long way from Chumash country, and the two groups are not related culturally or linguistically. It is possible the foreman's mother had some contact with local Chumash, but the account doesn't sound like Chumash tradition. Two details are interesting, however. Angel emphasizes that the Painted Rock was a sacred place of public assembly—an open-air temple—where a large group of Chumash met periodically in ceremony and deliberation. Angel also identifies the High Priest who presided at these meetings as the Dreamer. This is a reasonable title for a Chumash shaman, who entered a trance with the assistance of *datura,* a dangerous hallucinogenic plant used to induce visions throughout southern California. Chumash rock art is closely associated with shamanic activity. The billboard scale of the pictograph panels and the uterine amphitheater endorse the idea that the function of the Painted Rock involved public ceremony. In recent years, the northern Chumash band has revived traditional use of the place.

With a natural configuration that advertises female fertility and birth, the Painted Rock may have been regarded by the Chumash as the place of Creation, but it was not the center of the Chumash universe. One of Harrington's sources located that spot closer to Mount Pinos, the highest mountain in Chumash territory. With layers connected by a central axis, the Chumash cosmos followed the usual pattern for the universe in a traditional belief system. Powerful supernatural celestial beings, including Sun, Moon, Morning Star, and Sky Coyote resided in the sky, which the Chumash called *'Alapay,* or "upper world." Below the earth, *C'oyinashup,* or "lower world," was haunted by dangerous and malicious spirits. Two titanic snakes also occupied that zone and supported *'Itiashup,* or the "middle world," upon their vast backs. We, of course, live in *'Itiashup,* which the Chumash judged to be flat, wreathed by ocean, and round at the horizon.

Maria Solares, an Ineseño Chumash, confirmed that Mount Pinos was the most sacred place anywhere in the Chumash homelands, and another source described its connection with the center of the Chumash world:

> The place the *'antaps* [Chumash shamans] are is a lagunita [pond] in a cañada [canyon] near *'Iwihinmu* mountain [Mount Pinos] and beyond Cuddy's [Cuddy Valley]. The water flows not *pacá* but *payá* [away from here]. You hear bullroarers and *tocar* [probably whistles and flutes, which were used ceremonially with bullroarers by Chumash shamans] and *gritar* [shouting]—dogs barking—many people there—it is like a fiesta. And in a cave, *suterrano* [underground], the *'ichunash* [sacred deer tibia whistles used by shamans] are kept—four of them. Ancients never went near that vicinity. The wind blows strong, the earth quakes. If you get in there, you never get back again.

Like the painted vulva of Mother Earth on the Carrizo, the sacred mountains and the world's center seemed to be places where Chumash shamans ceremonially activated supernatural power they acquired from potent cosmic umbilicals.

In what is now San Diego County, rock formations resembling female genitals probably were shown to Kumeyaay girls participating in coming-of-age rituals. Although the ethnography is not as complete as we would like, the circumstantial evidence is compelling. Granitic boulders carrying X-rated fissures are present where such a ceremony took place. During their initiation, the girls ". . . were then led away to a hillside where they were shown the sacred stone, which it was said was to protect them." The anthropologist who collected this information was not permitted to see the stone and possibly confused it with a much smaller rock that was warmed and placed between the girl's legs during the ceremony, but he reported that there was a stone that ". . . was said to symbolize or have reference to the female organ of generation."

A ranch owner also provided information about power boulders, about which he had been told by a Kumeyaay shaman:

> . . . he told us about these fertility stones up there. He said that when a young girl acquired a man and she didn't have any children right away, they would take them up and show them the magic stones. . . .

The Kumeyaay regarded the sky as male and the earth as female, and concrete expressions of Mother Earth's generative capacity make sense in any magical effort to enhance fertility and to promote successful delivery.

Charlotte McGowan, an archaeology professor in southern California, finds *yoni*, the Hindu word for vulva, a convenient name for these sexually explicit landmarks. In 1982, after locating numerous examples in Kumeyaay territory, collecting pertinent

Natural female genitalia in the boulders near Jamul, California, deprive the landscape of a G rating. Features like this reveal the procreative power of Mother Earth with explicit geology. The message was not lost on the Kumeyaay, who exploited such places in puberty rituals and fertility magic. (photograph E. C. Krupp)

ethnography, and finding analogous symbolism from other parts of the world, she published her analysis in a monograph, *Ceremonial Fertility Sites in Southern California.*

At the Peterborough Petroglyphs in Ontario, Canada, there are several unambiguous petroglyphs of women or vulvas that exploit natural cavities and splits in the rock to guarantee no one misses the message. Other examples of this imagery are known in the rock art of Inyo County, California, and in the American Southwest, the Pacific Northwest, and in the Great Basin.

McGowan discovered that many of the southern California yonis are supplemented with pigment or grooves to enhance the resemblance to the real thing. John Rafter has studied similar features in Luiseño territory, particularly at the Bernasconi Hills in Riverside County. Pictographs, cupules, bedrock mortars, and grinding slicks share the neighborhood with the Bernasconi yonis. Another natural feature resembles a sun disk, and its rays seem to be cut deeper and straighter than expected from normal erosion. The largest of the vulvas also has sunburst rays and is on a flat tabletop of rock that slopes down gently toward the east, and Rafter has noted that the long axis of the yoni, which is extended by a pair of carved grooves, points east to equinox sunrise over Mount San Jacinto. Known to the Luiseño as Yamiwa, this peak is where Taakwic, a dangerous supernatural celestial spirit, roosts. Sometimes associated with

Where the cliff face is perforated by a natural (but perhaps enhanced) hole, the prehistoric artist of Clear Creek Canyon, in what is now Fremont Indian State Park, Utah, saw fertility in the rock and pecked the rest of the female figure into a petroglyph. (photograph E. C. Krupp)

the stunning light of a meteoric fireball, or perhaps more correctly with the eerie and rare appearance of ball lightning, Taakwic is also said to have visited one of the tall, rocky summits in the Bernasconi Hills. All of this connects Taakwic with the sky, for lightning and mountaintops jointly deliver displays of celestial power. Not far from the sunbursts and yonis, another boulder possesses a petroglyph traditionally known as Taakwic's Genitals. Harrington's unpublished field notes name the location of this carving as "takwic" and explain how it got there: "One of the first people, a man leaned against this rock and left imprint of penis and testicles."

Examined collectively, all of the elements of the Bernasconi Hills seem to tell a story about the procreative power of the earth and the fertilizing power of the sky. These are the same themes that structure the Luiseño Creation myth, and the ritual landscape of the yoni terrain put that creative, reproductive power of nature to work. Whether it operated on behalf of Luiseño women looking for children, on behalf of the world itself as it seasonally revived, or on behalf of shamans returning from their trances like spirits reborn, we cannot say, but all three are appropriate clients of cosmic power.

Symbolic reenactment of birth in initiation ritual may also explain the female imagery of tunnels and womblike cavities in other California Indian rock art sites. At

Counsel Rocks, near the Providence Mountains in the eastern Mojave Desert, one of the large and massive boulders is known as Womb Rock. Its naturally scooped east face provides the suggestive hollow responsible for the rock's name. The overhang is shallow. The lower face of the rock is outfitted with numerous petroglyphs and has too steep a slope to provide a comfortable foothold, but the stone is penetrated by an oval tunnel about 2 feet in diameter. From the back, this tunnel offers a window to the eastern horizon that accommodates a view of equinox sunrise in a notch formed by the horizon profile. The hole is just large enough to permit a person to scoot through. When John Rafter and the late Beverly Trupe studied this site, they noticed that the floor of the little passage and its entrance lip were remarkably smooth and glossy and unlike the unworn volcanic tuff surfaces on the rest of the boulder. They also observed a dark stain penetrating well into the upper surface of the lip and realized they were looking at the effects of person after person, year after year, crawling through the short tunnel. The stone had been made slick through repeated contact, and the stain was body oil transferred to the stone every time someone completed the passage.

Many of the petroglyphs resemble male and female genitalia. A pair of "arms" extends from a carved vulva on the east face of Womb Rock. Seven lines hang like rib-

Counsel Rocks, in southern California's East Mojave Desert, includes a set of boulders on which pictographs and petroglyphs have been inscribed. Womb Rock is penetrated by a short tunnel, and its smooth, stained floor provides evidence of frequent passage through the aperture, possibly as a reenactment of birth in initiation ritual. An oval split by a vertical line and with seven more lines dangling from it is just below the "window" on the sheltered face of the rock. With two small "legs" extending from the sides, it appears to distill a compact symbol from the Chemehuevi story of the woman who opened her legs to the rising sun. (photograph E. C. Krupp)

bons from the bottom of the same feature. There is another grooved oval above this vulva and small disk with seven rays to the right. In the high cavity of another boulder, just a few feet northeast of Womb Rock, the petroglyphs include another seven-rayed disk, framed with a ring, and a set of zigzags separate this "sun" symbol from a symmetric wiggly form that looks a little like the letter *w* in the hands of an expansive calligrapher.

Although we don't know the cultural identity of those who carved the petroglyphs, Counsel Rocks is now in Chemehuevi territory. The Chemehuevi are a Southern Paiute group, Great Basin desert dwellers whose Southern Numic language, like Kawaiisu, belongs to the Uto-Aztecan family. For the Chemehuevi, east and west were the primary sacred directions, and shamanic ritual began with east. The Chemehuevi word for it means "toward the dawn," and it carries, of course, the connotations of beginnings and births.

The Counsel Rocks petroglyphs may be too old to be attributed to the ancestors of the Chemehuevi, but Trupe and Rafter recalled a Chemehuevi myth that seemed to be illustrated by the drawings on the rocks. The story has versions throughout the greater Southwest and so may be relevant in any case. Carobeth Laird, who once had married and then divorced John Peabody Harrington, heard the tale from her second husband, George Laird, who was Chemehuevi. She published the story in her first book on Chemehuevi culture, where it is called "The Twin Sons of the Sun." The story begins with a woman who lived by herself in a cave. Apparently secure in her isolation, she habitually parked herself in front of the entrance to her cave and reclined with legs spread and her own dark grotto open to the sunrise. She must have improved the landscape, for Sun poured his rays into her and lit the darkness. Inseminated by sunlight, she conceived two sons of the Sun. In Chemehuevi, the word for Sun's rays is actually "whiskers," and the seven lines dangling from vulva on Womb Rock numerically match the seven rays on the sun disks at Counsel Rocks. Trupe and Rafter saw the vulva on the east face of Womb Rock as a portrait of the sunstruck woman of the cave at the moment of the mating. The two "arms" are actually her outstretched legs, and the ornate *w* up on the neighboring rock is a persuasive stylization of an east-facing vulva with legs spread and bent at the knee, the kind of view that is normally reserved for hard-core pornography. There is nothing indecent in the sacred, however, and the sacred process of creation is the agenda at Counsel Rocks. Here, too, the transcendent power of nature's cyclical renewal has something to do with ritual rebirth. Shamans may have passed through the birth canal of Womb Rock as part of a seasonal effort to renew the world. Youth on the edge of adulthood may have crossed the frontier with the passport of rebirth. Womb Rock may also have been a portal for the symbolic transformation of shamanic candidates into genuine dreamers. We can see what kind of power is transferred from these wombs of Mother Earth even if we can't be certain who were the beneficiaries.

Petroglyphs on another massive boulder at Counsel Rocks include a rayed disk that could stand for the sun. The zigzags that frame it have no conclusive meaning but could refer to irregularities of the mountainous horizon to the east. The odd design below the zigzag can be read as a vulva (oval bisected by vertical line) with legs, bent at the knee. (photograph E. C. Krupp)

There is a similar birth tunnel at a pictograph site in Joshua Tree National Monument that was compromised by motion picture production. On the Sierra Madre Ridge, a painted rock shelter known to the Chumash as the House of the Sun is entered through a short, tight squeeze that conveys the sense that entering the small domed chamber is an entry into another world. Reaching this shamanic sanctuary presumably involved a spiritual transformation symbolized by the process of birth.

Field research in the southern half of California's San Joaquin Valley by Mary and Jim Gorden has linked a Yokuts rock art site with initiation, birth, and the fertile womb of Mother Earth. They have emphasized the significance of the access route to one of the painted shelters in Round Valley. The entry is guarded by a painted figure who wears a distinctive, striped hat, split at the top and almost as tall as he is. One of his arms, bent at the elbow, puts his right hand behind his back. His left arm is extended with an open palm, as if conducting arrivals to their next destination. From this "sergeant at arms" a very narrow corridor conducts the visitor between high rock walls to a small roofless chamber. Water draining from the pilgrim's final destination runs across the stony floor, slippery with moss. Negotiating the jumbled boulders that obstruct the way, and crawling low beneath massive rocks and past small, isolated paintings, conveys you farther uphill and deeper into the rocks. After reaching another vestibule open to the sky, a damp passageway leads a little higher into the last room, a rattlesnake retreat and a gallery of painted women. At least some of them look like stylized females, although whether it's breasts or arms protruding from each side of the torso is hard to say. There are other figures here as well, but the ones with spread legs and upturned feet look like they have something to say about sex. We can reasonably guess that they are all spirit beings and that they have something to do with the visionary transformations experienced by shamans. Whoever went after that power had to reenter the body of Mother Earth before emerging with new knowledge.

WITHIN THE WOMB OF MOTHER EARTH

If suggestive natural clefts in the rocks are genital apertures of Mother Earth, caves must be her womb. Deep, dark, convoluted, and unfathomable, the cave conveys physically the character people have attributed to the generative interior of a woman. Just the fact that the cave is an interior space loads it with female symbolism. It is also a watery realm. As the home of underground pools, springs, lakes, streams, and rivers it seems to be the source of life-sustaining water. In limestone caves, dripping ceilings and damp walls help create the extraordinary features that turn the underworld into an uncanny kingdom of stalactites and stalagmites. Without electricity or modern lanterns, most of a cave is invisible. In prehistoric times, modest lamps with burning animal fat would reveal what was close, but only sound could disclose what might be beyond the reach of the light. Caves, like radio dramas, act upon the imagination, and for that reason alone they would be recognized as places of power.

Studies of generic symbolism make predictable assertions about the meaning of caves. They are the inner body of Mother Earth, and birth imagery links the cave with the navel that nourishes life. Mother Earth's vital rhythm is seasonal, however, and involves both life and death. She not only gives birth, she accepts the dead. Neandertal burials, the oldest ritual interments we know, returned the dead to Mother Earth in caves and rock shelters. One of these graves, discovered in Israel, is 90,000 years old. Detection of pollen sprinkled over the skeleton found in Iraq's famous Shanidar Cave led some, but not all, prehistorians to conclude that flowers were left as part of the funeral rites performed in 50,000 B.C. Often the body was left in a crouched position with the head pillowed on one of the arms. Stains of red ochre introduced the color of blood, life, and the monthly renewal of female fertility into funereal symbolism.

Caves, then, are not only the place of creation, they are the destination of the deceased. Seasonal renewal rolls away the stone and also makes the cave a center of rebirth. For that reason some people see the cave as the heart of the world and the center of the cosmos.

Burrowing animals make their homes inside the earth, and some of them hibernate in earth's cavities and caves. Because the earth's interior is a seasonal refuge for animals, the cave becomes the temporary retreat of the spirits that vivify the plants and animals. After a bear withdraws to the sanctuary of a cave for the winter, the young are born, and Mother Earth delivers more bears to the world the following spring.

Bears were especially effective symbols of Mother Earth's capacity for rebirth, for the bear reminded people of themselves. Bears are omnivorous. They can stand upright, and their bones resemble those of humans. People began incorporating bear symbolism into burials and other ritual arrangements in the paleolithic era. A Neandertal burial at Regourdou, in southwest France, was accompanied by a bear bone, and in a lower level in the cave, massive stone slabs formed a container for the jumbled skeleton of a bear. Paleolithic displays of bear skulls found in high altitude caves in Switzerland and Germany imply systematic storage with a symbolic purpose.

With the perspective of an historian of religion, Mircea Eliade acknowledged the importance of cave symbolism in the rebirth imagery of ritual initiation. The spirit that controls initiation into the Dancing Societies of the Kwakiutl, in the Pacific Northwest resides in a cave. Among Australian Aborigines, the Aranda shaman candidate must fall asleep at the entrance to a cave, and while unconscious, he is attacked by a spirit who kills him, takes him into the earth, replaces his internal organs with new, magical viscera, and lets him awaken as a new being. Mithraic mysteries of spiritual transformation were performed in a natural cave or an architectural replica.

If the interior of the earth is a sacred chamber where the soul gestates and is reborn, is the temporary residence of the creatures that seasonally counterfeit death, and is the final home of those who are truly dead, then the cave is the territory of spirits. That is how shamans regard caves. They see them as places of power. Whatever else may have been involved with the prehistoric use of caverns, these chambers of Mother Earth permitted access to the spirits and their power.

For the Lacandon Maya of Chiapas, Mexico, caves are sacred shrines associated with the dead. They operate, however, as homes for gods. The Lacandon will pray, burn copal incense, and leave other offerings for the divine husband and wife who reside in the cave, and through that interaction, they access the gods and their power.

The ritual significance of caves to the Maya of Yucatán was highlighted in 1959, when a guide, José Humberto Gómez, discovered deeper passages and chambers in the Cave of Balamkanché and also found extraordinary groups of jars and burners for offerings. This large limestone cave is just four miles west of Chichén Itzá, the ancient Maya ceremonial center that is world famous for its monumental pyramid, its sacred well of sacrifice, and its militaristic ornamentation. Although some of the ceramics were put in the cave almost 2,000 years ago, others coincide with the era just before the arrival of the Spanish. Most, however, belong to the centuries of Chichén Itzá's greatest political and economic strength, between 900 and 1200 A.D. They include incense burners decorated to portray the face of Chaac, the god to whom the Yucateco Maya assigned the power of control over the rain. Central Mexican influence, however, prompted the Maya to depict Chaac with the fangs and goggle eyes of Tlaloc, the god who presided over rain and storm on Mexico's altiplano.

Humberto Gómez took me on my first visit to Balamkanché in 1982 and explained how he found the sealed entrance to the inner corridor. Sitting alone against the back wall, where he and everyone else believed the cave's main passage ended, he was idly chipping at it with his knife when he realized the surface was not rock but stucco. In a moment, he broke through the masonry and realized there was more cave to be explored. Gómez found a massive limestone column about 500 feet from his new entrance to the underworld. Forged by the fusion of a stalactite from the ceiling and a stalagmite on the floor, it actually resembles a tree. Branches of stone all around the top hang like leafy boughs. Roots seem to bulge from the rocky floor. Humberto's flashlight caught a pair of red handprints on the pillar's "trunk." After walking up to the base of the "tree" for a closer look at the hands, he pointed his light toward the floor and discovered dozens of delicate incense burners and many more ceramic vases and bowls deliberately placed all around the column of limestone. There were also many miniature metates. The Maya use the full-size version of these flat stones to

In Mesoamerica, caves are portals to the otherworld, and a world axis makes commuting from one spirit realm to another easier. A massive tree trunk of limestone in Balamkanché Cave spreads into hanging branches of stalactites on the ceiling and widens at the base in a convincing incarnation of the world-axis tree that connected the underworld with its caves and waters to the middle world of human habitation and sky world of the celestial gods. About a thousand years ago, the Maya of northern Yucatán recognized the significance of this natural feature and made a shrine out of it with many ceramic vessels for offerings to the Maya equivalent of the god Tlaloc. (photograph E. C. Krupp)

grind maize, and the small ones left in the cave reinforced the agricultural connections of the offerings. Somehow, as Humberto Gómez approached the natural column, oblivious of everything the darkness concealed on the floor, he managed to avoid stepping on any of the fragile offerings for rain made on behalf of the maize.

Maya symbolism primarily associated the inner recesses of the earth with rain, water, fertility, and the cultivation of maize. Soon after archaeological investigation of the new grotto shrines was begun, Romualdo Hoyil, a prominent *h-men,* or Maya shaman-priest, told E. Wyllys Andrews, who was in charge of the work, that the tradition of a concealed sanctuary for Chaac in Balamkanché had persisted for generations among the local Maya. Now that it had been located, he wanted to organize a ceremony inside the cave to mollify the jaguar spirits who protected the interests of the rain gods. Their cave had been compromised, and immediate ritual reparations were required. A 24-hour ceremony of chants, offerings, and symbolic gesture began the next morning. When everyone—anthropologists, Maya priests, and frog-impersonating boys—emerged out of the smoking incense a day later, they were greeted by a downpour, an approval from Chaac.

James E. Brady, the foremost authority on Mesoamerican cave archaeology, emphasizes the role of the "Earth God" in Maya belief. The earth lord stabled the world's wild animals in his hollow, cavelike fortress within the mountain. There he also hoarded wealth, water, and grain.

Remarkable evidence of prehistoric ritual burial continues to be found in Mesoamerican caves. In 1994, a team of Honduran and American cave explorers found numerous bones from single and multiple burials accompanied by ceramics and jade in Talgua Cave in the Olancho Valley of northeast Honduras. Placed in the cave about 3,000 years ago, some of the bones and skulls were coated with calcite. They acquired these mineral deposits through the same process that produces the picturesque formations of limestone caves. Sparkling in the light of spelunkers' headlamps, the glittering bones prompted them to call their find "the cave of the glowing skulls."

In Mesoamerica, the earth's appetite for the dead was also transferred to monuments and temples that borrowed oral imagery from the mouth of the cave. This was evident early in the first millennium B.C., when Chalcatzingo was established as a highland ceremonial center with an Olmec overlay. A huge face carved in relief on a thin, flat slab of stone, about six feet tall, is believed to portray the "earth-monster." Its open mouth is actually a hole through the rock. In the hillside carving described earlier in this book, we think we see the same divinity in profile as the outline of the cave in which Chalcatzingo's rain king is enthroned. The Maya sometimes incorporated similar facial symbolism into the facades of their temples. During the Late Classic era (about 650–900 A.D.), the doorways of temples like Structure 2 at Hochob, in the Chenes region of east-central Campeche, and Structure II at Chicanná (which means "serpent-mouth house" in Yucatec Maya), in Campeche's Río Bec zone, are really fanged mouths that allowed a priest or king to enter a symbolic underworld to communicate with the spirits. Malinalco's temples for elite Aztec military organizations

Death is at home in the body of the earth, and caves provided underworld chambers for the remains of the dead. The "Cave of Glowing Skulls" in Honduras contained the bones and funeral remains of many burials. (photograph Gary Wirth)

A monumental door to the underworld was carved in relief on Monument 9, a freestanding stone at Chalcatzingo. The slab is six feet high and five feet wide and portrays the face of the earth monster that devours the dead. Living shamans who wish to enter that spirit realm must also pass through the monster's jaws. This seems to be the same mouth that frames the cave where the Rain King thunders out the storm like the caimans who roar at the sky and bring down the rain. In this context, too, the cave is the source of world fertility. (photograph E. C. Krupp)

were cut into the rocky mass of the mountainside, high above the valley floor, late in the fifteenth century A.D. The entrance of its Temple of Eagle and Jaguar Knights is a serpent's face, and the snake's forked tongue protrudes from the door like a unrolled carpet. Bernal Díaz del Castillo, who was with Hernán Cortés for the Conquest of Mexico, described the entrance of one of the temples he toured: "... one of its doors was in the shape of a terrible mouth, such as they paint to depict the jaws of hell. This mouth was opened and contained great fangs to devour souls." In Aztec ritual, as elsewhere, sacrifices of the living sustained the gods. They symbolized the death that precedes, and seems to make possible, the subsequent rebirth of life, the renewal of the seasonal cycle, and the reestablishment of the world order.

Providing admission to the domain of the lord of the earth, caves were regarded as portals to the supernatural underworld and the realm of the dead. For this reason, the cave had the symbolic power of the sacred center of the world, where contact between

The entire western facade of Structure II at Chicanná, a Late Classic (600–830 A.D.) Maya site in the Río Bec region of southeastern Campeche, Mexico, is the stylized face of a reptilian monster. In fact, the name *Chicanná* means "serpent-mouth house," and the doorway is the monster's maw. The nose is right above the entrance, and eyes are on either side of the nose. Ear flares buttress the eyes. Fangs hang over the top of the door and continue all around it in an upside-down *T.* According to Linda Schele, buildings like this are images of the *Witz* Monster, "the symbol of the living mountain." If Maya rulers performed shamanic rituals and sought visions in these chambers, they had to walk through the monster's muzzle to reach the spirits within the mountain cave. The king personified the World Tree that allowed the supernatural to come in contact with our part of the cosmos. Each Witz Monster temple was Grand Central Station where shaman-kings caught the spirit express to the underworld and to the sky (photograph E. C. Krupp)

The entrance to the interior of Malinalco's Temple of Eagles and Jaguars splits the face of snake and lets its tongue roll out like a red carpet for diplomats to the Otherworld. (photograph E. C. Krupp)

the levels of the universe is possible. Some experts are persuaded that the Maya saw this kind of cosmology in the treelike pillar of Balamkanché. According to the Maya, a cosmic ceiba tree grows at the center of the world. It is a transcendental axis erected by the Maize God, also known as First Father, and it connects the underworld, the earth's surface, and the heavens.

Brady sees caves as "sacred space that imbued their surroundings with the qualities of power, prestige, abundance, and sacredness." For that reason, Maya elite incorporated caves into the symbolism of political power. Affiliating themselves with the supernatural power of caves, Maya kings anchored their authority to the center of the world and the source of its vitality. Although these are all concerns of a settled, agricultural community in which power is centralized, the broad themes inspired by the natural architecture of the cave—hidden shelter, the female capacity for life, renewal of nature, and contact with the spirit world—very likely also seized the interest of the paleolithic hunters of Ice Age Europe.

MOTHER EARTH IN THE ICE AGE

The discovery of upper paleolithic cave art was first publicized in 1879. Although images of animals had been seen in the cave of Niaux in France 13 years earlier, no one seriously attempted to evaluate their age. At Altamira, however, in the Pyrenees of northern Spain, the daughter of Don Marcelino de Sautuola spotted large pictures on a low chamber ceiling while the pair was searching for Old Stone Age tools, bones, and hearths on the cave floor. Because Altamira's elegant, 12,000-year-old paintings were not primitive enough to satisfy preconceptions about prehistoric art, the experts decided they were a hoax. We are, after all, talking cavemen here, people still too low on the ladder of social evolution to carry on respectable agriculture and to live in proper houses. Altamira's refined polychrome pictures of bison, horses, and other beasts were eventually accepted as genuine, however, by the beginning of the twentieth century, after several similar painted caves had been discovered in France. Since then, more than 200 illustrated paleolithic caves have been revealed, primarily in the Dordogne of France and on both flanks of the Pyrenees. Whether or not the artists thought of caves as the womb of Mother Earth, most of the images painted on those interior walls are animals, and the artists did have a stake in the seasonal return—and rebirth—of the creatures that defined their world.

Astounding sanctuaries of prehistoric art continue to be found. In July 1991, Henri Cosquer, a French professional scuba diver, spotted stenciled handprints, incised markings, and paintings of Ice Age animals, on his fifth visit to a cave now known as Grotte Cosquer. The painted grotto can only be seen by negotiating a long narrow and dangerous water-filled passage, but in the Old Stone Age, when the sea level on the French Riviera was lower, the entrance was dry. Now starting 120 feet below the surface of the Mediterranean, a diver must swim 525 feet and 12 minutes (in a tunnel that scarcely admits the oxygen tanks) before reaching the chamber. Not long after Cosquer found the paintings, an unauthorized entry attempt ended with an accident that killed three divers.

Cosquer Cave's 145 prehistoric pictures of bison, horses, the extinct great auk, and other creatures, drawn in black charcoal, have been dated as 18,000 years old. The handprints were outlined even earlier, in 25,000 B.C. Neither era left any evidence of habitation in the cave, just its use as a painted sanctuary within the earth.

Cosquer Cave could have kept prehistorians satisfied for awhile with unprecedented images, but in December, 1994, Jean-Marie Chauvet, an official of France's Culture Ministry, found another new cave, in the valley of the Ardèche River of southeastern France, with more than 300 paleolithic paintings and engravings. Along with the usual bison, mammoths, horses, and woolly rhinoceri, Chauvet Cave has pictures of bears, lions, aurochs (wild prehistoric cattle), reindeer, a leopard, owls, and geometric signs. Someone in the Old Stone Age left a bear skull upon a rocky slab, probably as another nod to the principle of renewal embodied in bears and incubated in caves.

From the first recognition of prehistoric paintings at Altamira to the stunning discoveries in Grotte Cosquer and Chauvet Cave, we have tried to understand the meaning of this Ice Age art and the function of the caves. Initially, it was argued that the animal paintings were part of imitative hunting magic. Before undertaking an expedition for game, the hunters would symbolically kill the animal in the painting to help ensure a successful hunt. More systematic study of the caves showed, however, that most of the paintings don't include wound marks. We also know that the animals illustrated in the caves played only a small part in the prehistoric diet.

Assuming that the proprietors of the art recognized an ancestral bond with particular beasts, and avoided eating them, some interpreters saw a totemic function in Ice Age art. The paintings, then, would be clan emblems that identified territorial rights within the cave. The actual distribution of species through underground chambers and corridors, however, doesn't seem to delineate unambiguous prerogatives of lineage. The animals are more mixed than we would expect.

The modern tradition of museums and galleries proudly possessing collections of work by acclaimed artists prompted some to endorse the idea that the cave paintings represent art for the sake of art. The display value of art has created a cult of status

Artists of the Old Stone Age herded bison and other animals into the body of Mother Earth as paintings on the walls and ceilings of French and Spanish caves. Various scripts have been suggested to account for the presence of well-crafted images like this bison from Font-de-Gaume near the French town of Les Éyzies-de-Tayac. We don't know if the paleolithic hunters of Ice Age Europe saw themselves as dependents of Mother Earth or if they believed they had to interact with a supernatural Master of the Animals that governed the animals' seasonal release, but their understanding of cyclical renewal and the generational continuity of life very probably prompted them to imagine the interior of the earth as a womblike corral from which the animals would come and to which they would seasonally return. (from *The Rock Pictures of Europe* by Herbert Kühn, Essential Books, Inc.)

that qualifies today's art museum as a ceremonial center. Social power is exerted and sustained there through objects of prestige and aesthetic appeal. While this principle of public exhibition is at work in our time, the painted caves of France and Spain were too poorly lit to function as museums and too redundant to be interpreted as an expressive artist's personal statement.

Because almost all of the caves reveal no sign of domestic use, they are now almost universally regarded as sanctuaries for religious rites. Some see the caves as dark retreats within Mother Earth where youths were transformed through initiation ceremonies into adults. Others recognize the shamanic dimension of the caves. Visions experienced in the darkness allowed the shamans to accost the spirits and acquire power to perform cures and secure game. In fact, most of the spaces in most of the caves can't really accommodate a crowd. So little evidence of traffic exists in some of them that we can believe they may have been entered only once or twice, when they acquired their art. Even those that were used repetitively probably hosted only small groups or individuals. Certainly there are some exceptions. Lascaux's Hall of Bulls is about 55 feet long, 22 feet wide, 19 feet high, and fairly accessible. It is a large room, and with enough lamps, the largest paintings—bulls between 16 and 18 feet long—certainly could have been viewed by a few dozen people assembled at one time. That doesn't mean, however, that the horses, aurochs, and stags in the main room ever enjoyed a large

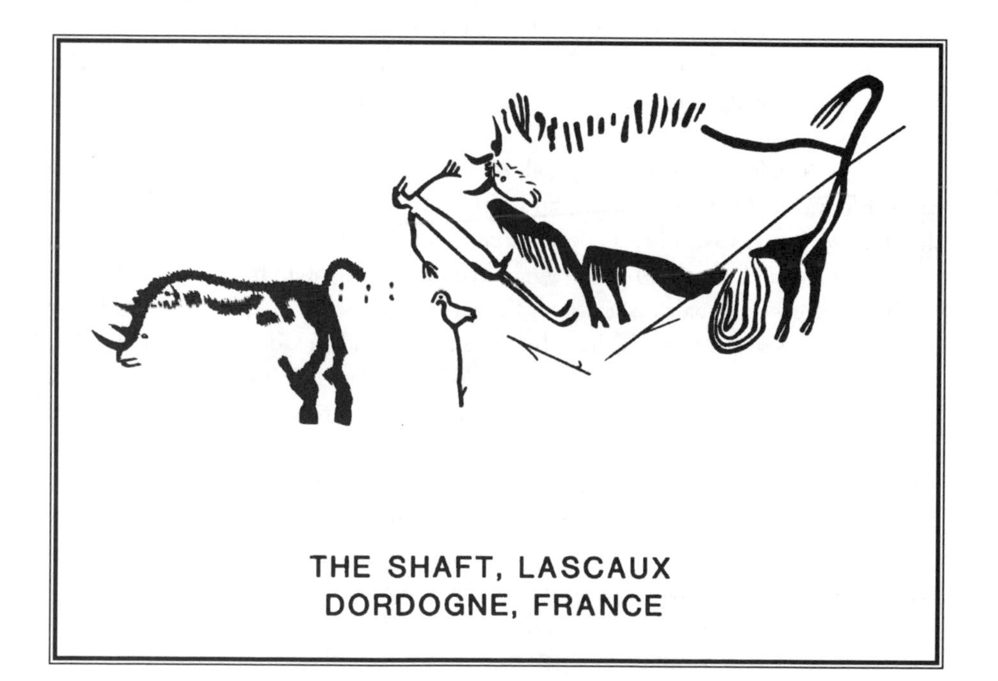

Shamanic trance in the Old Stone Age may be portrayed by the stupefied figure in front of the mortally wounded bison in this image from the Shaft, a hard-to-reach vestibule in the cave of Lascaux. (Griffith Observatory drawing, Joseph Bieniasz)

audience, and other zones of Lascaux wouldn't permit it. Only a few persons at one time could have ever reached the Shaft, a deep pit at the end of a side chamber off one of Lascaux's main passageways. There, a 23-foot descent through the dark vault, presumably by rope, seems to have been the only way the Lascaux artists dropped to the cave floor. Some assert this hard-to-reach pit is the true heart of the cave. Its primary painting has provoked a great deal of commentary, and at first glance it seems to qualify as hunting magic. On the right, a wounded bison endures the loss of intestinal loops from its belly. The spear that disemboweled the animal is still there, and some argue the bison was struck in its genitals. A bird-headed man with an erection appears to have dropped dead or unconscious in front of the bison. Near him is a vertical rod on which a bird is perched, and there is perhaps another dart on the ground. A woolly rhinoceros departs to the left, and there are six dots arranged in two rows, below its tail. In an alternative approach the scene was interpreted as a narrative, the story of a hunter killed by a bison and the bison subsequently gored by the rhinoceros.

There is, of course, no way to know what the artist really had in mind, but the painting also has shamanic potential. The man could be laid out in a trance. The bird imagery is consistent with the symbolism of the flight of the soul, and the battle with the bison actually may be a contest with the spirits or a metaphor for shamanic trance. If so, a large game animal in the upper paleolithic had the same kind of meaning the San see in the eland. Petroglyphs of bighorns in California's Coso Range and the antelope and deer in the great murals of Baja California's Sierra de San Francisco may have fulfilled the same purpose for prehistoric hunters who left the art on the rocks. In any case, the enigmatic painting in the Shaft was never intended for public consumption.

Professor André Leroi-Gourhan analyzed the distribution of animal types in many caves. He was looking for patterns, and different animals, he concluded, dominated different zones in a cave. He also thought he could see gender opposition in the bison and the horses and in other figures. The horses he judged were male elements, while the bison expressed female associations. Sexual connotations could be intended in the use of these two quadrupeds, but the system also had broader implications. Any dynamic and complementary opposition could be indicated with such symbolic pairs. Critics have challenged the concept of clearly defined cave topography and also regard Leroi-Gourhan's gender assignments as arbitrary. These are valid complaints, but intriguing differences in the painted wall surfaces have been verified. Bison and cattle almost invariably appear on bulging, convex walls and ceilings. Horses, on the other hand, and most deer occupy concave surfaces. In addition, the notion of complementary opposition is a natural way to organize the process of cyclical change. It is a common structuring principle of myth, and the concept is also reflected in seasonal ritual and ceremonies of renewal. Although the evidence for polarization in paleolithic art is inconclusive, the idea is plausible. If present at all, it may have something to do with an Old Stone Age wish for the seasonal rebirth of the world and for the transformation of the soul at life's key transitions.

By this point, perhaps, the profusion of interpretations and the difficulty of testing any of them may seem discouraging and confusing. It helps to stick with the basics. Upper paleolithic cave art is mostly animals, and most of them are big game herbivores. Bison and horses dominate the underground menageries. For example, Font-de-Gaume, a cave just 1½ miles southeast of the village of Les Éyzies-de-Tayac, the "capital of prehistory," on the left bank of the Beune River, is fundamentally an underground parade of bison. Discovered in 1901, the 16,000-year-old paintings and engravings of Font-de-Gaume were among the first in France to be accepted as authentic. After walking about ¼ mile from the parking area, visitors reach the entry. Taller than it is wide, the female entrance conveys in natural rock the mystery of birth. It faces west and is about 66 feet above the valley floor, halfway up a limestone palisade. From the door, it's almost 200 feet of straight, unadorned, single-file passageway. Its width varies between 6 and 9 feet, and it is about 25 feet high. Where the tunnel then narrows and descends, the paintings begin. Bison march in file on both sides of the corridor. Many are headed back toward the entrance, but others are advancing deeper into the cave. There are other animals as well—including a short line of mammoths and a male and female reindeer facing each other—on this subterranean thoroughfare and in a passage that branches to the side. Still flanked by bison as you penetrate farther, you reach a small "chapel," the Cabinet des Bisons, with

Like San pictographs of the dying eland, the wounded bison in Lascaux and the hunting of bighorn sheep in the prehistoric petroglyphs of California's Coso Range may be local expressions of shamanic metaphor and not the wishful thinking of hungry hunters. (photograph E. C. Krupp)

bison swelling from its ceiling. Beyond that, the main corridor dead-ends a little over 400 feet from the entrance. With a fairly simple layout, Font-de-Gaume is not a labyrinth, but it could be a vagina and a womb.

Although the individual paintings of bison and other creatures in the cave are what first attract the eye, they accumulate in the mind and become a great herd on the move. There aren't just animals here. There are a lot of animals here in the womb of Mother Earth, and it's as if nature is getting ready for its own cattle drive. The body of the earth is full with animals that migrate seasonally and reproduce seasonally. Font-de-Gaume seems to be animated by those expressions of seasonal change.

Of course, the place where we find the paintings is as important as the paintings themselves. It is a cave, and caves are dark, protected cavities in the earth. They are interior space but not architecture. Their configuration is an aspect of nature, found not engineered. Although caves are embellished symbolically, they are not structurally altered. People start with what nature provides, not a clean slate. The boundaries of a cave cannot be seen in the darkness, and so caves have itineraries rather than floor plans. They are not territory but environment. Even though we can't always verify specific functions of the caves and their art, we can identify the primary themes—transformation and renewal. Caves are places of creation and places of power. They induce emotional reactions that can be channeled through ritual to reinforce beliefs that perpetuate the community's values and identity. The art we see in them and the activity that took place in them reflect access to the power of cyclic renewal.

Alexander Marshack detects signs of ritual renewal in the Ice Age art of Europe. As the author of *The Roots of Civilization,* he is probably best known for assembling evidence for paleolithic use of time-factored symbols. He has identified what he believes to be notations modulated by the phases of the moon and figurative references of seasonality. One of two rump-to-rump, galloping bison painted in Lascaux, he observed, is in its summer molt. The other wears the full black coat of winter. A bone knife found at La Vache, a rock shelter dwelling in the Pyrenean foothills of southern France, is, for Marshack, a lesson in seasonal ecology. One face is engraved with symbols he identifies with spring—a doe, wavy "water" symbols for the runoff, and flowering plants. On the other side of the blade, the message is autumn—a bellowing bison in rut, bare branches or evergreens, a drooping flower, and assorted nuts and seeds. In a similar way, mating seals, snakes with conspicuous genitalia, a sprouting plant, a flower, leafy branches, water bugs, and a salmon carefully drawn to include the jaw hook the males develop when spawning all endow an engraved bâton from Montgaudier Cave (near the Le Placard rock shelter) with the burst of annual renewal delivered by the spring thaw. Marshack's seasonal analysis of the details seems convincing now, but to prehistorians nursed on hunting magic and decorative intent, his ideas were unprecedented.

Through seasonal symbolism and patterned marks, Marshack seems to be able to enter the Ice Age mind. He finds evidence for the theme of cyclic renewal everywhere in upper paleolithic art, including the new discoveries of Chauvet Cave. Drawings of woolly rhinoceri there were later reoutlined on several separate occasions. New horns

Alexander Marshack recognized paleolithic imagery of two different seasons on opposite sides of an engraved bone knife found at La Vache. The doe and flowering plants on the left say, "spring." On the right, a bison announcing the intentions of his autumn rut, seeds, evergreens, and a flower at the end of its seasonal rope all refer to fall. (from *The Roots of Civilization* by Alexander Marshack)

were added to one rhinoceros portrait at least a half-dozen times. What looks at first like a tight column of rhinos in profile is actually a single rhinoceros ritually and repetitively renewed. The rhino's midbody skin folds, represented in the drawing by black vertical lines, are exposed in summer when the animal's coat is thin. Other sites provide similar rhinoceri of summer with symbolic wounds. According to Marshack, these are not signs of hunting magic but ritual killings of an image of an animal that probably was not normally hunted for food. Instead, he thinks these pictures allude to death and to the inevitable period of dormancy that comes with seasonal change. The multiple horse heads of Chauvet Cave portray the animal with its summer muzzle and also project seasonal references onto the concept of renewal.

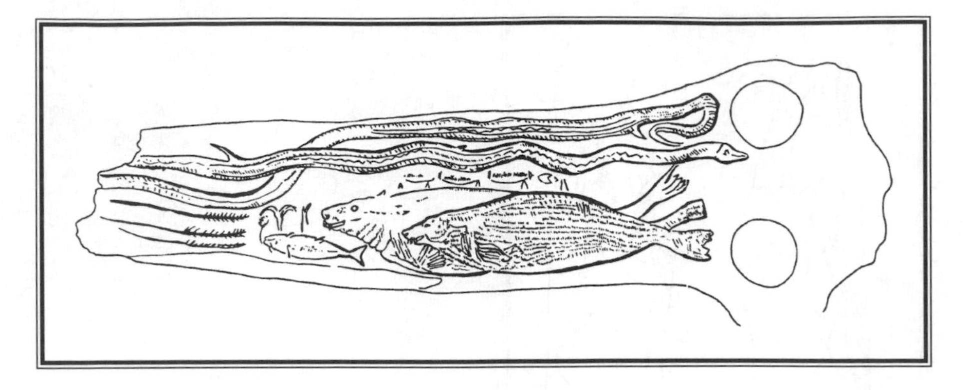

Seasonal references on the engraved bâton from Montgaudier are detailed and unmistakable. In this drawing, the object is "unwrapped" to reveal the entire design, and it includes genitally explicit snakes, a male and female seal, and a spawning salmon—all emblems of springtime in the Charente Valley. (from *The Roots of Civilization* by Alexander Marshack)

We don't know exactly what kind of renewal the Solutrean and Magdalenian hunters of Ice Age Europe had in mind when they transformed the inner body of the earth into subterranean wildlife reserves. Perhaps they were just concerned with the annual return of the game. Or, anxious about the seasonal continuity of the world itself, they may have used the animals to symbolize the earth's cyclic rebirth. If the caves were connected with the idea of birth, the resurrection of the animals could have had something to do with human reproduction. The continuity of culture, reinforced by initiatory rites, could have been another goal of symbols related to seasonal transformation. And, finally, the alchemy of the soul, achieved through ritual, visions, and trance, may have been expressed with the imagery of cyclic renewal. In analogical thought, one cycle of ordered change is like every cycle of ordered change, and the cave art of the upper paleolithic may have embraced not one, but all, of these functions. We shall never really know exactly what took place in the caves and why it occurred. We can identify some of the fundamental themes implied by the images and the environment, and the presence of the concept of cyclical renewal confirms that even in the Old Stone Age it was an organizing principle in our understanding of the cosmos.

To the Ice Age hunters, it wasn't so much structure that defined the universe as process. The world was not a place but action. This may be why we see so little celestial and topographic imagery in paleolithic art. Certainly seasonal change drew Old Stone Age eyes to the sky. They may have seen the stars as great hunters, as the !Kung do today. They may have marked time according to the phases of the moon and sensed power in the rising sun. But when they looked for a metaphor of the cosmos, they found it in the animals, whose behavior conformed to the rhythms of the rest of nature. They hitched themselves to the universe by forging a mystic rapport with the animals, a rapport negotiated by their shamans and cemented by their paintings.

For the nomadic hunter, the world is a fluid, changing landscape, soft around the edges, everywhere in contact, and centered in the mind. Supernatural power is dispersed, not concentrated at the center, and the shaman encounters the spirits around every corner. It isn't just the sky that is cosmic. Everything is cosmic. Every hunter is an off-road vehicle. For herders and farmers, on the other hand, the land is mapped like a system of freeways. The useful territory is a landscape of routes and destinations. The surrounding territory is wilderness—interesting, perhaps, and evocative, but unpacified, unpredictable, and dangerous.

Roberte N. Hamayon, a French ethnologist and expert on central Asian shamanism, sees a difference between hunting shamanism, which emphasizes supernatural exchange with the animal spirits, and pastoral shamanism, in which the spirits are more like humans. According to Hamayon, ancestors, spirits, and the mythical founder of the tribe converge as sources of transcendental power. Land becomes linked with lineage, and the geography begins to have a center and an edge. Social hierarchy remolds the egalitarianism of the hunting band, and the vertical axis of the cosmos accumulates greater meaning.

Hamayon believes Old Stone Age hunters moved horizontally through the universe. For nomadic hunters, landmarks indicate where things happen. In a society of hunter-herders, landmarks also define territory. They encourage centralization of power and impose a social template on the land. The sky no longer blankets the earth but instead makes contact with it at significant locations that impose a structure onto space. Sacrifice and reciprocation are still necessary, but the sacrifice is not the animal killed in the hunt. It is instead the animal raised domestically. Fertility, measured in rain for grazing land and in the growth of flocks and herds, rather than sustenance, measured in productive foraging and successful hunts, is what the spirits provide.

Aspects of hunting shamanism persist, of course, in pastoral shamanism and even in agriculturally based chiefdoms, in sacred kingships, and in imperial states, but as people become anchored more firmly in place, as village-based foragers, as seasonally migrating herders, and as landbound farmers, their concept of the universe and the functions they attribute to the sky evolve. The locomotive of cyclic renewal still pulls the seasonal train, but the destinations on the timetable depend on the economy and the complexity of the culture making the trip.

CHAPTER FIVE

AGENTS OF RENEWAL

Just because the upper paleolithic artists carved sexually-charged images of abundant women and emphasized their procreative possibilities, that doesn't mean they imagined the earth itself as a Great Mother. They did understand, however, the indispensable role of women and female animals in replenishing their world with life. For that reason, we are not surprised to find symbolic projections of female power (the maternal power to bear life) and of seasonal change (the most immediate evidence of death and rebirth) onto the art. The art reflects observation and belief.

When the way of life shifted from nomadic band to permanent village, the transition did not diminish the divine and cosmic power of the earth. The underlying principle, after all, is constant: Life is given, and life is withdrawn. But herders and farmers, peasants and kings, did reformulate its imagery to match the landscape of their lives. They retained the vocabulary of seasonality and of female sexuality in one form or another even though artistic conventions and human interaction with nature evolved. Caves, graves, and womblike temples sheltered the spirit of Mother Earth in her journey from the world of the Old Stone Age hunters into the world of the New Stone Age farmers, even as she abandoned her old wardrobes. Whether the power of the earth is attributed to cosmos at large, recognized in the land itself, or personified in a god, we continue to encounter the belief that it sustains the rhythmic renewal of life. We find that belief in references to female fertility and in allusions to sacrifice and death.

FROM CAVE TO TEMPLE WITH MOTHER EARTH

Settled on the land, then, with an altered vision of the cosmos, people interact differently with the earth through their symbolic systems. Familiar themes like Mother Earth are not abandoned but modified. The sanctity of caves persists. For example, two different caves on Crete are claimed to be the place where Zeus was born. One is a low, dark, damp, and slippery sanctuary among the thick oaks on the hillside of Mount Dikte above the village of Psykhro. The Minoans, who used this shrine long before the Greeks introduced the Olympian gods to Crete, dedicated its worship to their mother goddess. Votive offerings were found in the chamber now said to be the birthing room of the earth goddess Rhea, mother of Zeus and his Olympian siblings. The other alleged birthplace of Zeus is a cave high on Mount Ida, near the center of Crete. It was in ritual use as early as the Bronze Age of Minoan Crete. Also illustrating the maternal power of the earth's interior, the Cave of Eileithyia, at Amnisos and about six miles east of Iraklion, was dedicated to the goddess of childbirth. It overlooks the Mediterranean and was mentioned in the *Odyssey* by Homer.

Long before the Minoans were offering sacrifices to goddesses in otherworldly caves, the people of Malta were fashioning limestone slab temples whose rounded chambers are reminiscent of the inner body of Mother Earth. These are believed to be the first freestanding stone temples in the world, and the earliest were built around 3500 B.C. by prehistoric farming communities. There is no evidence of paleolithic occupation on Malta. The archipelago was first colonized at 5000 B.C. by immigrants who probably sailed from Sicily, about 50 miles to the north. They cultivated wheat and barley and also kept goats, sheep, cattle, and pigs. The Mediterranean provided fish and other maritime resources.

There is nothing quite like the megalithic temples of Malta anywhere else in the world. Massive, high, corbelled, and probably roofed with timbers, these monumental sanctuaries had concave facades that created an exterior forecourt for the entrance. Crossing the threshold, you continue on a short passage that opens into an inner central court. Doorways connect the court with semicircular chambers on each side. From the far end of the court, the central passage, depending on the design, ends in a final apse or accumulates additional lobes on the sides. Subsequent stages of construction complicate the layouts of temples with shared walls and odd intersecting angles, as at Tarxien and Mnajdra. The conventional four-chambered core of Hagar Qim was transformed by making a passage out of one of the side apses to provide access to an architectural swelling larger than the primary temple. It encloses a radiating set of four oval rooms.

No one really has a clue to explain the anomalies of each Maltese temple. A couple of them, Mnajdra and Hagar Qim, have been shown to accommodate astronomical alignments with the rising sun at solstices or equinoxes, but taken as a group, the 23

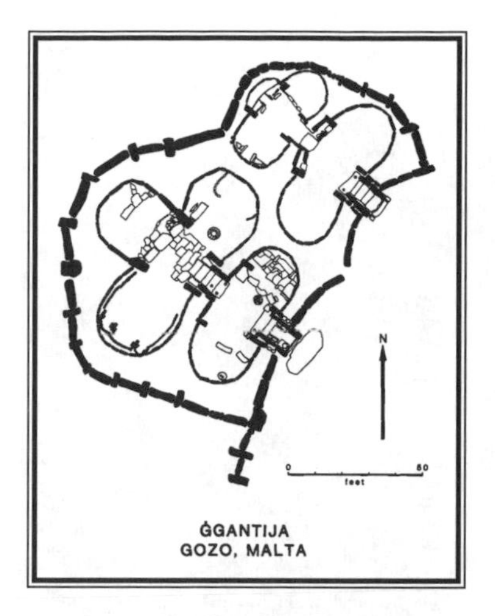

The Ggantija (giants' tower) temples on Gozo, the second largest island in the Maltese archipelago, share the same megalithic walls. Their multilobed interiors have been likened to the stylized outline of full-breasted, wide-hipped women and to the cavernous womb a woman's body contains. Neither image is supported by the actual outline of the chambers, but the dark and enclosed temple very likely symbolized the fertile interior of Mother Earth or the goddess of cyclical renewal. If the first pair of chambers really represents the thighs of the goddess, sexual accessibility is what permits entry into the temple. (Griffith Observatory drawing, Joseph Bieniasz, after J. D. Evans, *The Prehistoric Antiquities of the Maltese Islands*)

known temples don't adhere to any clear pattern of orientation. The British archaeologist Colin Renfrew believes their geographical distribution reflects the territorial prerogatives of the separate communities that shared and subdivided the islands.

What really strikes you about the Maltese temples is their curves. Although they have some rectilinear features—doorways, for example—it's hard to find an ordinary corner anywhere. The perimeters are round. The chambers are round, and all those curves probably stimulate a subconscious intimation of women. Hallie Iglehart Austen voices this subjective impression in her book *The Heart of the Goddess—Art, Myth and Meditations of the World's Sacred Feminine:*

> . . . they were made in the shape of the Goddess's body, with her womb as the ritual inner chamber and her vaginal opening the entrance and the exit.

A little overconfident perhaps, but many see hips and breasts in the outline of the inner floor plan and equate the interior volumes of the chambers with an internal cavity of procreation.

Women, we know, had something to do with religion in prehistoric Malta, for many female figurines have been found in the temples. Exaggerated hips, breasts, buttocks, and thighs are sometimes supplemented with exposed pubic triangles and explicit genitals. Although the women are usually completely nude, they sometimes wear a flounced skirt. To call them full figured would be courteous. Most interpreters regard them as an earth-mother-goddess of one sort or another, and a few offhandedly refer to them as the "fat lady." Her corpulence has been accepted by most as indicative of her talent for fertility and renewal. A massive statue of her, originally nine feet high but now intact only to the waist, dominates the first apse on the right as you enter the South Temple at Tarxien. David H. Trump, a professional archaeol-

Although the roof is gone and the upper walls have fallen, it is still possible to sense what the environment inside the larger (south) Ggantija temple was like. From high on the back wall, the view is southeast and toward the entrance. (photograph E. C. Krupp)

ogist who has worked in Malta and elsewhere in the Mediterranean, is convinced this statue acknowledges the ancient mother goddess of Malta and writes about her in *Malta: An Archaeological Guide:*

> She wears a very full pleated skirt. It would be ungentlemanly to quote her hip measurements, and her calves are in proportion. She is supported, however, on small, elegant, but seriously overworked feet.

The "Goddess Movement," in part a by-product of recent efforts to advance the status of women, identifies her as the primordial Great Goddess worshipped in matriarchal societies, run by women in a golden age of peace and a time of harmony with the natural rhythms of the earth. This, of course, may all be true, but it is also a story that for some serves an agenda of our time. The story in prehistoric Malta may not have been so simple. In some of these carvings, the gender is ambiguous, and the apparently female character of the temples and of most of the statues could as easily indicate male preoccupation with sex and procreation in a system over which priests, and not priestesses, presided. In this case, gender preference is probably arbitrary, and every person may, in ignorance, indulge personal prejudice.

Whoever it was who performed the rites and managed the temple, the circumstantial evidence of architecture and female imagery in Malta favors the theme of cyclic renewal in neolithic belief, just as caves and animal seasonality imply a similar princi-

Further evidence of a female divinity in the prehistoric temples of Malta is provided by the surviving skirted lower torso of a goddess built for durability in the temple of Tarxien. (photograph Robin Rector Krupp)

ple in paleolithic religion and concepts of power. Temple design in Malta was traced by archaeologist J. D. Evans to earlier rock-cut tombs with similar multilobed layouts. Traditional associations with tombs and possible evidence for animal sacrifice, especially at Tarxien, are understandable features of Malta's temples because death is an inevitable element in the myth of rebirth. Some kind of ritual sacrifice often accompanies ceremonies for renewal, and spiritual contact with the dead is sometimes thought to provide guidance for the living.

In Malta, a complicated system of subterranean rooms, the Hypogeum of Hal Saflieni, seems to have brought the living and the dead together in an unusual combination of temple and tomb within the body of the earth. The Hypogeum was discovered accidentally in 1902 through excavation for residential development on a hill just a few hundred yards to the west of the temples of Tarxien. The uppermost room is nine feet below the surface of the ground, and a three-story honeycomb of staircases, corridors, and 30 rooms was tunneled into the limestone to a depth of 35 feet. The bones of perhaps 7000 individuals had been interred in the Hypogeum, but it has other features that make it more than a cemetery. Some of its rooms were cut to mimic the megalithic uprights and lintels of the temples. The facade of the so-called "Holy of Holies" has four false support blocks and a pseudo crossbeam all carved to a delicate curve. Other details, including side chambers, pits, and altars, imply ritual use of various rooms, and the famous Sleeping Lady terracotta statuettes were found here. In these, the prodigiously endowed woman, bare-breasted in

This formal chamber and sanctuary deep within the Hypogeum is sculpted from the inner bedrock to resemble the megalithic construction of Maltese temples. The vaulted ceiling may indicate how the temple chambers were actually roofed. (photograph E. C. Krupp)

an ankle-length fringed skirt and with eyes closed, reclines on a simple couch. We cannot really tell if she is the Goddess, the mother in the earth, who sleeps in seasonal dormancy and pregnant with the world's life grows huge. Perhaps, instead, she is a pilgrim or an oracle who descended into the earth for a dream chat with the spirits in Mother Earth's condominium for the dead. Asleep, entranced, or dead, her soul is on the move.

Further confirmation of affiliation between the temples, the spirits of dead, and the interior of Mother Earth is provided by recent excavation on Malta's largest companion, the island of Gozo, which is separated from the main island's northwest shore by three miles of Mediterranean and the small island Comino. About six thousand years ago, the inhabitants of Gozo cut a two-chamber tomb into the rocky plateau to bury the dead and periodically removed and rearranged the bones as others were deposited there. In time, additional pits and tombs were extended to a small system of caves. The cave chambers were architecturally enhanced with the kind of megalithic features we see in the later temples, and this burial ground, now known as the Brochtorff Circle, took on some of the character of a shrine. Along with more human remains, bundled in some cases and burned in others, small sculptures of the familiar mother goddess and some animal burials were ritually left. Because the underground chambers are not very deep, they were probably roofed. In fact, the site looks

The Hypogeum's Sleeping Lady may be dreaming sweet dreams till sunbeams find her or picturing tangerine trees with marmalade skies. It's hard to tell, but she is probably sleeping with the ancestors to birth visions in the womb of Mother Earth. (photograph E. C. Krupp, object displayed in the National Museum of Malta in Valletta)

a little like the footprint of a typical chambered temple, only here the foundation is below the surface of the rock, and a temple actually vaulted the collapsed cave at the center of the megalithic circular enclosure. The temple and the tomb were in direct contact. In neolithic Malta, then, the continuity from tomb to temple seems to have served rituals of renewal that required the participation of the living and the dead.

THE SEPULCHRE OF THE SUN

More than five thousand years ago, a little after the time the prehistoric people of Malta started erecting their first cavelike temples in stone, the Irish were burning their dead and depositing the remains in monumental passage graves that also imitated the birth chambers of the earth. These neolithic tombs appear to incorporate the same themes of death and renewal that we find in Malta, but in some of them the sky may be playing a more active role in fertility. In alignment with the sunrise or sunset at a seasonally significant station in the sun's annual itinerary, they admit warm—and perhaps revivifying—sunlight into the tombs of the dead. The artificial caves housed their bones—and perhaps offered sanctuary for their spirits—like the womb of mother earth.

Irish passage graves are described as megalithic structures because the skeletal frames that comprise their corridors, chambers, capstones, and curbs are large (*mega*) stones (*lith*). Probably the best known of the 300 or so passage graves cataloged in the Irish countryside is Newgrange, located in the so-called Bend of the Boyne River, about 26 miles north of Dublin and just east of Slane. The whole area is a large prehistoric cemetery dominated by Newgrange and two other nearby passage graves known as Knowth and Dowth. Written references to Newgrange appear as early as 656 A.D. in medieval Irish annals. It was already a ruin in the Roman era, and the ancient Celts before that probably regarded it as an abode of spirits and a former home of their predecessors in Ireland. The Irish Celts called these mythological ancestors "the peoples of the goddess Danu" (*Tuatha Dé Danann*). As a divine ancestress, Danu filled the role of mother goddess, and there is reason to link her with Anu, another Celtic mother goddess who sponsored fertility and prosperity in the land. In any case, the Celts regarded the founding fathers and mothers of Ireland as a supernatural presence in the landscape. According to some Celtic traditions, Newgrange belonged to the Daghda ("the good god"). He was a divine tribal patriarch, and he mated with Boann, who was not only the spirit of the Boyne but the nourishing river itself. One end of the Daghda's club could slay the living while the other brought the dead back to life.

Newgrange is a megalithic surprise. The white facing of the prehistoric chambered passage grave is architecturally austere but flamboyant in its contrast with the surrounding terrain. It is not integrated into the landscape with the sensibilities of Frank Lloyd Wright. Instead, it is imposed on the countryside like a highway tourist attraction. It is a tuxedo button on fields not used to formal attire, and its insistent quartz facade redirects all attention its way. (photograph Robin Rector Krupp)

To the Celts, the abandoned prehistoric tombs were portals to the dangerous Otherworld within the hills and beneath the earth. They called these gates to the underworld the *sídh.* The ancestral spirits, who now occupied them, returned back into this world at certain seasonal junctions when the border with the supernatural was frayed.

Although modern archaeologists recognized that Newgrange was not the work of Iron Age Celts or their spirit ancestors and belonged instead to New Stone Age farmers, no one knew how old it really was until the late Michael J. O'Kelly excavated and restored the site between 1962 and 1975. His organic samples, according to recalibrated radiocarbon dating, place the construction of Newgrange in the last centuries of the fourth millennium B.C., probably around 3200 B.C. It is then older than the Great Pyramid (about 2600 B.C.) and the earliest phase of Stonehenge (about 2800 B.C.).

By Stonehenge standards, Newgrange is huge. It is really a small hill in the Irish farmlands. The mound is 36 feet high at the center and about 280 feet in diameter. Stonehenge, at least its central stonework, could have been buried within Newgrange and we would be none the wiser.

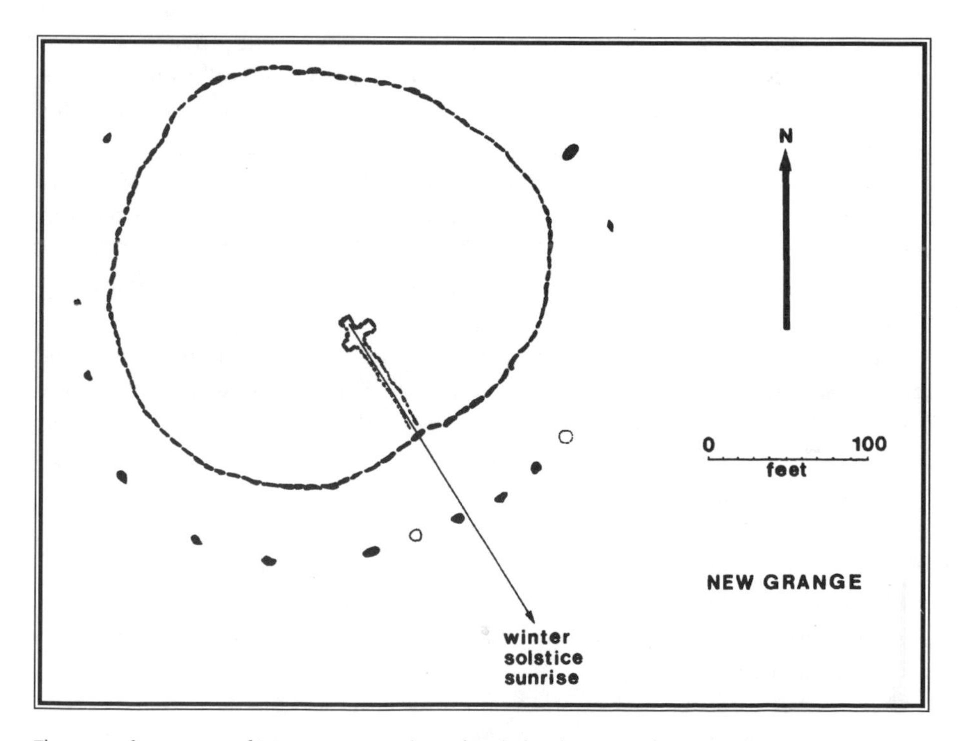

The axis of symmetry of Newgrange is aligned with the direction of winter solstice sunrise and coincides more or less with the megalithic entrance passage. (Griffith Observatory drawing, Joseph Bieniasz)

O'Kelly's discovery, in 1969, of the convincing celestial alignment designed into Newgrange, turns the site into the oldest astronomically oriented monument in the world. Its architecture admits the light of the rising winter solstice sun. Its warm golden rays penetrate the mound and reach into its most intimate compartment. According to *The Oxford Illustrated Prehistory of Europe,* Newgrange's encounter with the winter sun reflects "the esoteric interest in astronomy and the calendar of the seasons." Actually, there is nothing esoteric at all in noticing the southern limit of the sunrise in the dead of winter. The winter solstice is a key turning point of the year, a conspicuous signal of seasonal change. Otherwise a perfectly respectable source of information, the Oxford reference could take a cue from prehistorian Aubrey Burl, who suggested that sunbeams interacting with the prehistoric sepulchre may have been intended to "reanimate the souls of the dead."

The dead were deposited in Newgrange, but probably not too many of them. At most, the remains of five individuals were found inside, and the ashes of three of them were left mostly under and around the great stone basins in the two side chambers. Portions of two unburnt skeletons were found as well. Some other passage graves contain the remains of a much larger number of persons, but most of the evidence indicates the bones and ashes were interred collectively. In any case, the dead in the tombs are still many fewer than a normal community would produce over centuries of use. Reaching the same conclusion about the astronomically aligned Clava cairns, in Scotland near Inverness, Burl writes, ". . . there are simply too few bodies for these to be cemeteries."

Rather than operating continuously as a cemetery, the passage graves seem to have been consecrated by the remains of honored dead. Their status is not reflected in elaborate grave goods or individual handling. They are elite because they, and not others, made their home in the tomb. If the tomb is a home for special spirits, it may be more accurate to understand it as a temple or shrine than as a grave.

Appreciating the astronomy-merged-with-funerary dimension of Newgrange requires an understanding of its architecture. Before its reconstruction, it actually used to look like a small hill. A doorway on the southeast side of the heart-shaped mound provides access to 62 feet of stone corridor. No mortar was used to stabilize the structure. The uprights and lintels are kept in place by their own weight. At the end of the tunnel, a central chamber is roofed by a corbelled vault capped by a massive slab 20 feet above the floor. Three smaller side chambers open on the central room to give the passage and compartments the overall shape of a cross. Antiquarians who opened the entrance of Newgrange in 1699 saw its resemblance to a cave and for convenience called it a cave.

Restoration brought more surprises. The megalithic chassis of Newgrange was covered with stones to build the mound, but the relaxed slope of its grassy turf was the product of age and prehistoric collapse. Originally the exterior wall was vertical to a height of 13 feet and surfaced with quartz to create a brilliant white facade that glared in the Irish sunshine. Probably more symbolic than ornamental, quartz is often encountered in the prehistoric tombs and stone rings of western Europe. Just beyond the decorated curbstone in front of the entrance, O'Kelly also uncovered an enigmatic oval scattering of smaller quartz pebbles fenced by low thin slabs of stone.

With its magnificent wall of white quartz, Newgrange looks abstract and incongruous, with all of the charm of a Los Angeles bank. It is not beguiling. It is powerful and disorienting. Newgrange is a pearly hockey puck, capped in green velvet and grown gigantic on the sloping fields.

The real surprise was the roof box. This is a small window, about three feet wide, just above the entry. Until O'Kelly cleared the roof box, no one knew there was an aperture over the door. He found a block of quartz still in place within the roof box, and it sealed half of the opening. Another had originally closed the other side, and scratches on the roof-box floor demonstrated that the quartz blocks had been removed and returned on several occasions, as if someone had deliberately opened the window for some special event. With winter solstice intuition, Michael J. O'Kelly, decided to watch the sunrise inside Newgrange on December 21, 1969. Four minutes after the sun eased itself above the local hills, it was in position to beam through the roof box and splash a pool of sunshine on the floor in front of the rear compartment. It took the sun 17 minutes to move across the window and leave the chamber once

A vertical groove cuts the design on the entry stone at Newgrange in half and seems to mark the line that leads to winter solstice sunrise. The door to the passage is just behind the decorated curbstone, and a slab leaning against the wall, to the right of the door, once sealed the passage. Winter solstice sunrise light still reached the inner chambers, however, through the roof box, the "window" above the door. (photograph Robin Rector Krupp)

again in darkness. O'Kelly was persuaded the passage grave builders of prehistoric Ireland knew something about astronomy.

It had been known for some time that Newgrange faced southeast, more or less in the direction of the rising sun at winter solstice. There was even an unsubstantiated tradition for the entry of sunlight to the inner sanctum, but no one took that seriously. In writing about this in 1964, the late Glyn Daniel, an expert on the chambered tombs, quoted a caption from a 1960 pictorial calendar,

> The rays of the rising sun at certain times of the year penetrate the opening and rest on a remarkable triple spiral carving in the central chamber. . . .

and then foreclosed any speculation in Newgrange astronomy by dismissing the claim, and the rest of what he called a "wildcat account" as a "jumble of nonsense and wishful thinking." Actually, Daniel was right. The passage in Newgrange bends a little here and there and slopes uphill. Any sunlight that made it through the front door would hit the ground somewhere in midpassage. Winter solstice sunrise had no home in Newgrange until O'Kelly reopened the window for it.

Newgrange was deliberately designed, thoughtfully engineered, and carefully constructed. It has, after all, not fallen down in over five thousand years. Its roof slabs were contrived to drain water away from the passage and chambers to keep the interior dry for the dead. Its astronomical orientation is restated by its general outline. Because it is heart-shaped, it is bilaterally symmetric, and its axis of symmetry coincides reasonably well with the line of the passage and with the astronomical target.

The roof box has a decorated lintel. Its facing edge is carved in relief with a chain of framed Xs. Other abstract symbols were engraved on the upper surface of the interior stone that supported the lintel, but once the builders put that stone into position, its carvings would have been hidden. Several other big stones, including three in the curb, are elaborately carved with geometric designs and are considered trophies of neolithic art. O'Kelly found, however, many other stones with intentionally concealed art. Their presence implies the purpose of the art was not decoration but meaning. The meaning must elude us, for Newgrange is a *prehistoric* monument, which means the people who built it did not have a written language. That means the monument didn't come with the megalithic manual of style.

There are some details, however, that reinforce the astronomical alignment. A deep, conspicuous groove divides the design of spirals and lozenges on the entrance stone in two. There is another curbstone, on the northwest side of the mound, just opposite the entrance stone, and its decorated surface is also divided in half by a prominent vertical groove. Martin Brennan, an independent and sometimes freewheeling researcher and the author of *The Stars and the Stones,* spotlighted the significance of these grooved curbstones. They define a line that for all practical purposes coincides with the axis of symmetry and the direction of winter solstice sunrise. The megalith that reclines in front of the west entrance of the Knowth passage grave—just a little less than a mile to the northwest of Newgrange and comparable in size—is also bisected with an upright groove. Perhaps it confirms the idea that vertical lines on curbstones have an astronomical connotation. George Eogan, the Irish archaeologist who has excavated and restored Knowth, has pointed out that the west passage faces equinox sunset, and another passage, on the east side of the mound, faces equinox sunrise.

Lingering doubts about the astronomical intentions of Ireland's Stone Age farmers evaporated in 1989 when T. P. Ray, an astronomer-physicist at the Dublin Institute for Advanced Studies reevaluated Newgrange and concluded its solstitial alignment was better than we thought. Accounting for the slow change in the tilt of the earth's rotation and reconsidering details of earlier calculations, Ray demonstrated that the sun would have shined into the roof box at the moment of sunrise and actually entered the far chamber as a narrow beam, about two feet wide. For a moment, half of the compartment would have glowed in stunning sunlight while the other half remained dramatically dark. Reflected from the upright stones, the light would have fallen on the "triple spiral" mentioned in the misinformed calendar. Ray showed the sun would have started its trip across the window in the lower corner on the east. By the time it crossed to the other jamb, it would have also been close to the

On the northwest, and back, side of Newgrange, there is another elaborately carved curbstone that is bisected by a vertical groove. Together with the grooved curbstone at the entrance, this stone restates the winter solstice sunrise axis of the monument and endorses the idea that astronomical orientation was something the builders had in mind. (photograph E. C. Krupp)

upper corner on the west. To Ray, this was evidence of intricate planning that not only got the direction right but also the dimension and proportion of the opening.

None of this means that Newgrange was an astronomical observatory. It was a place of the dead. But interaction of winter solstice sunrise with the bones and ashes of the dead was important to someone. Its builders understood the basic cyclic behavior of celestial objects. Beyond the passage graves themselves, we don't have much information about the scale of their society. It took, however, organization and effective mobilization of resources to plan and build these monuments. The tomb-building tradition began a thousand years earlier as a family enterprise and gradually evolved into a clan or tribal endeavor, which produced a community shrine that probably announced territorial control as much as respect for the dead. In the last half of the fourth millennium B.C., cultivation of wheat and barley and pasturing livestock must have created surpluses that could subsidize these larger monuments. Knowledge of the sun's seasonal shifts along the horizon is not especially esoteric, but the ability to incorporate it into monumental architecture implies some specialization. Although there is no obvious sign of social stratification in the remains of the dead, it seems likely that there was some centralization of power, enough to commit 800,000 hours of effort to the construction of a passage grave like Newgrange.

In a discussion of the nature of chiefdoms and the ideologies that sustain them, anthropologist Timothy Earle recognized the symbolic use of the heavens, through celestial alignments of monumental architecture, in neolithic "group-oriented" chiefdoms. In these, he suggests, "leaders served group rather than individual interests." If so, the chiefs of neolithic Ireland operated in a sacred landscape defined by principles of cosmic order and cyclic renewal. Ceremonies were staged in special places constructed to promote and provide communication with forces and spirits that seem to move and control the world. The power of these chiefs is conferred through their ability to make contact with the gods, and they do that, in part, through the celestial references in their monumental architecture. This means that the themes we have already encountered are the themes that should be present in Newgrange, just packaged differently.

At Newgrange, the dead and the winter solstice sunrise seem to be telling us something about fertility and renewal. Winter is the death of the earth, but when the sun changes course at winter solstice, death is on the way out and making room for the springtime revival that always emerges from the body of Mother Earth. In that sense, the year is reborn at winter solstice, and the day is certainly reborn at sunrise. When that sun comes up in front of Newgrange, it pumps life back into the tomb as if it were fertilizing Mother Earth. The sunlight and the tomb may have had something to do with revivifying the world, or perhaps the light just put life into the spirits of the dead and made them available for consultation. In another thread of Celtic legend, Oengus, the son of the Daghda, brought the body of Diarmaid, one of the great heroes of Irish folklore, back to Newgrange, where he "put an aeriel life into him so that he will talk to me every day." Oengus was the god of love, and Diarmaid had

The sloping floor of the entrance corridor prevents light that passes through the front door from reaching the back chamber. The roof box, however, is another story. It was apparently designed to permit the first light of the winter solstice sun to reach the inner sanctuary and to accommodate the rising path of the sun, which exited from the diagonal corner of the window. This winter solstice photograph demonstrates that the sun still performs in the roof box despite a slight change in the tilt of the earth's axis over the last five millennia. (© 1996 Yorkshire Television Limited)

connections with the god of death. Perhaps the ancient Celts managed to converse with the spirits of the children of the goddess Danu at Newgrange after all. In any case, the Celts laminated a tale of fertility and death onto a monument that held the same message for its neolithic owners.

If the passage grave builders woke up the spirits at Newgrange, they may have wanted to talk to them about the revival of the earth, the rebirth of the year, the rejuvenation of the sun, the transformation of the soul, or the renewal of chiefly power. We don't know what was discussed, but these are not mutually exclusive concerns. The language of the tomb, however, is fertility, death, and the sun. These are all chapters in the book of cyclical renewal. That is the story told by Mother Earth, sometimes with a little help from the sky, and that is what makes the passage grave one more womb of Mother Earth.

MOTHER LODE COUNTRY

Mother Earth keeps reentering her own parade. We spot her through the centuries, not by the details of her costume, which must change like fashion, but by her anatomy, which is usually obvious, and by her function, which is cyclic renewal. Even when she is surgically separated from the land and humanized into a goddess—when she is the Mother Goddess instead of Mother Earth—she fulfills the same role. We find her everywhere, but her encores in prehistoric Turkey offer an appealing series of performances by a woman who continually changes into what she just was.

In Turkish, the traditional name for Asia Minor is *Anadolu.* Europeanized, it is Anatolia, the name still given to the plateau that dominates most of the country, and it means "land of mothers" or even "full of mothers." The mothers first start showing up in Turkey about 9500 years ago, in Çatal Höyük, a neolithic village about 32 miles southeast of the modern city of Konya, in the south-central part of the peninsula. The absence of streets in the New Stone Age settlement may explain the absence of doors, or vice versa, and either would probably make Çatal Höyük an interesting study in the archaeology of urbanization, but Çatal Höyük is probably more striking for its highly developed and decorated religious sanctuaries and elaborate birth and funeral imagery. The people entered their temples and homes through an opening in the flat roof. Small, baked clay figurines of ample women were found in the ruins of almost every home. Sometimes seated, sometimes standing, they all have hips that abhor the vacuum of space, vast buttocks, enormous thighs, and fully tanked breasts. One of the most famous of these naked matrons is enthroned with a pair of leopards (or other feline predator) for armrests as she gives birth. The great cats make her a master of the wild animals and reflect paleolithic ideas. Her motherhood obviously connects her with fertility and procreation. She was collected from the household grain bin, which may hint at her neolithic transformation as the goddess that renews

An abundant supply of female figurines from Çatal Höyük, a neolithic settlement in southern Turkey, argues in favor of long tradition of goddess worship in Anatolia. This is probably the most famous depiction of her. Midwifed by leopards, she is bringing new life into the world. (photograph E. C. Krupp, object displayed at the Museum of Anatolian Civilizations, Ankara)

the productivity of the cultivated land. Some have argued that the shift to agriculture elevated the importance of women and that the evidence is the proliferation of mother goddess figures. Actually, the economic importance of women is already high in the life of nomadic hunters. Although the men occasionally bring home the bacon, the women forage the fruit and vegetables and are the ones who put most of the food, if not the protein, on the table.

We encounter the goddess again in the early Bronze Age at Alaça Höyük, where she was cast in silver with gold inlay capping her breasts and covering her feet like go-go boots. Her hips are wide. Her waist is thin. Her arms are fins, and from the neck up she is a hammerhead. Nevertheless, in the middle of the third millennium B.C., the message was still clear. When they baked her in clay at Kalinkaya at the end of the third millennium B.C., they omitted her arms and legs but left the essential curves and a groove between her hips. At Hasanoglan, Bronze Age believers contrived her from electrum and gold at the end of the second millennium B.C. Gold straps cross her chest between her tiny gold breasts. Long-waisted, thin, and flat, her gender is announced by the oversize, stippled triangle between her thighs. In the nineteenth century B.C., Assyrian colonists at Kültepe influenced the ivory cutter to give her wide eyes, a vulnerable smile, tiny hands that hold her modest breasts up for display, hips broader than her waist would suggest, and a dark triangular inlay in the pubic zone. The Hittites turned her into a skirted breast-grabbing jug in the sixteenth century B.C., and the Phrygians found in her Kybele, the mother and fertility goddess that eventually made her way to Rome. They eyed her in Isis after the Hellenistic mystery cults reached Asia Minor. Equating her with Fortune, they equipped her with a cornucopia that could choke a supermarket and a rudder for steering lives. When the Romans got ahold of her at Aphrodisias, she became Venus, the divine embodiment of all erotic love, fertility, and life in the cosmos. She competed with the Christians at Ephesus as the multibosomed Diana. She shelved a chestload of bulls' testicles, eggs, or enough breasts to nurse a child-care center. No one knows for certain what the round and firm and fully packed forms really are, but on Diana they convey world-class fertility.

MADE IN HEAVEN, CONSUMMATED ON EARTH

In this litany of mother goddesses, we can feel the female rhythm of the earth and sense the power of renewal attributed to women, but there is little sign so far of their male partners. The men, in fact, are usually on the premises, despite the claims of ardent feminists and the chants of goddess revivalists. Even the voluminous and self-assured mother goddess of Çatal Höyük received her consort. There is reason to think he is symbolized in the bulls that dominated their own shrines in the village. Small figurines of men standing on bulls have also been retrieved from the rubble.

If seven thousand years of Anatolian goddesses are going to culminate in anything, it ought to be this. Known as the Beautiful Diana of Ephesus (from the period of Roman Emperor Hadrian, 117–138 A.D.), she projects fertility by the breastload and wears the zodiac around her neck. In other statues of Diana, the egglike forms on her chest have nipples and certainly are meant to attire her in a multibreasted suit, but in other sculptures, the message isn't so clear. Many alternative explanations have been suggested, but they all rely on the theme of fertile renewal. (photograph E. C. Krupp, object displayed in the Ephesus Museum)

These little statues superficially resemble the bull-riding, fertilizing storm gods we later encounter in the Near East.

Two bulls towed the chariot of the high father god of the Hittite empire, the "Weather God of Heaven," as he barnstormed the mountaintops and surged through the skies. With chariots and bronze swords, the Hittites forged an empire that reached into Mesopotamia, Syria, and Lebanon and made a cosmic family out of their king and queen. The royal couple partly operated as counterparts of the mother goddess of the earth and the father god of the sky. It is believed they perpetuated the Sumerian ceremony of sacred marriage, which ensured the renewed fertility of the land through the consummation of ritual sex. The king was the real mate of Mother Earth, and his contact with her took place through seasonal nuptials with her high priestess, the queen.

This seasonal celebration of royal ceremonial sex originated in Mesopotamia in the fourth millennium B.C. and was perpetuated in the *akitu*, or New Year festival, in Babylon, more than 1,200 years later. Scheduled to coincide with the vernal equinox,

The male role in the cosmic renewal of world fertility is usually fulfilled by a celestial god that rides the storm. To the Hurrians of Anatolia, he was Teshub, and his Hittite name has been reconstructed as "Tarhunda of the Sky." In this Hittite relief from Til Barsip (Tell Ahmar) in northern Syria and the eighth century B.C., we see him with the ax that releases thunder every time it is struck and with the forked lightning that is the most unnerving evidence of his power. (photograph E. C. Krupp, object displayed in the Louvre, Paris)

the time in spring when the sun rises due east and sets due west and when the hours of daylight equal the hours of the night, the *akitu* ceremony was intended to revive the world from its season of dormancy. The 12-day ceremony relied on the fertilizing power of the king and included a reenactment, through ritual drama, of the world's creation. In Mesopotamian tradition, Creation was really the establishment of the cosmic order, and cosmic order naturally included the process of seasonal renewal. Magically reinvigorating the land with life was, in fact, a duty of the king. Bedding with the priestess of the great goddess of renewal was an affair of the state not of the heart. High responsibility, however, did not pull the plug on great sex.

In her role as the receptacle for divine male vigor, the Mesopotamian fertility goddess was an amorous partner whose power was her reproductive equipment. In ancient Sumer, where she was Inanna, the Queen of Heaven, she was sometimes associated with the planet Venus and strictly speaking not a goddess of the earth.

According to the ancient hymns, however, Inanna has a "holy lap." With it she draws her consort Dumuzi to the marriage bed and so promotes the abundance of the fields and the proliferation of the flocks. Were some of Inanna's songs adapted to MTV, their salacious detail would probably create a whole new audience for cuneiform. The triangular clearing above her thighs is an "untilled plot." In "Plough My Vulva," she informs the world, "My vulva is a well-watered field," and a "piled high hillock," and she asks, "Who will plough it?" In another text, her breast is a "wide, wide field which pours out plants" and grain. She may not be Mother Earth, but she is earthy enough.

Hittite scholars believe the same kind of transcendental matrimony described in the Sumerian liturgy is portrayed in the reliefs cut in the thirteenth century B.C. onto the rocky outcrops at Yazilikaya, an open-air Hittite sanctuary just two miles northeast of Hattusha. Hattusha was the Hittite capital, and its ruins are near Bogazköy, a town in central Turkey about 124 miles east of Ankara. In Yazilikaya's main chamber, the reliefs include two long processions of gods that converge on a large rock face at center stage, where the Weather God of Heaven faces his consort. Goddesses queue up on one side of the rock chamber. The male deities advance along the other. Only a single exception in each file defies the supernal symmetry of the entire assembly. Sharruma, the son of the two great gods renewing their marriage vows, stands right behind his mother and leads the line of divine ladies. On the west wall of the sanctuary, Shaushga, the Hittite version of the planet Venus as the goddess of war, marches with the men.

Rock reliefs are known from other Hittite sites, but none is as large and as elaborate as the galleries of gods at Yazilikaya. This unusual shrine in the natural rock was connected to the north end of Hattusha by a road that began at Temple I, the weather god's main shrine in the city. As the route ascended the hillside, it passed through one of the capital's major cemeteries and then continued to a small complex of temples that controlled traffic into the carved chambers at Yazilikaya.

Although Hattusha was the center of power of the Hittite empire, archaeologists were surprised to discover that the carved hieroglyphics that accompany the gods at Yazilikaya identify them with Hurrian names. The Hurrians established the kingdom of Mitanni southeast of the Anatolian plateau—in the upper Tigris, in northern Mesopotamia, and in eastern Syria. With a language unrelated to the Semitic peoples of Babylon or to the Indo-European lineage of the Hittites, the Hurrians were actually agents of cultural transmission between those two independent traditions. Puduhepa, the Hittite queen who appears to have worked out the marching orders for the gods on parade at Yazilikaya, came from Kizzuwatna, a Hurrian province in the southern lands of the Hittite empire later known as Cilicia. She was the daughter of a priest, and as a priestess herself, she brought tablets inscribed with the names of the gods to her Hittite husband. She was married to Hattushili III (1275–1250 B.C.), who along with their son, Tudhaliya IV (1250–1220 B.C.), commissioned the sculpture at Yazilikaya.

Yazilikaya, an open-air shrine on the slopes above Hattusha, the capital of the Hittite empire, puts the gods on parade in the wedding procession for Teshub, the weather god, and Hepat, the goddess of cyclical renewal. Here, we are looking north, into the main gallery. The gods are sculpted in relief on the natural faces of the rock. Farther in and to the right, the central panel illustrates the Sacred Marriage which the gods performed and the theocratic ruling couple imitated. (photograph E. C. Krupp)

Careful study of the style, symbolic details, and engraved names of gods in the reliefs, supported by lists of gods included in international treaties that have survived as Hittite texts, allows us to determine at least part of the content and purpose of the reliefs. The groom in the sacred marriage is identified as Teshub, the Hurrian "weather" god. His home is the mountaintop, but he moves through the sky as the power of the storm. Thunder booms from his hammer or mace. He fires lightning from his fist. He inseminates the earth with rain. Although his Hittite name has slipped through the cracks of time, we encounter him throughout the ancient Near East. Elsewhere in Asia Minor, he was Tarhun, and Babylon saluted him as Marduk. The Canaanites called him Baal. To the Assyrians he was Adad. He had the power of a wild bull, bellowing thunder and mounting the earth.

Teshub's bride at Yazilikaya is identified as Hepat. She was the great Hurrian Mother Goddess, but in adopting her into their system, the Hittites also equated her with the Sun Goddess affiliated with the city of Arinna. She, like many other Hittite gods, seems to be a legacy of the Hatti, non-Indo-European predecessors of the Hittites in central Anatolia. To the Hatti, she was known as Wurusemu. Through steps of logic that now elude us, the Hittites added her solar dimension to the Hurrian incarnation of divine female power. In that sense, Hepat's duties for the Hittites paralleled the role of Inanna. In each, a celestial luster accompanied the talent for sex and procreation. Both were queens of heaven who mated with the storm king on behalf of the arbors, orchards, and fields.

These are not, however, simple earth goddesses of village farmers. Inanna and Hepat are high status deities. They belong to societies of rank and privilege. They, like a royal dynasty on the social hierarchy, are frosting on the layer cake of divine power. Their celestial attributes generically reflect the lofty status of the sky, but there is an additional reason for equipping mother goddesses with the houseware of heaven. Celestial cycles are utensils that help bring the earth to term. Time seemed to induce all of the significant transformations on the earth, and the movements of the sun, moon, planets, and stars were used all over the world to track the passage of time. Recognition of the annual character of the growth of vegetation and the cultivation of produce and grain made the sun an appropriate partner for the rain.

On the surface, then, the ceremony of Sacred Marriage seems intended to conjure the domesticated body of the earth back to fruition. Through royal copulation, the king and his paramour impersonate the gods, and the gods they mimic personify natural principles responsible for renewal. The king and the Weather God are allied with the rain and are licensed to fertilize the world. The priestess-queen and the mother goddess-sun goddess are vessels ready to incubate new life. Libido, however, is propelling more than procreation among the gods and kings. What really makes a meaningful relationship out of a one-night stand is the renewal of sovereignty through divine conjugal union. In the body of the king's sacramental courtesan, the great goddess of order and renewal embraced the king, and her willing embrace publicized her endorsement of his authority. This is bedroom politics in the service of the social order.

Although the Sacred Marriage ritual looks at first like a magical effort by the king to renew the agricultural capacity of the land, we realize that all of the symbolic activity in the Babylonian New Year rites seasonally renewed the king's license to rule. By rededicating the legitimacy of the king's privilege and power, the *akitu* maintained the structure of political and social order through ideology and state religion. That was useful, for a functioning government also helped ensure another cycle of prosperity.

There was, then, more to the Hittite mother goddess than reproduction and sex. When they invoked her as the Sun Goddess of Arinna in prayer, the Hittites credited her with "righteous judgment" and acknowledged her "royal authority" over "heaven and earth." When "the door of heaven" opened for her, as it did for the sun, she stepped through. Like the movements of the sun, which establish order in space and time, and like the will of a king, which orders human affairs, her survey from heaven ordered the world by marking out "the borders of the land." She also crossed through the boundaries of gender, for she was "father and mother of every land."

The wife of the Weather God of Heaven was still a lover and a mother, but we begin to see how the demands of kingdom and empire complicated her character. In the Old Stone Age and in the New Stone Age, she could have operated like the earth and delivered babies, young animals, and another crop in season. But when a society like that of the Bronze Age Hittites gets complex enough to install a king and centralizes regional power to ensure its health and continuity, the gods that drive the cosmos have to act as sponsors for the ruler. In fact, the Hittite Queen of Earth and Heaven was also the queen of the land. Her hymns confirm that she protected the king and queen, her deputies on earth.

The Hittite economy, like that of any respectable empire, was supported by agricultural prosperity, and that they understood in terms of timely precipitation, celestial cycles of time, and seasonal renewal. But both partners at the head of the Hittite pantheon were more than personified elements of nature. Instead, they were modeled on aspects of nature and cosmic forces to illustrate their character and their power. They were really divine incarnations of sovereignty. A prayer recited by the king at a palace dedication affirms their interest in his fortunes.

> To me, the King, have the gods—the Sun God and the Storm God—entrusted my land and my house.... (The gods) have taken care of the kings. They have renewed his strength and set no limits to his years.

Through the imagery of renewing the economic basis of kingship—the prosperity of the land—the union of Teshub and Hepat renewed the political foundation of kingship. Power on earth required congruence with the gods. Just as procreation was the ancient model for cyclical renewal, the family was the prototype for leadership. The high gods were the ruling family in heaven, and the king and the queen followed heaven's lead as the ruling family on earth.

Sacred kings and weather gods are not just semen donors. They are fathers, and paternity means more than generational continuity. Their real responsibility is mobilizing, organizing, and protecting all of the endeavors that enhance social cohesion, promote cultural continuity, and ensure survival. At the family level, whoever is head of the household has to exercise some kind of leadership. It may involve negotiating priorities, encouraging participation, distributing resources, or any other action that supports the family's mutual endeavor. In the social and political arenas, leadership means the same thing but operates on a different scale and obviously requires different tools. Ideology that explains and ratifies the institutions of leadership is part of the kit.

First shamans, and then chiefs, rely on sacred topography and on myth to advertise and explain the origin and acquisition of power. In a chiefdom, the ruler calls upon celestial forces with magic and practical observation. He manipulates celestial symbols to publicize his powerful sponsors. He shares, then, some of the insights, obligations, and perquisites of a shaman.

In kingdoms and empires, power is more centralized than ever, and that invites more explicit identification with the center of the world. The sacred king occupies

In this detail of the central panel in the main gallery of Yazilikaya, Teshub, on the left, faces Hepat, his divine consort and partner in the renewal of sovereignty. A detailed analysis of the character and function of this shrine tell us it was the center of the Hittite universe. (photograph E. C. Krupp)

the center. His reach extends to the borders. He taps the power of earth and sky. The scale of government requires an orderly delegation of power, and the center helps delineate the social order. The elite are no longer necessarily relatives by blood or marriage. Specialization creates a military, an administrative bureaucracy, a system of law, and religious institutions to protect the interests and the sovereignty of the state. That includes the symbolic placement of the ruler at the center of the world.

Somehow, the power invested in the ruler must be explained and justified. It is a human invention, and without some doctrine or belief that cements it to the real world of season and landscape, it looks artificial. If it is artificial, it can be challenged, and so ideology is enlisted to legitimize political power. Ideology links the ruler's franchise with the natural order of the world.

Ideology, however, emerges from nature, and the Hittites, like everyone else, were inspired by the natural environment when they went looking for gods. They called their territory the "land of a thousand gods," and they apparently tried to make room for everybody. Most of these gods were local spirits, and as incarnations of the divine power of their neighborhoods, they all followed the same pattern. Despite the empire's ethnic diversity, each community—a city and its countryside—did most of the same things. They cultivated wheat, barley, fruit, and vines. They herded cattle and sheep. Craftsmen worked in wood, leather, metal, and clay. Governors, nobles, and bureaucrats organized the life of the cities, and over them all the king ruled with political power, military power, judicial power, and religious power. Each community had its own personalized storm and weather god, a male master of the local mountain and lord of the bull. He was a warrior and a conqueror and was inevitably paired with the fertility goddess, the nourishing, life-renewing deity of the local river or spring. Everywhere, their conjugal partnership was evident in the well-being of people and in the health of the land.

At first glance, Yazilikaya looks like a prelude to the same kind of intimacy of Mother Earth and Father Sky. We know, however, that their pillow wrestling was an exercise in political science, and at Yazilikaya they summoned an audience for an inauguration. An ancient text confirms the Hittite kings staged a great festival of Heaven and Earth at the beginning of the year, when

> All the gods assembled and entered into the
> House of the Weather God.

The Hittite New Year occurred in spring and probably coincided with the vernal equinox, as it did in Mesopotamia. Other sources document Hittite celebrations in the spring and the fall, with the springtime event observed beyond the city walls in the natural landscape. It is reasonable to argue, as Kurt Bittel, one of Yazilikaya's excavators did, that the open rock sanctuary was the site of the royal New Year rites. Unlike the multistory buildings within Hattusha and other Hittite strongholds, the structures that front Yazilikaya were not fabricated for heavy public use. Bittel believes their rubble foundations, casual construction, and one-story design all imply limited intermittent use. That, he adds, is what you would expect at the site of seasonal performance of royal ritual. As far as the rest of the empire's population was concerned, the Sacred Marriage was a private affair, but for the crowd of gods at Yazilikaya, it looks like it was standing room only.

There was something new in the way the Hittites handled their gods at Yazilikaya. Presented as members of the wedding party, they became an organized system of gods instead of an empire of dispersed powers. Their allegiance was focused on the ruling family of the gods. Because Teshub and Hepat were personifications of sovereignty, the rocky sanctuary's pictorial pantheon performed like propaganda. Monumental like the mountainside, the message was clear. The king ruled Hattusha with the divine sanction of Teshub and Hepat. By the time they appear on the wall at Yazilikaya, their domain is no longer local. From every subordinate city, their local epiphanies have come to acknowledge their supremacy. Because the divine couple are cosmic sponsors of empire, celestial gods with cosmic prerogatives also get in line. Along with Shaushga for the planet Venus, the moon god and the sun god are present. The link between celestial power and earthly affairs is emphasized by two bull-men who stand upon a sunken symbol for the earth and hold the sky's emblem above their heads.

Regularizing the Hittite gods at Yazilikaya, most likely according to Queen Puduhepa's sacerdotal recipe, was a conscious application of theology to the process of centralizing power. With approval and support from Teshab and Hepat, power over so broad a land was permitted to be so concentrated in the king and his queen. From the small imperial family, power flowed through the social pyramid. The same trickle-down theory of empire franchised all of the subsidiary gods and goddesses.

If the main chamber at Yazilikaya really was the House of the Weather God, it was probably the original home of the local Weather God of Hattusha. The stream that runs by it could have been the property of the local Fertility Goddess. Perched upon a slope above the capital, its natural chambers and exposed rock bring the earth into direct contact with the sky, like the mountaintop where Teshub alights. The attention lavished upon it by the Hittite kings suggests they saw it as a place of numinous power, a site where nature's mystery and divine majesty seemed revealed in the awe it inspired. Such places have a theatric presence that beckons the imagination to connect with the landscape. As the sacred boudoir where the gods mated and royalty followed suit, it was the place of creation, where sex ignited life in the land through fertility magic, activated the seasonal renewal of the world in the New Year, and made the king a companion of the gods. As the place where the king could communicate with the gods, it was a cosmic modem. As the place that nurtured the empire through the king's contact with the gods, it was a cosmic navel.

Yazilikaya was like the bull-men in its own New Year's Day parade, a junction between earth and sky, but its second chamber gives it depth. The reliefs in the smaller vestibule to the east seem to refer to a Hittite equivalent of Nergal, the underworld god in the era of Old Assyria and Old Babylonia. Evidence of burial and bird sacrifice were found in the passage. Bittel speculates that the room served as the funerary temple for the dead king, who was deified on his demise. There was also a mortuary component to the New Year ceremony that this gallery could have satisfied.

If any of the funerary connections of the second room are true, they place Yazilikaya in touch with the underworld. It has, then, the properties of a world axis, for it links heaven with hell via the the world we know.

The bridge between worlds, the paradigm mountain and the paradigm water, the complementary opposition of male and female, the New Year recapitulation of Creation and the power of cyclical renewal, the earth's navel and the uplink to the cosmic gods, the sacred open sky and the earth's fertile womb—they are all here, and to our ancient ancestors they were the vehicles of revelation that resided at the center of the world. This synthesis of the evidence tells us what texts have yet to reveal. Yazilikaya was the center of the Hittite universe, a shrine of cosmic kingship at the heart of the Hittite empire.

CHAPTER SIX

Shamans, Chiefs, and Sacred Kings

We know we can trace our use of the sky's power back to small bands of hunters and gatherers. Their shamans and medicine men sensed the presence of celestial power in seasonal change and in the architecture of the natural world. They mastered that detailed knowledge of the world around them. Through their charismatic affiliation with the spirits they believed populated the cosmos, they attracted the support of their families and villages. With these tools, they forged a system of symbols and rituals that mobilized community belief.

If we all lived by hunting and gathering, we would not have much more to say about the cultural applications of cosmic power. Human society, however, can be much more complex.

As a society grows more complex, and regional power is consolidated into the hands of chiefs, symbols of power are crafted to clarify who possesses the power. These symbols also tell people how the power is acquired, exercised, and transferred, and what the power is capable of doing.

Although we usually associate the power of chiefs with physical strength and with success in raids and warfare, the true source of their power must transcend human affairs. The ruler's power must be sanctioned by ideology. Ideology is the symbolic framework that allows society to believe it is operating in harmony with the order and rhythm of the world. Because the sky seems to drive those rhythms and establish that order, there is a celestial dimension to the power of chiefs.

A chief, and those that social stratification places closest to him, control resources like agricultural surpluses and the water supply. The chief's power is political and economic. Decision making and the management of risk are centralized. His ability to perform these responsibilities are explicit expressions of his power. The seasonal dimension of chiefly power may, then, be reflected in the symbols and ceremonies of office. When French explorers encountered Natchez settlements in Mississippi in the seventeenth and eighteenth centuries, they learned that nobles of the highest rank were known as Suns, and the highest chief was called Great Sun. His house was built upon a large earthen platform. From his door each morning he faced east. At the first sight of the sun, he yelled three times and bowed to it. The sun was his elder brother, and from the summit of his mound, which placed him closer to his celestial sibling, the chief smoked tobacco in the sun's honor from a special pipe and blew the smoke toward the rising disk. Then he acknowledged the four sacred corners of the world in the same way.

When the size of the territory, the diversity of the population, and the complexity of society demand the integrating power of a true state—a kingdom or an empire—the ruler still invokes a charter from heaven. It is, in fact, what allows him to channel his power to others and what indemnifies his right to call it back. This ruler is still the primary lobbyist with the spirits and the highest gods and calls upon them personally. The sacred king and the divine emperor trace their lineage to the celestial realm. In this way, they justify their monopoly of power and institutionalize astronomy in the bargain.

Because systems of rank and prerogative vary with social complexity, we find that the way in which power is accumulated, consolidated, manipulated, and employed is in part linked to cultural evolution. Different societies have different needs, and so the purpose of power depends on the social environment. Celestial symbolism, world view, and astronomical knowledge are components in the mechanism of power. So, what people do with the sky is also related to the scale and complexity of their social and political organization. An emperor can afford a bureau of professional astronomers. A nomadic forest hunter doesn't need one.

POWER AND THE CHUMASH SHAMANS

At the time of the first European contact in the sixteenth century, California Indian society in some parts of the state hovered between independent villages and the development of chiefdoms. Anthropologists described the system as *tribelet organization.* The term was devised in the first part of the twentieth century by Alfred L. Kroeber, a pioneering anthropologist, to describe a system in which several villages were affiliated together. The largest, usually, was the home of the tribelet chief. A typical village of the coastal Chumash of southern California was occupied by a few hundred persons and had about 30 houses. The largest house belonged to the village leader, or "captain."

It may seem strange to illustrate the relationship between social complexity and astronomy with evidence from California. Most people are unaware of the California Indians. Even those who know something about them recognize that they don't fulfill the American Indian stereotype. They didn't wear bison robes and trailing war bonnets. They didn't build apartment villages like high-rises on the desert mesas of the Southwest. They didn't teach the Pilgrims how to farm the New World just in the nick of time for the first Thanksgiving. There is no Disney animated feature film about a California Indian princess. With the possible exception of the Mississippi Valley, however, California was the most populous region north of Mexico at the time of European contact. For all practical purposes then, the indigenous tradition of California is the mainstream for North America and not an obscure and marginal development.

In the sixteenth century, California's Indian population totaled about 300,000. Roughly 18,000 of them were Chumash, whose territory reached from San Luis Obispo to Malibu. The Chumash occupied the Santa Barbara Channel Islands, the mainland coast, and inland mountains and valleys to the western edge of the San Joaquin, California's great central valley. Essentially a maritime people, the Chumash took advantage of ocean resources, including fish, marine mammals, and shellfish. Wild acorns were the primary component of the Chumash menu, but other wild seeds and plants supplemented their diet. Like almost all of the rest of the California Indians, the Chumash were hunters and gatherers. They did not farm. Some hastily conclude that the absence of agriculture indicates less technical advancement. That may be so as far as cultivating plants is concerned, but the California Indians didn't farm the land because they didn't have to grow food. In California, nature was usually generous, and hunting-and-gathering was an effective strategy for survival. Settled villages, tribelet organization, and high population distinguish these California Indians, however, from small, nomadic hunting groups. It would be more accurate to say that the California Indians systematically harvested the landscape. They did it with the seasonal sophistication of farmers and herders but without the capital investment. In California, no down payment was required, but people there, like everywhere else, watched each month carefully and managed their interaction with the environment to keep food on the table.

The power of a Chumash chief was limited. Although wealthy by village standards, the chief did not control all of the wealth. Responsible for hosting guests, arranging ceremonies, planning intervillage skirmishes, and arbitrating disputes, the chief operated as an officer, not as a ruler, and shared power as a member of an elite group of known as the *'antap.* Craft specialization and a shell money economy contributed to social stratification in Chumash territory. At the top of the social pyramid, the *'antap* collectively managed most of the wealth and power.

Members of the *'antap* identified themselves with powerful aspects of nature, including the sun, the earth, the eagle, the condor, the swordfish, and the dangerous hallucinogenic beverage prepared from the datura plant. Visions induced by the consumption of *datura,* or jimsonweed, provided encounters with divine sources of super-

The triangular skirt around this stick figure painted at Ven-51, a Chumash pictograph site in southern California's Los Padres National Forest, resembles the ceremonial dance costume worn by Rafael Solares. Solares, it is known, was a member of the shamanic elite, and this image probably portrays a Chumash shaman. (photograph E. C. Krupp)

natural power such as the sun. Drug-inspired trances also facilitated the acquisition of a guardian spirit helper, an ally no shaman would want to do without. Chumash shamans used this power to cure disease, to defeat enemies, and to control the weather. They also presided over seasonal community ceremonies, instructed and initiated the youth, kept the calendar, and coordinated most aspects of village life. The shamans were themselves specialists. The *'alchuklash* was the skywatcher. He counted the cycles of the moon, established the times of the solstices, observed the seasonal appearances of stars, participated in the naming of newborn children, and read their destinies in the sky.

Chumash religion was rooted in shamanic concepts of power. Spirits provided or withheld food. They inflicted disease or subdued it. They inhabited a traditional, layered cosmos. Each level was believed to be circular, more or less flat, and skewered by a central cosmic axis that reached to the pole of the sky. Celestial beings, such as Sun, Moon, and Morning Star, lived in this upper realm. The sky was ruled by Eagle, who possessed all of the eagle's attributes. He was a high-flying, sharp-sighted predator. The surface of the earth was the middle world. Oriented by the cardinal directions, its center was associated with a high, sacred mountain. This middle world was supported by two gigantic snakes, and their occasional movements caused earthquakes. The underworld where they coiled and turned also accommodated dangerous spirit creatures of the night.

Power permeated the entire Chumash universe, but it was not distributed uniformly. Concentrated in one place but not in another, power could be acquired through informed interaction with the supernatural landscape. The ability to access it conferred social, political, and economic status. Inequitable dispersion of supernatural power was, then, reflected in the other inequalities of life, including a society of rank, privilege, and elite power. That is where the *'antap* enter the picture. They stabilized Chumash life with expenditures of the power they managed to accumulate through ritual, sacred objects, visions, and esoteric knowledge. Some of this was cosmic power, for it was extracted from the fundamental order and rhythm of the world and from the celestial beings that made those rhythms visible. Targeted doses of ritual power were also believed to influence the elemental forces of the cosmos. By doing the right thing the right way at the right time, the shamans and the people they served helped restore equilibrium to the world at those junctions of time and circumstance when the natural order of things seemed to be threatened by some crisis.

Winter solstice brought a crisis every year. It was a time of scarcity, colder weather, and a sun headed south. It was also the time when the Sky People—Sun, Sky Coyote, and the others—tallied up a year's worth of their nightly gambling scores to see what was in store in the year to come for the folks down there on earth. They played as teams, and when Sun's team won, there was grief ahead for the Chumash. Sun was powerful and demanding. If he prevailed, lives would be lost and food would be scarce. Sky Coyote, on the other hand, extended a helping hand from the north celestial pole, the stable hub of the Chumash cosmos. He turned in his chips for birds, acorns, seed, deer, fish, and everything else people needed and expedited their delivery through the hole in the sky.

In California, the universe was essentially stable, but the uneven allocation of power and the fluid nature of power made things uncertain. The balance of power was forever shifting. The winter solstice windup of the yearlong gambling tournament was the real California Lottery, and it was in the best interest of the Chumash to influence the outcome. So the *'antap* watched the phases of the moon and the daily shift of sunrise along the horizon to determine when the sun would reach the end of its tether and mobilized the community in public activities intended to honor the sun and harness its power on their behalf.

The Chumash called their winter solstice ceremony *Kakunupmawa,* a name the event shared with the sun itself. *Kakunupmawa* means "the radiance of the child born on the winter solstice," and it refers to the yearly rebirth of the sun. Most of what we know about this event was told by Fernando Librado *Kitsepawit,* a Ventureño Chumash from Santa Cruz Island. He died in 1915, but just a few years before his death he shared his knowledge of Chumash belief with John Peabody Harrington, an energetic ethnographer and linguist. According to Fernando Librado, a yearly meeting was called by the 12 members of the *'antap* to plan the winter solstice ceremony. Village captains also attended, and 13 were selected to assist in the planning. One of them won the responsibility to supervise the planning, and for the duration he held the title *slo'w,* or eagle. Prior to the solstice they constructed poles topped with feathers. They were erected during the festival, and there was one to honor each settlement—and the captain or some other worthy person—and one for Sun. Debts were publicly cleared, an appropriate move at the time of a new beginning. Then, on the day determined by solstitial protocol, the "Shadow of the Sun," one of the 13 ceremonial officials, placed a sunstick in the ground in a hole that had been dug into the ceremonial plaza the day before. His 12 assistants were known as "the Splendor of the Sun," and they symbolized the sun's rays.

Of modest size, the sunstick was a wood rod about 18 inches long. On it was mounted a stone disk with a central hole. The stone was tipped at an angle to the wand. Those that still survive are painted or inscribed with radial lines. The sunstick was part of a ritual Fernando Librado said was intended "to make the sun return for

In Chumash society, ritual specialists at the top of the social ladder organized public seasonal ceremonies. Here the *'antap* are conducting the *Kakunupmawa* activities during the winter solstice season. (Griffith Observatory painting, Lois Cohen)

another year." He explained that the name of the stick means "to divide" or "separate in the middle." That is especially interesting here, for that is also a name given to the North Star, which is associated with Sky Coyote, the other celestial captain involved with earth's prospects for the coming year. Librado also said the sunstick was under the four cardinal directions. All of these attributes tell us the sunstick was a stand-in for the center of the world.

After fixing the sunstick in the ground in the middle of the afternoon, the 12 Rays of the Sun tossed up eagle down and said to the sun, "It is raining! You must go in the house!" The sunstick was hit twice as this occurred, and then the Shadow of the Sun spoke briefly. After a few words concerning the weather and the food supply, he

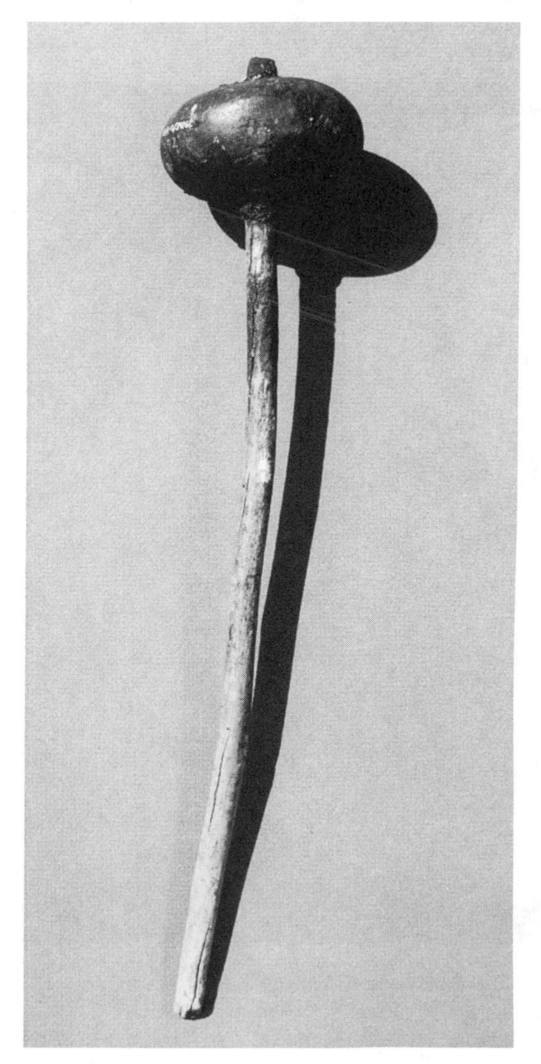

Solstice ritual in winter and summer included planting a sunstick in the open ceremonial area to tether the sun and keep it from migrating past its seasonal limits. This is the Bowers Cave sunstick. The perforated stone on the stick resembles a doughnut, and its upper surface has painted rays for icing. (photograph E. C. Krupp)

then made his point to the crowd assembled around the sunstick: "... Here is the force of the Sun—see how it drives this into the earth. ..."

Announcing the rain was intended to persuade the sun to go easy on the earth. From time to time, California experiences multiyear droughts, and the power of the sun is all too evident. It was also important for the vivifying energy of the sun to penetrate the female earth and revive it from its season of dormancy with the fertilizing power of the sun.

Fernando Librado also likened the sunstick stone to the sand dollar. The esoteric name for the sand dollar is "shadow (or image) of the radiance of the child born on the winter solstice," a title that clearly tells us the Chumash were interested in the resemblance between the sun and the sand dollar. Its ordinary name meant "stuck to the sand." Through this name, the sand dollar is also bonded to the earth. The Chu-

Sun visits Sand Dollar periodically, but when he does, he parks his rays outside. You can always see the sunbeams on the ventral side of a sand dollar. (photograph Anthony Cook, Griffith Observatory)

Some rayed disks at Chumash pictograph sites are equipped with forked rays. Their resemblance to Sun's forked rays on the sand dollar inform us we are looking at an image of the sun. (photograph E. C. Krupp, SBa-1380, Los Padres National Forest)

mash said that when the sun goes down, it enters the hole in the bottom of the sand dollar. Through analogy, the sun enters the earth through the sunstick stone just as it enters the earth by passing through the door of sand dollar's house.

As the sun goes inside the sand dollar, it leaves its rays outside. We do see forked rays emerging from the ventral opening of the sand dollar. These, of course, are the sun's rays, and that may explain the radial lines on the sunstick stone. There are, as well, disks in Chumash rock paintings that have the same forked rays, and those forks in the rays tell us those symbols must be emblems of the sun.

All of these solar and solstitial connotations of Chumash symbol and ritual belonged to the shamanic elite who governed and guided Chumash life. The Chumash occupy a threshold in social complexity and concentration of power, but even in the transitional zone, celestial imagery and celestial power are key elements in the exercise of power.

POWER AND THE DOGON PATRIARCHS

Even if the economy of a group shifts from foraging to herding and agriculture, the seasonal dimension of their enterprise keeps them in contact with the sky. The sun, the moon, the stars, the clouds, and the rain take on new meanings pertinent to the goals of farmers. For example, hunter-gatherers of the upper Amazon encounter a period of scarcity during the tropical jungle's rainy season, but the Hopi desert farmers welcome the rains as the renewers of life. Understandably, the celestial events that herald the rains in each territory differ profoundly according to their relationship with the environment. The vocabulary of celestial symbols and seasonal ritual shifts to accommodate a society's primary concerns. In any case, the traditions of hunters, foragers, herders, and the first farmers tell us that even the earliest roots of society reached into the sky to tap the power that abides there.

Rain, for the Dogon of west Africa, is associated with a primordial sacrifice in heaven that was required at the time of the world's creation to establish cosmic order on an earth contaminated by confusion. Supervising this enterprise, the high god Amma established vital links between the underlying plan of the universe, the behavior of celestial objects, and the grains the Dogon cultivate in the arid savanna and the mountainous plateau and cliff country south of the bend of the upper Niger, in what is now the landlocked nation of Mali. Today there are about 250,000 Dogon adhering to traditional life in Mali and Upper Volta. They are farmers who rely on seasonal rains to grow millet, sorghum, and *po* (the minute white grain of *Digitaria exilis*), or fonio, a wild grass the Dogon cultivate.

Timbuctu, the fabled medieval trade-city on the edge of the Sahara, is about 200 miles north of the present Dogon homeland on the Bandiagara Plateau. The Dogon have occupied their rocky terraces and cliffside villages for at least five centuries, but originally, as peasant warriors, they lived farther to the west, in the area known as the Mande. The Dogon left their original territory in the twelfth century to preserve their own ancient religious traditions as Islam embraced northern Africa.

Related through language, belief, and culture, the Dogon are politically dispersed into village settlements. Leadership and power belong to each village's elders and are particularly focused in the *hogon*, the oldest man of the primary lineage. This patriarch and his colleagues are "men of learning" who have mastered the complexity of Dogon philosophical and religious thought and who understand the elaborate symbolic system of correspondences in which it is expressed.

Dogon villages include houses, granaries, altars, and other sacred places, and the layout of the village components is likened to human anatomy. Important public spaces correspond to the head, and some of the altars are the genitals. In the Dogon world view, human anatomy, human physiology, human behavior, and human society conform to the patterns and processes of nature and the cosmos. The present state of the world and the character of people's lives are understood by the Dogon to be direct consequences of the original conduct and policies of Amma.

The Dogon elders rely on their specialized knowledge to make decisions that guide and protect the community. This esoteric lore is complicated and abundant, but it is not restricted. Age, experience, and personal application permit almost anyone to acquire the deep understanding possessed by the village leaders. Prestige is conferred through the acquisition of knowledge, and prestige is what permits the community to place decision-making power in the elders' hands.

Ceremonial responsibility is naturally assigned to the village hogon. This includes the care and ritual operation of the village altars, which, like other sacred places in the Dogon landscape, maintain the people's connections with their ancestors, their gods, and the sky. The first altar of the world was dedicated to the first ancestor of the Dogon, and a terrace of altars in the village operates as a stage for reenactment, through ritual, of the ancestor's sacrifice and the reorganization of the world. Only the authorized elders, or priests, are permitted to go there. This terrace is said to be the sky, which is where the first sacrifice occurred, and it is also called the "navel of the world." Every altar houses a concentrated dose of the vital force, or power, that animates living things, and every altar communicates with the others through the sacrifice of blood. Ritual sacrifices reenergize the altars as their power is tapped.

The importance of granaries in a society that depends on the cultivation of grain is easy to understand, and among the Dogon, the granary was used as a model of the cosmos. Ogotemmêli, the blind and respected elder of the Dogon village of Upper Ogol, explained the universe in those terms to Marcel Griaule, the French anthropologist who contributed so much to our understanding of Dogon belief. Village elders and priests had made a deliberate decision to share their tradition with Griaule and assigned Ogotemmêli the difficult task of communicating its principles and subtleties to an outsider.

According to Ogotemmêli, the world's structure was established when a pair of divine celestial spirits the Dogon call the Nummo descended from heaven in the Granary of the Master of Pure Earth. This transcendental celestial granary had a circular base that melted into four sides that met at the top in a square roof. In shape it resembled the *tazu* millet stalk basket in which the Dogon carry harvested ears of grain. A spindle whorl emerged from a circular hole in the center of the granary roof, and each side of the granary supported a stairway to that upper platform. Ogotemmêli's account of the symbolism was explicit. The circular base is the sun, and the square roof is the sky. The circle where the spindle extends from the roof is the moon. Each side of the granary faces a cardinal direction. The stairway on the north side belongs to the Pleiades (Grouped Stars). The south stairway is associated with the Belt of Orion (Stars Three). Venus (The Morning) is affiliated with the east set of stairs, and the western stairs are linked with Orion's Sword (Tail Stars). This granary of the gods landed on a piece of heaven that was also a piece of celestial earth, oriented as well to the world's four directions. Although earth was corrupted and disordered, heaven was pure, and heaven landed on ground consecrated by its touch. Much more had to happen before the world became what it is today, but the process could not begin without seeding the earth with celestial order from the granary from heaven.

That spindle whorl in the granary roof is the counterpart of Amma's Pillar, the world axis that reaches out of the flat disk that is the earth and points toward the Pole

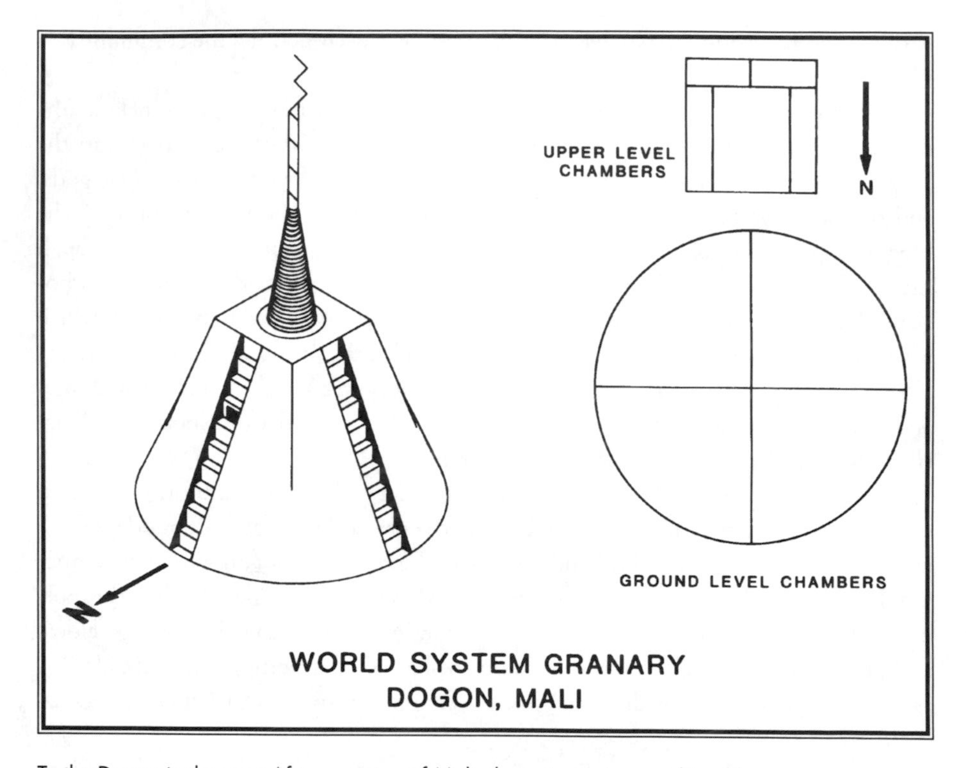

To the Dogon in the west African nation of Mali, the universe is something like a granary, a storehouse of potential and the source of all that is. Translated into the kind of rendering that is at home in our system of graphic representation, the Dogon world system has a circular base that tapers with four cardinal sides into a cosmic polar axis. (Griffith Observatory drawing, Joseph Bieniasz, after Marcel Griaule)

Star, which the Dogon call the "Star of the North." It also extends south to the Southern Cross, the "Star of the South," and both ends of the cosmic axis are known as the Eyes of Amma. Connected by the Milky Way, they watch over the world and hold it in place. Fundamental celestial events accompanied the establishment of the world's order. The first sunlight glowed in the east. The stars were shifted into rotational gear. The first eclipse of the sun blinked over the earth. Since then, the moon, the sun, and the stars have continued to "spiral" around Amma's Pillar and establish the rhythms of time.

Dogon astronomy punctuates the rhythm of time with the disappearance in June of Sirius, the brightest star in the sky, and the start of the rainy season. The Dogon also keep calendars by the sun and the moon, and winter solstice starts the solar year. Someone has to determine when the solstice is. At Ogol, that task was performed by one of the family heads at a set of three altars west of the village. The sunrise upon a distant horizon feature was aligned with a small rod lodged upright on one of the altars. Calling this observation the "measure of the direction of the moment," the Dogon sun-

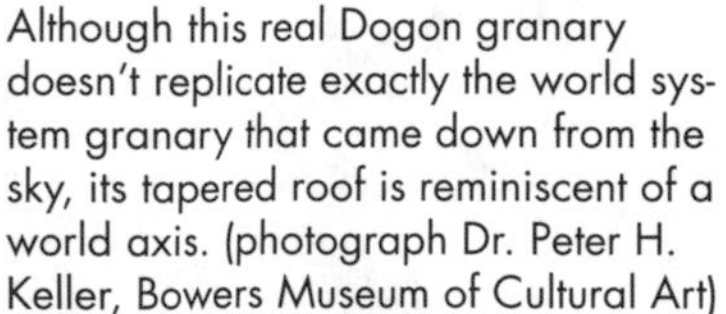
Although this real Dogon granary doesn't replicate exactly the world system granary that came down from the sky, its tapered roof is reminiscent of a world axis. (photograph Dr. Peter H. Keller, Bowers Museum of Cultural Art)

watcher performed the task four times a year, at the solstices and the equinoxes. These four yearly positions of the rising sun were associated with the four Dogon tribes and the four pairs of heavenly ancestors—four male and four female—that descended to earth before the sacrifice of Creation. Each ancestor corresponds to one of the eight Dogon family lineages. Seasonal rituals accompany the four "moments" measured by the family head.

Cardinal directions and the number four restate the cosmological character of the *toguna,* a ramada-like post-and-beam structure in the village plaza where the men gather to discuss community issues and plan events. Its four open sides ventilate the roof-shaded shelter and are oriented to the cardinal directions. It replicates the anthill on earth where the eight ancestors first sheltered themselves and decided how to proceed with the world reordering Amma had assigned to them, and the carved anthropomorphic support posts of the *toguna* represent the ancestors. They were "working the work of Amma" in "Amma's place." Now, long after the initial restructuring was done, the village elders continue to fulfill Amma's intention through symbols and ceremonies that insure cyclic renewal of society and nature.

World order is also reflected in the instructional diagrams Dogon diviners draw in the sand to explain how the footprints of the pale fox of Dogon country (*Vulpes pallidus*) reveal destiny on the divination tables they also draw on the ground. The fox is

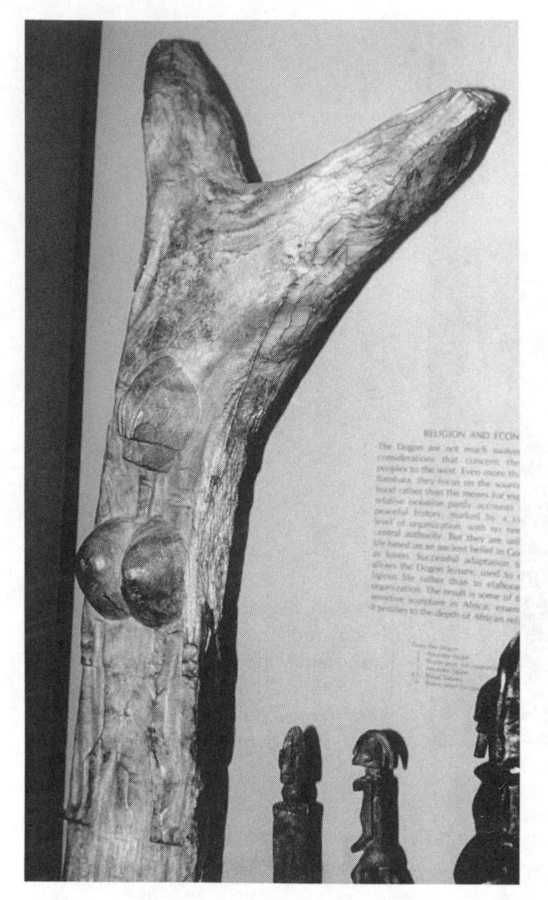

This Dogon house pole represents an ancestor. Primordial ancestors are accorded great respect, for it was they who established the circumstances of the present world order. (photograph E. C. Krupp, object in American Museum of Natural History)

the terrestrial incarnation of the primeval celestial being whose descent to earth put the world in disarray and required the reorganization described in Dogon myth. Attracted to the tables by peanuts left for it overnight by the diviner, the fox steps in various compartments of the drawing. Its tread is read the following day. Patterned somewhat like the divination tables, the diviner's instructional drawing is rectangular. Its corners reference the cardinal directions. It has 12 compartments that represent the year's 12 lunar cycles. The Dogon name for this diagram is "the place of the sun, the path that it walked the year of twelve moons."

Counting the phases of the moon, keeping track of the configurations of Venus, and watching the arrivals and departures of the stars are all part of Dogon practical astronomy. The Dogon recognize 266 stars or asterisms and associate them with 266 pictographic symbols that are painted on rock shelters, altars, and other shrines. At the village of Kangaba, every seven years, the Malinke, who are related to the Dogon, rebuild a shrine they name "the Vestibule of the Master of the Sky" and repaint all 266 sacred signs on its walls. The Dogon call the 266 signs the "signs of creation,"

Dogon village elders meet in the *toguna* to organize the community endeavors. The mythic ancestors are symbolized in the anthropomorphic poles that support the canopy, and the shed itself copies the place where the ancestors first met to discuss the future course of the world. The structure is rectilinear and cardinal. (photograph Dr. Peter H. Keller, Bowers Museum of Cultural Art)

and the 266 celestial objects they stand for are "the stars of the support of the foundation of the world." These symbols are regarded as the first way in which Amma conveyed his thoughts and are believed to be "the seed of all knowledge." Their number equals the length in days of human gestations and so connects the heavens and Creation with people and procreation. A comprehensive understanding of the signs and their meaning is part of what it takes to be an informed Dogon elder and a member of the decision-making team.

HIGH CHIEFS IN PARADISE

As village-based societies, neither the Dogon nor the Chumash engaged in monumental architecture, but both recognized the existence of sacred places where the supernatural was present and could be tapped. They marked those places symbolically with carved and painted symbols and performed rituals there to insert themselves into the supernatural and mythic landscape, a domain defined by celestial order and cosmological landmarks. Neither the Dogon nor the Chumash concentrated regional power in the grasp of a chief, but elders, officers, shamans, and priests acted together to mobilize, stabilize, and regenerate the community. In a more complex society governed by real chiefs, the manipulation of celestial knowledge and imagery continues, but their scale and applications are altered. Chiefdoms extract new functions from astronomical expertise. The underlying principle of leadership is the same, however. Whoever holds power maintains a bond with nature by observing and interacting with the forces at work in it. This alliance with the great forces that move the universe enlists the ruler in the effort to uphold the natural order. It also authorizes the transfer of power, through the ruler, into the structural integrity of society. To the extent that our world possesses a share of fundamental order of the world, it remains in touch with the Paradise, the era of Creation when the pattern of natural order was first installed. Any ruler bonded with beginning of time and able to create a coalition with the powers of order and creation clearly has to enjoy heaven's blessing.

Most of us don't carry the burdens of power and possess a far narrower concept of Paradise. Tropical islands—with palms that hula in the trade winds, surf that bubbles onto black sand beaches, and sunrises that break first upon the rims of high volcanoes—probably come closer to our everyday image of Eden. Hawaii probably looked a little like Paradise to Captain James Cook's crew when Europeans first made contact with its indigenous Polynesians in 1778.

Timothy Earle and other anthropologists who have studied Hawaii's highly stratified traditional society regard it as a complex chiefdom. Its native population, organized into four competing chiefdoms at the time of Cook's voyage, totaled two or three hundred thousand, but by 1795 Kamehameha I, the high chief of the big island of Hawaii, had subdued the other six main islands and united the archipelago for the first time under his rule. According to Earle, the development of the Hawaiian elite, culminating in Kamehameha's complete conquest, was driven by population growth and the need to centralize power and put it to work in solving problems of subsistence on a regional scale.

Chiefs understand that productive agriculture and military success are the foundations of their economic power. Nature's inequitable allocation of useful land and other assets, an entrenched Polynesian tradition of rank, and intense competition by the elite for land, resources, products, and status set the stage for expansionist ambitions. Because the lifestyle of the elite was supported by the portion of the goods and produce they claimed from the territory to which they held title, they had a financial interest in maintaining and increasing their power. This sounds a lot like feudalism, the economic and political system of Medieval Europe's kings, lords, and vassals. In fact, some of these feudal institutions are now recognized as properties of chiefdoms.

The chief's edge in these conflicts is spiritual power, what the Hawaiians and other Polynesians called *mana.* It was transmitted to the chief through the bloodline, a consequence of the fact that the gods in Polynesia were ancestor spirits. The chief's power was divine mana inherited from the gods to whom the chief's lineage could be traced. Under these circumstances, investment in genealogy was essential, for the chief's ancestry established the chief's divinity. He was, in Hawaiian terms, "a god that could be seen," and when the chief died, he acquired even more power through funeral rituals and became a full-fledged god. His descendants, then, could petition him for direction and help.

While still alive, the chief applied his supernatural muscle to the challenges of cultural continuity. His primary effort involved enhancing the agricultural productivity and reproductive fertility of his domain. Directing his mana to influence the divine forces of nature was, then, a perfectly practical endeavor. He was intimately bonded with land. The well-being of its animals, plants, and people all depended on the effectiveness of his mana. That mana was visible in the ritual activity over which he presided in the *heiau,* the temple or shrine where worship took place and where sacrifices were offered. The chief was, then, the agent who helped keep the world fit and on track.

The Hawaiians saw the world as a multilayered universe and counted nine levels in the sky alone. The earth on which they stood had four subterranean floors and at least eight belts of the sea encircling Hawaii, which was at the center of the cosmos. The islands, they imagined, were like floating mountains and not attached to the deep, silty, and gelatinous bed of creation at the bottom of the ocean and the cosmos. Similar, if not identical, beliefs about the structure of the universe were held throughout Polynesia. The world was likened to a coconut in Mangaia, one of the Cook Islands, in central Polynesia. Ten concentric celestial domes comprised the Mangaian sky. The Tuamotuans, who occupy a group of islands in French Polynesia south of the Marquesas, pictured a universe with nine heavens, each populated by ancestors who became immortal as they migrated through the layers of the world and lifted the skies into place. Through their effort, the sun, moon, and stars were able to rise to their celestial stations and then travel daily between the sky and the underworld.

Ancestors, gods, dynastic power, and ritual energy all ultimately originate from a dark and all-encompassing invisible realm beyond the horizon and the nine heavens known to the Hawaiians as Kahiki. Mystical and transcending, Kahiki is the garden where the root of power grows in a pure, formless, and unseparated state. It is the undifferentiated, immaterial paradise of the beginning of time, ready for Creation before the curtain is raised. This place in the Hawaiian cosmography is also the home of the god Lono, whose talismanic color is black. Lono is one of the four primary gods of the ancient Hawaiians, and he is primarily responsible for the healthy return of the crops that are watered by rain. The black rain clouds and their winds belong to him, as does the last day of the lunar month, when the moon goes dark. He influences fertility, encourages births, and sponsors medicine. Each of the four main gods is assigned a season, and Lono chairs the committee during the time of winter rains. When Lono's voice was first heard in the thunder, the *Makahiki* festival began. As a New Year festival, it celebrated the harvest's first fruits and coincided with the beginning of the year.

In old Hawaii, the high chief was also the supreme mediator between the people and their gods. Divine mana gave him the right and the responsibility to meet Lono

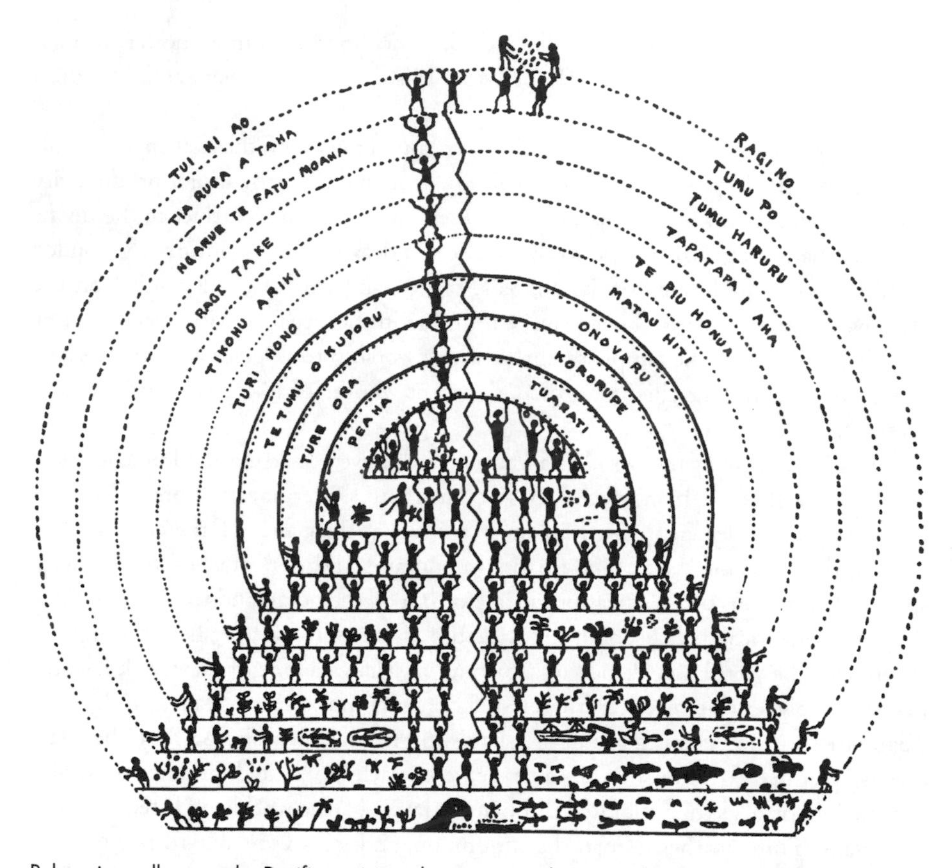

Polynesians all across the Pacific envisioned a cosmos with structural hierarchy. There were multiple layers for the earth and multiple levels in the sky. This sketch of the universe described by the people of Tuamotu has nine floors for any celestial elevator that might be operating. Each of these celestial shells merged with one of the terrestrial realms. Originally the layers of the earth were collapsed together, but overcrowding drove the primeval ancestors to jack up each level of earth and sky with their own muscle. (from *The Morning Star Rises* by Maud Makemson, Yale University Press)

when he returned to this world at the New Year. For that reason, the high chief had to know when Lono would be paying a call. How did the Hawaiian chiefs anticipate his arrival?

The answer was provided by Kamehameha I's last astronomer, Hoapili. Simply knowing that an astronomer officially reported to the Kamehameha tells us that astronomy was institutionalized in the service of the chief. The Hawaiians kept a lunar calendar counted by the phases of the moon, and according to Hoapili, as reported by John Papa I'i in late-nineteenth-century newspaper articles about Hawaiian customs, the arrival of Lono's rains was announced in mid-November by the first rising of the Pleiades after sunset. Celebration of Makahiki actually began, then, in the last lunar month of the dry season and continued through the three lunar months of winter precipitation.

On the first night of the first month of Makahiki, the high chief of Hawaii visited the *luakini,* a special temple only he and his delegates could enter, and left a rag, a symbol of the passing year, in front of it. Inside, the high chief sacrificed a pig and also left bananas and coconuts as an offering that said goodby to the gods that were leaving with the old year. Waiting with the high chief at one of the temple's altars until they saw the Pleiades appear in the eastern sky, the priest then called out the name of each lunar month, recited an invocation, and identified the star that signaled that lunation. This ceremony was said to "feed" those months and their stars and was itself timed by the risings of stars, including the Belt of Orion. Green coconuts were cracked, and their milk, which symbolized new life, renewed and purified the temple. When the last timing star rose and the last coconut was broken, the ceremony concluded and the altar was closed. In the days that followed, the high chief and his priests took part in a series of additional rites related to the primary purpose of Makahiki.

In *Kingship and Sacrifice—Ritual and Society in Ancient Hawaii,* anthropologist Valerio Valeri, identifies every aspect of Makahiki as a celebration of time's renewing return. The Pleiades return to the evening sky. The southwest winds return to the leeward coasts. The rains return, with rough seas, thunder, and lightning, and favor the growth of sweet potatoes, as well as vegetation in general. The first sacrifices and prayers offered in honor of Lono's return trigger the next year's systematic round of the moon and stars. The New Year returns to restart the seasons. The season of renewal returns to reinvigorate the earth. It is really Lono that returns, and he brings the wind, the rain, the stars, and the New Year with him. He mates with the land and impregnates it with another cycle of life. The sexual connotation of his role is preserved in other details, too, for he is called both Lono the Father (Lonomakua) and Pele's Fire Keeper. Firemaking sticks symbolized fertility, and firemaking was a metaphor for male sexual entry. Lono is the male sky that penetrates Mother Earth and imseminates her with rain. His seasonal reproductive re-creation of the world is a cyclical replay of the original Creation.

We are told by the *Kumulipo,* the Hawaiian "Hymn of Creation," that the cosmos itself began with the same event that begins the new year, the "time of the rise of the Pleiades." In Rubellite Kawena Johnson's translation and commentary on that chant, she uses the phrase, "When the Pleiades are small eyes in the night." It was the time the earth first grew warm to the touch. It was the time when the starry sky first began to turn. The moon was first seen, thanks to the shadowed light of the sun. It was the time of cosmic night, and from that night the first couple was born.

In the myth that narrates the origin of the Makahiki festival, Lonomakua dispatches his brothers from his home in the sky to the earth in search of a mate for him. They run into a stunning woman named Kaikilani on the island of Hawaii, and Lono hikes down a rainbow to see her. His route to reach her already references the rain. After they are married, Lono is persuaded she has betrayed him through intimacy with another man. Although she has remained faithful to her husband, he kills her.

Then, profoundly sorry, he exiles himself from the Islands, but promises to return. His return at Makahiki revives the earth, who is really his lost wife.

There is a curious historic footnote to this mythic tale. After discovering Hawaii on January 18, 1778, Captain Cook left and then came back to Hawaii a year later, on January 17, 1779. He was recognized by the Hawaiians then as the god Lono returned. The Makahiki festival was well underway at the time. At one of the stone temple platforms, the priests chanted the *Kumulipo* to him. Cook's plans to leave Hawaii again became, in time, evident to the natives. Disappointed and irritated, they interfered with Cook's effort, and he started to retaliate against the population. The theft of his ship's boat on the night of February 13 fired Cook into a kidnap attempt on the high chief. His action prompted a violent response from the crowd who followed him to the beach, and in the fight, Captain Cook was killed. In the days that followed, the people of Hawaii asked their chief if their god would be returning again.

Lono is the complete embodiment of the reawakening of life, and the high chief interacts with him because the high chief is obliged to exert his mana on behalf of the fecundity of the land he rules. In doing so, the chief mimics the gods. He participates in the annual restatement of the Creation in the New Year and in the renewed world. He is actually linked to Creation through his genealogy, for his family line begins with

When Captain Cook reached the Sandwich Islands in 1778, the Hawaiians decided he was the living incarnation of Lono, the divine force behind fertilizing rain and well-watered fields. His festival was timed by the first evening rising of the Pleiades and the New Year. This scene places Cook in front of Lono's shrine, where he is being given offerings. A tall effigy of Lono, closer to the fence, wears a tall miter. ("An offering before Captain Cook, in the Sandwich Islands," by R. C. Barnfield. Photo courtesy of Bishop Museum Archives, Honolulu)

primary gods who were there at the beginning. To fulfill his role, the high chief also interacts with the sky in seasonal calendric ritual. His behavior is symbolic and practical. It also reinforces his mana, for it exhibits the prerogatives of high status. Through celestially tempered ceremony, the chief aligns himself with supernatural, natural, and social power. His judgment verges on royal privilege, for kings serve as heads of state and hold broad powers to order society. They, like the high chiefs of old Hawaii, also borrowed symbols and knowledge from the celestial agencies of cosmic order and put them to use in the governance of their domains.

THE LION KING

Despite the fact that we are no longer ruled by chiefs and kings who have a hotline to heaven, the celestial connotations of kingship reemerge again from time to time in popular culture. That's what happened in 1994 with the nationwide release of the Walt Disney feature-length animated film *The Lion King.* Strength and the lion's place as an aggressive predator at the top of the food chain have conferred on this creature the familiar title of "King of the Beasts." That is how the lion was "justly stiled by all writers," according to Edward Topsell's encyclopedia of Elizabethan animal lore, *Historie of Foure-Footed Beasts.* Published in 1607, it ostensibly synthesized the opinions of the ancient writers of Greece and Rome. The lion, we know, was affiliated with kings from medieval India to ancient Greece. Legend tells us that Philip of Macedonia's dream of a lion was regarded as an auspicious sign of the imminent birth and high destiny of Alexander the Great. Anchoring his lineage to King Solomon, the emperor of Ethiopia called himself "The Lion of Judah." In the twelfth century, Richard, the Crusader King of England, was nicknamed "the Lionheart."

The lion's mane conveyed majesty, like the crown of a king, and was equated with the sun's rays. In Egypt, the lion was an emblem of the sun, at least by the time of the New Kingdom (1570–1070 B.C.). When the Persian Shahs ruled Iran, the solar lion in the royal coat of arms also appeared on the flag. The sun rising over the lion's back on the old Persian flag reinforced the themes of royal power, the light of justice, and adherence to cosmic order. Completing the equation between kingship and the lion, a crown floated above the sun. The scimitar the lion sometimes brandished in its raised paw was the sword of Ali, son-in-law and cousin of the Prophet Mohammed. Ali was the fourth caliph, or ruling successor of Mohammed, of the early Islamic empire. From Kufa, his capital in what is now Iraq, he initiated many successful expansionist campaigns. He is known as a soldier and ruler, and his sword symbolizes the king's military strength.

Ample tradition documents the connections people have drawn between the lion, the king, power, and the sun. With no surprise, then, *The Lion King* begins with an absolutely majestic sunrise over the African veldt. It is a triumphant fanfare of light that summons all of the animals of the lion's kingdom to a great assembly for the first

To find the mythology of our age, look at popular culture. That is where the universal mythic themes ever reemerge. The old celestial connotations of kingship revived in the Walt Disney animated feature film *The Lion King* were accessible enough to advertise the movie. The spirit of kingship is enthroned in the sky, which simultaneously sparkles with the night's stars and also beams sunlight to earth, where it spotlights the place where the Lion King rules and surveys his domain. (© 1994 The Walt Disney Company)

public presentation of the lion cub, the king's rightful heir. It takes place on a special day and at a special place associated with the king, Pride Rock. A baboon shaman officiates the event by marking the cub and so conferring on him his special identity. Dust scooped from the ground and sprinkled over him bonds the lion prince with the land he will someday rule. Then, the old baboon carries the cub to the edge of Pride Rock and holds him aloft for all to see. The sky opens, and crepuscular rays of sunlight spotlight the heir. In acting as witnesses to this ceremony of heir designation, the animals ratify the selection of their future king and so reinforce the continuity of the community through a process of lawful successsion.

The diverse and heavy burdens of kingship promote the role of subordinate officers who assist in the management of the country. They serve the king, but their power, like his, is also obtained from supernatural beings, forces of nature, and the order of the world. Magicians and court shamans, like the Lion King's baboon, operate on the community's behalf, but they do so within the political institutions of kingship. In this case, the baboon performs rituals, paints pictographs, knows what others do not, sees the future, and influences the outcome of events through informed action.

The sky continues to tell the story of kingship in *The Lion King* as the father and the cub watch the sun come up on another day. With the horizon as a classroom for the future Lion King, his father explains "the circle of life." "A king's time as ruler," he says, "rises and falls like the sun. One day the sun will set on my time here. It will rise with you as the new king." In the night, the Lion King shows his son the stars and tells him, "The great kings of the past look down on us from those stars." They guide the living king on the earth.

Later, the Lion King's ambitious and treacherous brother murders the king, condemns the heir to what he thinks will be death at the hands of the hyenas in the wilderness, and installs himself as the new king. The land suffers under his self-serving judgment, his disregard for his people, and his unlawful reign. His offense is more than a crime against another individual. In supplanting the legitimate king, he has damaged the world. Parched, desolate, leafless, and starved, the kingdom of the lion is abandoned by all who can leave. It cannot be renewed until the real king returns.

At this point, the baboon reads in the wind the ripening of another stage in the cycle of time. The missing lion cub has grown into an adult. Prompted by a visit from the old shaman, the future king, longing for his dead father, experiences a vision. He sees his father in his own reflection in a pool that also mirrors the twinkling stars. Although he rejects the image, the baboon insists it is the old king and tells him, "You see, he lives in you." Still unsure, the new Lion King looks into the night sky. Clouds condense out of the night and glow, and in them the old king materializes. As real as the stars, he speaks to his son and reminds him of his royal destiny. Transformed by this supernatural encounter, the young king returns to his homeland to challenge his uncle.

The king's return to Pride Rock is accompanied by the crack of lightning and the roll of thunder. The wind rises, and the fight begins. Engulfed by flames ignited by the lightning strike, the battle of lions ends with the victory of the true king. Cleansed by the fire, the land of the Lion King is ready for the renewal that comes with the rain. The storm recarpets the ground with grass and refills the water holes. Leaves and flowers dress the once-dead branches in new life. In time, on another morning when the sun breaks over Pride Rock, the old baboon shaman lifts a new cub, the next Lion King, to the sky with time's promise for the continuity of life and for the health of the land through the lawful rule of the real king.

ENGINEERING THE ARCHITECTURE OF KINGSHIP

A large population—at the level of hundreds of thousands—and challenges that power at the local and regional levels cannot resolve offer opportunities for the consolidation of power in a king and favor the formation of a true state. The unifying local lineage of a village and the cohesion between related lineages of a tribe or of the territory of a chief don't operate at this scale. Family ties do not link all of the communities that are governed by kings. Like most chiefs, the king inherits his power, but his bloodline does not flow from the population he governs. In fact, the people may not even share the same ethnic background. Even the king's kinship to the elite is limited, and the elite is likewise unrelated through family lineage to the general population. The ways in which all of the various members of the kingdom make their livings may also be diverse.

Despite the diversity of the population and the absence in a large, mixed populace of allegiance inspired by consanguinity, a variety of problems may lead to engineering regional solutions with more centralized power. Water control and irrigation for agriculture may influence this decision if the agricultural enterprise and the mechanism of water distribution require capital investment—in canals and dams—and require administrative systems a chief cannot afford. The complexities of trade and the redistribution of commodities in a stratified society is another. Effective use of technology and labor specialization may be part of the picture. Defense against enemies acquisitive for land, resources, or products is also a powerful stimulus. Anthropologists, in fact, regard the process of state formation as a complex, significant issue in the study of cultural evolution and recognize that all of these elements contribute to one degree or another. Allen W. Johnson and Timothy Earle have jointly synthesized these studies in *The Evolution of Human Societies,* established the key factors, and outlined the ways in which the development proceeds.

More complicated societies sometimes evolve, then, in a response to the practical need for innovation, in response to the obvious need to defend their own interests, or in response to the desire to exert control over other people and territory. Once kingship is in place, the perennial concerns of the state encourage its perpetuation of itself. Its complex economy has to maintain the ability to redistribute commodities where they will do the most good, and there won't be any agricultural commodities to redistribute without control over water supply, an understanding of its allocation, and the ability to move it. Divisions in a mixed population are inevitable, and the central government has to be able to maintain and guarantee civic order. A universal calendar is not just a vehicle for getting fields seeded and grain harvested on time, it is an essential part of organizing and integrating all of the activities of a group. If you want the right things in the right places at the right times, you need an accurate calendar and the authority to impose it. Without economic control, public order, and successful cooperative endeavor, there is no state, and without the state, there is no king.

The identity of the ruler and membership in the elite are advertised by the luxuries and exotic items in their possession. Control of this symbol-laden merchandise is also part of the apparatus of kingship. Monumental palaces, temples, and tombs further endorse the power of the king, and their construction involves the resources of the state. Architectural proclamations of royal power are supplemented by public documentation of royal lineage and dynastic history. Control of this history and the ability to billboard it therefore command the interest of the king and the state. Finally, concentrated power, aware that its authority may be vulnerable to other sources of power, will consolidate its position through military strength and also assess the landscape beyond its borders for opportunities to expand its influence. That means conquest.

Several specialized tools for management and control allow a king to do things over a large and diverse region that elude a chief operating within a more limited district. Because these tools are ultimately at only the king's disposal, they are reflections of royal power. They include the military, the bureaucracy, and the priesthood. As permanent, full-time institutions, all are subsidized by surpluses. Usually this means agricultural surpluses. Surpluses are supplied by the general population but controlled by the king. And they mean the services of an army of soldiers, an army of accountants, an army of pontiffs, and maybe a platoon, or at least a squad, of astronomers. These functions overlap, of course. The skywatching calendar keeper likely performs liturgy and writes like a bureaucrat. The bureacracy manages information and maintains the economy. It collects tribute, directs trade, and puts surpluses to work. The priests develop and operate the ideology that sanctifies, and therefore legitimizes, the authority of the king and the rest of the structure of government. Through state religion, the king rules with approval of the gods and a handshake from heaven. Intimidation and direct force allow the military to pacify the countryside, turn back invaders, and conquer new lands.

All of these institutions distinguish the king and the state from less powerful rulers and less complicated societies. They are mechanisms that sustain the majesty of the king and the proficiency of the state. At the core of this system resides the genuine belief that the king is truly different and truly special. To perform the tasks required of royalty, the king must display and restate his status, and so the state exalts the king in rituals that establish his divine mandate. He is installed in office in public ceremonies that invoke cosmic order. He enjoys certain prerogatives and obeys certain restrictions that apply only to the king. He is assigned unique insignia that symbolize his place, purpose, and his power.

Celestial references uphold and enforce these expressions of state power, particularly as it is embodied in the king. The power is concentrated in him and symbolized by him. If Louis XIV, the Sun King of Versailles and absolute monarch of seventeenth-century France, were with us now, he would explain the principle as succinctly as he is said to have lectured to the Paris Parliament: "*I* am the State." And he glorified himself with solar emblems to prove it. Just like the ancient sun kings of antiquity, he was chartered by heaven and ruled with the order and power the sun gilded on the sky.

Cosmic kings hold hereditary titles and are tenured for life. The welfare of the people and the well-being of the land are believed to depend on their leadership and the effective and proper use of their power. Although kings do not usually talk to the spirits in a shaman's trance, like shamans, they are intermediaries with great forces that govern the universe. Nature in the cosmos of a sacred king still admits the presence of divine, supernatural power, and the king manipulates supernatural power through state religion and royal symbolism. As a participant in the divine regulation of the world, the king is the junction between the will of the gods and human affairs. Contact with divine will makes his endeavor sacred and makes the king divine. In the wake of the excesses of Europe's Age of Kings, we used to argue about the divine right of kings. In time, the secular power of the king was harmonized with the old habit of sacred power through a contract, or charter, that relied on divine approval to temper conflicting interests, and eventually modern democracies dispensed with the king altogether. Even the United States Declaration of Independence, however, justified separation from the British Crown with references to divine will. "The Laws of Nature and Nature's God" are mentioned in the first paragraph, and the last paragraph enlists the approval of the "Supreme Judge of the World."

The concept of divine kingship was first introduced as a compelling anthropological theme when it appeared in 1890 as the central idea in Sir James George Frazer's monumental synthesis of seasonal myth, ritual magic, and the lore of nature spirits. Frazer framed his research with the tradition of the dying god and royal sacrifice. His sacred king was also a magician and a priest, and the myths collected by Frazer emphasized the king's role as the source of the abundance and fertility of the land. This power could presumably be seen in the king's own vigor and sharpness of mind, and when his power waned, it had to be revived through magical ritual or through rites of succession. Frazer's work generated controversy and conflict, with many at the time doubting whether the ritual murder of a declining king was as commonplace and fundamental as Frazer argued. In fact, the periodic replacement of the king through orderly, authorized succession fulfilled the requirements for revival of the soul of the people and the fruitfulness of the land.

Cultural and social continuity are the goals, of course, and complex societies governed by sacred kings treated kingship as the connective tissue in the cosmic order that held their world together. The king's performance as a mediator with nature and heaven has led some to speculate that the office of sacred kingship evolved naturally from the magician chieftain who fulfilled so many of the same ritual responsibilities. A simple extrapolation would slip us back into the spirits and spells of the shaman and tempt us to reason that pedigree of power of the sacred king begins with the shaman's trance. Such a simple linear evolution of power seems unlikely, however. High population and large territory don't eliminate the shamans. Like what is magical and supernatural, they just show up under different auspices and in different costumes. Even in our era there are urban shamans channeling messages from beyond. Some are bogus. Some are sincere. And some, no doubt, are slouching toward cable television.

The similarities we see between shamans, chiefs, and sacred kings in acquisition and use of supernatural power are probably reflections of a process that is always in effect, a reaccessing of old, dependable tools that require refitting for new circumstances. Crises linked with population growth, resource limits, natural catastrophe, disease, and powerful enemies weaken the validity and resolve of the prevailing belief system. A resourceful and resilient society somehow manages to retool its ideology. With new meaning pertinent to changes of the times, it is put back in service. If successful, the new institutions that congeal around it promote adaptation, and the society survives. For that reason, the myth of cyclic renewal, with its dying gods, sacrificed kings, and regeneration of nature, always makes sense. When the process works, it illustrates the myth in action.

As a living character of myth, the sacred king possesses the supernatural powers and divine attributes that enable him personally to activate the process of world renewal. These attributes may be conferred on him because he is the terrestrial agent of a deity, the son of a deity, or is himself a deity. He rules by divine election. If he is not actually divine, he at least governs with divine authority. His personal identity and special status entitle him to the favor of the gods. His emblems of sovereignty set him apart. He has a throne only he can occupy and sits at the center of the world, the point of contact between heaven and earth. His radiant crown mirrors his majesty and his divinity. The scepter he holds in one hand is the symbol of his authority and power. The orb he holds in the other represents the world he rules and its order.

According to Frazer, evidence for his tradition of the divine king could be found in the Sudanic kingships of Africa. Secular political power and sacred religious duties both belonged in the jurisdiction of a single autocratic king. He owned the land and could authorize its use to others as he wished. His ceremonial activity included a vegetation rite in which he planted the first seed and gathered the first fruits. He was, like Louis XIV, the State.

DIVINE KINGS IN WEST AFRICA

Similar traditions of the sacred king were part of the royal protocol and foundation myth of Ife, the sacred spiritual capital of the Oyo empire in thirteenth-century Nigeria. As the center of Yoruba civilization in west Africa, Ife is most famous for the artfully and realistically modeled portrait heads discovered in the groves that grew over the ruins of the old city in 1910. These were excavated and publicized by Leo Frobenius, a German anthropologist, who thought that Ife had been colonized in southwest Nigeria by Bronze Age Greeks in the second millennium B.C. The graceful bronze and terracotta sculpture, no doubt, influenced his judgment, and more bronzes found in 1938 also fueled speculation about European inspiration. More careful study of the sculpture and Yoruba traditions, however, now permit us to recognize the antiquities of Ife as part of a completely independent African tradition that began to appear around 800 A.D. Most of the faces have somewhat idealized features. The expression is usually serene and dignified, and many of the heads bear crowns. Some of the sculpted heads were mounted on full-length funeral effigies. Most are thought to represent deceased Yoruba kings and members of the ruling families. Others appear to portray slaves and captives sacrificially killed.

Known traditionally as the holy seat of kings, Ife was believed in Yoruba myth to be the place where Olorun, the high celestial god, first touched down when he descended to earth. It is also the place of Creation and the place where the first king took office. To the Yoruba, Olorun, who personifies the sky, is known as the Owner of the Sky. An essential but indifferent source of divine power, he is remote from earth, unresponsive to prayer, and interested in his own agenda.

After roughing out the general shape of the world, Olorun dispatched Obatala to perform the finish work. After getting the job under way, Obatala consumed a little too much palm wine and fell asleep. Wondering why it was taking so long to get the world under way, the Owner of the Sky sent another god—Oduduwa—to earth. Oduduwa is the one who actually piled up sand upon the formless sea to create solid ground and made human beings to occupy it. Olorun filled them with the breath of life, and Oduduwa became their first king, at Ife. The rest of Ife's kings are his descendants. The primordial source of the royal lineage is therefore celestial and divine.

Yoruba mythic history actually credits Oduduwa with siring 16 sons, and each of them becomes a king of one of the 16 states that owed allegiance to the high king of Ife. The number 16 seems to mirror the belief that 16 gods subordinate to the divine sky made the first descent to Olorun's new world on a celestial chain and first set foot in what would become Ife's Olose Grove.

A slightly different tradition from northern Nigeria also documents the idea that heavenly power brought Ife to life. These northern Yoruba regarded Obatala as female. She was created by Olorun to be Oduduwa's wife, and she is the one who puts children into the wombs of women on earth. Obatala's own children included the lord of the soil and the lady of the water. Together, they parented the god of the upper air. He took his own mother by force, however, and incestuously fathered an entire family of cosmic gods, including the sun, the moon, the lightning, the spirit of vegetation, the god of cultivated plants, the master of the hunt, the divine blacksmith, and the god of smallpox. Yemaja, the mother goddess of the waters, exploded while dropping these deities at term and fell out of the sky. Where she struck the earth, Ife was founded.

Today, a collection of the portrait heads of Ife's *onis*, or sacred kings, and related royal sculpture is on display in the museum that occupies the ground where the *Afin*, the palace of king of Ife, once stood. Igbodio, the most important grove in Ife, is right next to the palace and is said to be where Oduduwa first molded the land. The palace, then, was the center of the center of Creation, the home of the primary king, and the legendary birthplace of the mythic ancestors of the other 16 kings. Ife was the center of the world, and Ife's king occupied the center of Ife. The rest of the old city radiated from the sacred royal center, and a series of more or less concentric walls established a geographic hierarchy focused upon the king and sealing him in the middle of it all. Even after Ife's decline in the sixteenth century and abandonment, more than once in the eighteenth and nineteenth centuries, the groves where so many of its dignified images of kings were buried retained their reputations as sanctuaries of

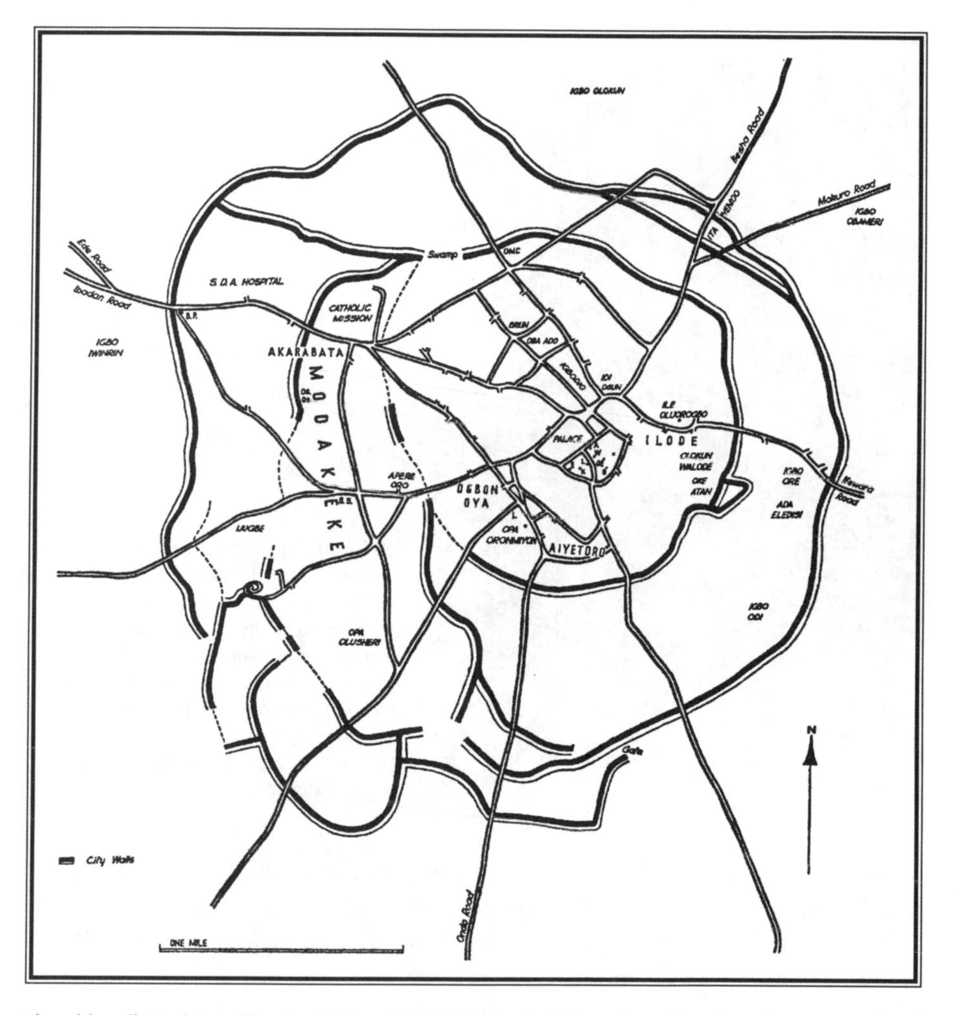

The old walls (indicated by the bold and doubled lines) of Ife enclosed its divine king in a central palace with a series of rings that emphasized his status as the living center of the Oyo empire. The radial lines in this city plan are more recent roads. (from *Ife in the History of West African Sculpture* by Frank Willett, Thames and Hudson Ltd.)

sacred power. One bronze head from Olokun Grove was retrieved from the earth, used in yearly ceremonies, and buried until needed again.

Centrally controlled tribute and trade brought wealth to Ife's king, and the king's power was perhaps preserved from one generation to the next through rituals in which Ife's celebrated heads of bronze and terracotta helped convey the message of dynastic continuity. Similar effigies of dead kings were contrived in Benin, the center of power in southern Nigeria that had superceded Ife by the time of European contact in the late fifteenth century. The *Oba,* or king, of Benin possessed the military, medical, and

Ife's habit of memorializing its sacred kings with portrait heads was perpetuated by the kingdom of Benin, which had succeeded Ife in power and influence by the fifteenth century. This casting honors an *Oba,* or king, of Benin with a bronze portrait. (photograph E. C. Krupp, object displayed in the National Museum of African Art, Smithsonian Institution, Washington, D.C.)

magical power, and his funeral statue was outfitted with the beads of kingship. Barechested, his successor publicly mourned him and then later reappeared in the same beads at his own coronation. It is possible the custom originated at Ife. Ife's continuing importance to the kings of Benin is reflected in their habit of sending the dead king's "head" (perhaps really just a small relic, like hair or nail clippings) to Orun Aba Ado, a zone of Ife next to Igbodio Grove. This last occurred in 1888. Like the Yoruba of Ife, the kings of Benin were plugged into the sky's power through an umbilical connected to the point of Creation, and they even traced their lineage to the youngest son of Oduduwa, born at Ife when the world was young.

CHAPTER SEVEN

CELESTIAL EMPIRES

By 1227 A.D., Chinggis Khan—better known in the West as Genghis Khan—had forged an empire that straddled the steppes of inner Asia from northeast China to the Caspian Sea. He stormed out of Asia's heartland with a unified horde of warring clans who hitched their destiny to the supremacy of Eternal Blue Heaven, their name for the highest god that embodied all of the power of the sky. Often the Mongols just called him Tengri, or "Heaven," and Tengri's army of horsemen provided the muscle for a mobile imperial state based on a shamanic belief in celestial sovereignty. It was twice as large as the Roman Empire at its height, and after the death of Genghis Khan, his descendants continued to expand in both directions. Khubilai Khan, the grandson of Genghis, ruled one-third of the world's land, from the tip of the Korean peninsula to the threshold of Europe on the Danube River.

Building on three thousand years of Chinese tradition, Khubilai Khan preserved the official character of astronomy. It was a tool in the service of the state, and he subsidized the best astronomy in the world in the thirteenth century. His Mongol dynasty later created a sacred, astronomically aligned cosmological design for his capital that can still be seen in the streets and historic monuments of Beijing today.

The Mongol empire was accumulated in a remarkably short time by Altaic nomads—herders of cattle, sheep, and horses—who established a new world order in less than half a century under a single family. They were at the gates of Cracow, the

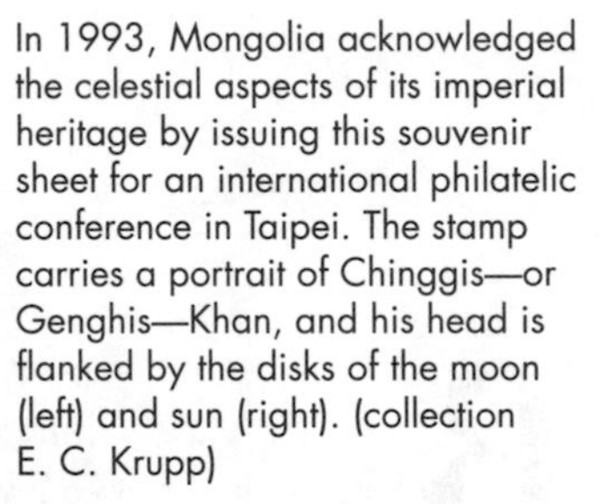
In 1993, Mongolia acknowledged the celestial aspects of its imperial heritage by issuing this souvenir sheet for an international philatelic conference in Taipei. The stamp carries a portrait of Chinggis—or Genghis—Khan, and his head is flanked by the disks of the moon (left) and sun (right). (collection E. C. Krupp)

capital of Poland, by 1241, and then they burned it. In the same year they controlled the entire Hungarian plain. They entered the lands of Austria and the Balkans, and smaller commands approached within 60 miles of Venice.

Ögödei, one of the four sons of Genghis Khan, was his chosen successor, and when Ögödei died in Karakorum, the Mongol capital, the political waters turned murky. Batu, who was the younger brother of Genghis Khan, withdrew from Hungary and retreated to Sarai, north of the Caspian, to consolidate his own position while the players at the center of Mongol power resolved the issue of succession. Confusion and intrigue in Karakorum, then, kept the Mongols from swallowing the rest of Europe when the armies of Europe could not.

IN STEP WITH ETERNAL BLUE HEAVEN

Genghis Khan, the real originator of Mongol military expansion, transformed a dispersed nomadic and pastoral people of central Asia into unified world conquerors with help from the sky. From its mythic accounts of its own origins to the institutions that defined its power as the ruling dynasty of China, Mongol ideology was cinched to the sky. *The Secret History of the Mongols,* a semilegendary chronicle of the Mongol dynasty and the earliest source of Mongol history, traces the lineage of Genghis Khan to the pairing of the "blue wolf" and the "fallow doe." For the Mongols, the color blue meant the sky. This celestial ancestor, the "blue wolf," was, in fact, credited by *The Secret History* with having been born with a "destiny from heaven above." The tradition of an ancestral azure sky wolf can be found in the older beliefs of Turkic-speaking peoples, to whom the Mongols are related. In the time of Genghis Khan, the Mongols called themselves the "Blue Mongols" to publicize their celestial bloodline.

Tengri, the lofty divine sponsor of Mongol imperialism, allowed Genghis Khan to justify his conquests with a celestial mandate, and other images from the sky also found their way into *The Secret History*'s story of his ascent to power. Alan Qo'a, one of the Khan's female ancestors, gave birth to three "sons of Heaven" after conjugal visits from a radiant yellow man who descended upon her through the smoke hole of her yurt and who returned to the sky via sunlight and moonbeams. The Khitans, a seminomadic people who spoke a language related to Mongol and who overwhelmed China in the tenth century A.D., claimed that their conquering king was fathered by a ray of sunlight that penetrated his mother's womb. Genghis Khan reinforced his own alliance with heaven by claiming to be the "son of the sky." By some later legends, he, too, was the fruit of the same kind of solar insemination.

After Genghis Khan died, he was deified and worshipped at several shrines. In one prayer he was said to be "heaven-born" and "born from the decision of sublime heaven." His "great high tent-pavilion" became "the hearth-circle of heaven." In the Ordos region of China's Inner Mongolia Autonomous Region his divinity is still recognized, and his cult is centered at a modern shrine in the Yejinhuoluo Banner. A *banner* is a traditional Mongolian administrative unit on the scale of a county. The neighborhood of this shrine—the hilly lands beyond Dongsheng, southwest of Baotou—is believed by local people to harbor the secret grave of Genghis Khan. There has been some kind of monument for Genghis Khan in this place since 1616. The cardinally oriented shrine now on Gande'er Hill, where the grasslands meet the Kubuqi Desert, is regarded as a mausoleum. It was built in 1956, and its design in part derives from the yurt, the traditional house of the steppe. The yurt was itself said to be based on the architecture of the cosmos, with the floor as the earth, the canopy as the sky, and fire pit and the smoke hole as the world axis. Ceremonies for the deified Genghis Khan are performed at the mausoleum four times a year, and Mongolian people gather there in a great crowd each spring. In 1962, the 800th anniversary of Genghis Khan's birth was sidestepped in Mongolia, which was closely allied with the former Soviet Union and wary of China, but 30,000 Mongolians in China made the pilgrimage to Gande'er Hill to honor the Great Khan.

Genghis Khan's mausoleum in Inner Mongolia, an autonomous region of the People's Republic of China, does not necessarily house the remains of the founder of the Mongol empire, but it does enshrine his reputation among the Mongolian people. It is cardinally oriented on a north-south axis. (photograph E. C. Krupp)

The name by which we know Genghis Khan is really a title that he took at a great assembly of the Mongol clans. His name was actually Temujin, and *The Secret History of the Mongols,* commissioned by Ögödei in 1240 A.D., details the career and murder of the chieftain Yesugei, Temujin's father. After Yesugei's death, the family's status deteriorated. At times Temujin was an outcast or a captive, but bold and ruthless, he gradually worked his way through his enemies and rivals to become, by the age of 20, the leader of his clan. It took him two more decades to unite the Mongols and claim the title by which we know him today. Kököchu, a shaman close to Genghis Khan, told him that the King of Heaven had informed him that Temujin should "hold the nation in his hands." Rashid-al-Din, a thirteenth-century Persian historian, reports that Kököchu insisted it was God's will that Temujin should become "the master of the world." The title "Genghis Khan," Kököchu added, belonged to Temujin. The translation of this title is disputed by linguists. Some read it as "universal khan." Others believe it means "chief khan." In any case, it refers to the Mongol high chief or king.

Genghis Khan actually took that title and asserted his authority over the Mongol tribes in 1206 A.D., at the source of the Onon River, traditionally the place where the blue wolf and the fallow doe had mated to engender the true line of Mongol rulers. Nearby, the mountain Burkhan Khaldun fulfilled the requirements of the traditional mountain cult of the peoples of inner Asia. This 7680-foot-high peak in the Hentei Mountains northeast of Ulaan Baatar, the present capital of Mongolia, was Genghis

Khan's "guardian mountain." It was his refuge from his enemies when he was still young and vulnerable. In gratitude, he prayed to it on his knees, with his hat in his hand and his belt over his neck. Later, in the territory of the same holy mountain, Genghis Khan summoned the Mongol tribes and declared his celestial patronage: "I was designated by Mighty Heaven."

Erketü Tengri, or "Mighty Heaven," belongs to the shamanic tradition of Mongol folk religion. As the Mongols' "Supreme Lord of Everything," he received their highest offerings, including incense burning and horse sacrifice. Tengri was worshipped in high places, because those places put the pilgrim closer to heaven. A sacred mountain like Burkhan Khaldun was a point of celestial contact, a place where the power of the world axis and the world's center seemed to be within reach. It was also the home for Lords of the Earth and mountain spirits and therefore an inevitable destination of shamans. Cairns of stones, known as *obos*, were built on mountains and other places recognized for their supernatural power. The vertical wand that emerged skyward from the stone and brush base of the *obo* was itself a miniature world axis. Shamanic offerings at an *obo* invited the presence and encouraged the cooperation of the spirits that gathered there. *Burkhan Khaldun* means "Mountain of the Shaman Spirit." Reflecting the significance of a mountain's power, the story about Genghis Khan's consultation with the shaman Kököchu places them on a mountainside. Genghis Khan honored Kököchu with the title Teb Tengri, or "Most Heavenly." It was an apt alias for a shaman who claimed, "God talks to me, and I visit the sky."

This shrine at the Genghis Khan Mausoleum collects prayer flags and other offerings and makes a cosmic mountain out of a modest cairn. (photograph E. C. Krupp)

Important ancestors were buried on the mountain slopes to add their spirits to the mountain's power. The same kind of ancestor spirit worship that nourished Mongol shamanism transformed the historical Genghis Khan after his death into the spiritual protector of the Mongol people. Another version of his last days locates his hidden tomb on the flank of Burkhan Khaldun. It is the paramount landmark in the traditional pasturelands and forests of his family line, and the *obo* on its summit is known locally as "Genghis Khan's Seat."

From the heights, Mongol shamans enjoyed uninterrupted audiences with Tengri. Because Tengri presided over all of creation, he was also known as Khan Eternal Heaven. The name reflects the political dimension of the sky. The high Khan controlled his empire through power applied by his political and military delegates. By analogy, the sky exerted pressure on all of nature, particularly through 99 subordinate powers in heaven, also called *tengri.* Among them were counted the four gods of the world's corners, the five gods of the world's winds, the sun, the moon, the planet Venus, the Pleiades, the stars of the Big Dipper, and more. Tengri and his officers controlled the order and the destiny of the universe, and with Tengri's sponsorship, the Mongol Khan and his chiefs did the same. Associated with the warrior aristocracy, the high-celestial ruling god protected the populace from enemies and was the patron of war.

David Morgan, a British historian and the author of *The Mongols,* explained the preoccupation of Mongol shamans and khans with the sky: "The Mongols believed in taking out as much celestial insurance as possible." Shamans also engaged in astral ritual. They offered sacrifices and prayers to the stars of the Big Dipper. When they made a circuit around a shrine, they mimicked the nightly circling of the stars around the north celestial pole, the "golden nail" that held the sky in place. A man buried in front of the tent of Genghis Khan inside the Ordos mausoleum personifies the unmoving "golden nail" of the sky, and the shamanic prayers recited during offerings direct him to remain still.

As steppe nomads, the movements of the Mongols were tuned to the sky by seasonal change. Grasslands that provided perfect pasturage in summer turned brown when winter approached. Each group then retreated to its own river valley territory to wait out the winter. Through seasonal ceremonies the shamans aligned their communities, in a familiar rhythm, with the cyclical order of the sky. The Mongols timed the "Day of the Red Disk," one of the most important religious festivals of the year, by the summer solstice and the northernmost limit of the sun's daily path. A summer solstice feast that involves offerings of fermented mare's milk was observed and described by Marco Polo on his trip to China in the last quarter of the thirteenth century. The fire god, who has an interest in the increase of wealth, in the growth of the herds, and in the healthy fertility of the land, is also called upon when the milk is presented. The horse is the talismanic animal of the summer solstice, and elsewhere in central Asia, Turkic peoples tint a horse blue and run it around in a circle to mimic the dynamic behavior of the sky.

To the Mongols, the north celestial pole was the "golden nail" that held the sky in place. The circumpolar motion of the northern stars provided the essential pattern of cosmic order attributed to Tengri, the highest power in heaven. (photograph E. C. Krupp, December 1, 1989, Baja California)

Nine, the special number of heaven, was incorporated into supernatural symbols of authority. Nine yak tails or horse tails, strung from a pole, symbolized the power of the Mongol military and the royal family. This standard was recognized as a protective spirit. When Genghis Khan bowed to the summit of Burkhan Khaldun, he dropped to his knees nine times. That celestial etiquette and the fermented mare's milk he scattered at the same time to honor the sun further linked the mountain with the sky.

Despite political changes in the beginning and the end of the twentieth century, the flag of Mongolia has retained celestial symbols and other emblems of traditional Mongolian belief. Mongolia became independent from China in 1921, and the basic design of the banner, the symbol of national identity and power, was adopted in 1924. It includes a vertical blue band for the eternal blue sky in the center, framed by panels of red. Components of the *soyonbo,* the yellow symbol near the hoist, include quadrilaterals that stand for the earth, triangles that represent fire, and the yin yang emblem of complementary opposition in a dynamically stable cosmos. In addition, a lunar crescent, a solar disk, and a flame symbolize the upper universe. Collectively, these insignia represent a magical dagger that possesses the power to disperse demons.

Today, the national flag of Mongolia still flies with celestial emblems. The symbols on the left include the sun, the moon, and the fiery light of the upper universe along with other magical insignia of cultural sovereignty. (from *The Observer's Book of Flags* by I. O. Evans, Frederick Warne & Company Ltd.)

Other peoples of inner Asia also introduced celestial elements into demonstrations of the sanctity and validity of royal power. When the Khitans informed the great celestial vault of power that a new ruler had assumed his place in the cosmic hierarchy, they lit a fire. Its smoke carried the news to heaven. In the sixth century A.D., the Tujue, or upper Orkhon Turks, seated their king in office by raising him on a rug in presentation to the divine sky. Genghis Khan put his license to rule on the line with Tengri in his campaigns of conquest. His preparations to move on the Manchurian Jin dynasty in Beijing included placing his case before Heaven—"O Eternal Tengri, I am armed to avenge the blood of my ancestors upon whom the Jin inflicted an ignominious death. If you approve of what I do, vouchsafe me the aid of your strength." He withdrew like a shaman for three days to communicate head-to-head with heaven, and when he finally reappeared, he reported publicly his successful negotiation of a celestial contract for victory. Of course, military triumph indicated Tengri's approval of Genghis Khan.

Belief in Tengri's personal selection of a sovereign and in the ruler's ability to draw upon celestial power is rooted in Altaic shamanism. Political power was actually conferred upon Genghis Khan by the sky through shamanic protocol. The shaman Kököchu, we know, was persuasive in announcing that Eternal Blue Heaven had already endorsed Genghis Khan.

In the Mongol tribes, the chief and the shaman were often the same person, and there was a special title for a khan, or chief, who also possessed supernatural talents. Like any shaman, it was his responsibility to travel as a spirit to the ancestors, to the underworld, and to the sky for knowledge and power. For the Mongol military, the shamans were an essential weapon. It was believed they could persuade the elements and the storm to assault the enemy. Genghis Khan himself was a shaman. He was said to be able to enter a trance and to read the future, like any inspired shaman, in the cracks of a sheep's scapula. His own shamanic powers validated his cosmic claim to the high command. "My strength," he explained, "was fortified by heaven and earth. Foreordained by Mighty Heaven, I was brought here by Mother Earth."

Allied with Eternal Blue Heaven, the sovereign lord of the entire universe, shamans used their influence in politics and war to share the responsibility of ruling on earth. Because the heavens were understood to be a zone of power, however, any court shaman who claimed to be able to travel to the sky was clearly engaged in political activity and could represent a challenge to the power of the Khan. Rashid-al-Din's history tells us that Kököchu, the influential shaman who lubricated the mechanism of Genghis Khan's political ascent, and who was himself elevated in rank by Genghis Khan, in time exceeded his own shamanic reach and found himself on a permanent trip to the spirit world.

Kököchu realized he was in position to acquire more personal power through intrigue, and he informed Genghis Khan that Eternal Blue Heaven was supplying conflicting messages about Genghis Khan's own brother Kasar. Believing that Kasar was scheming to replace him, the Khan made plans to assassinate him. His mother intervened, however. Kasar was spared, but Kököchu's ongoing manipulations resulted in factional desertions and the embarrassment of Genghis Khan. Realizing, at last, that the shaman was the source of his troubles, Genghis Khan had him quietly murdered and then announced that Kököchu had fallen afoul of Tengri. In accord with standard shamanic practice, he had risen through the smoke hole of a yurt, presumably a little more accountable to the will of Eternal Blue Heaven now that he had disappeared into the sky.

Through its political dimension and its imagery of celestial power, Mongol shamanism fueled the engine of Mongol territorial expansion and formulated effective additives to imperial power. Of course, to rule effectively with heaven's mandate, it was necessary to have the power to discern heaven's intentions. That is partly why kings and emperors put astronomers to work in keeping calendars and observing the activity in the sky. The Mongol Khans of China's Yuan dynasty may have started as nomads and practitioners of folk shamanism, but like all kings and emperors they had to govern. They needed a bureaucracy to administer their empire, and among their bureaucrats were shamans who handled the technical details of celestial ritual. Shamans of the Mongol court read the heavens, anticipated eclipses, and presided over ceremonies intended to diminish their supernatural effects. It was believed they could modify the weather and negotiate with the 99 spirits of the sky. Marco Polo's narration of the story of the conflict between Genghis Khan and Prester John, the legendary Christian prince of Asia, includes full participation by court astrologers who cast reeds, recited chants, and consulted books to read the will of heaven.

Caroline Humphrey, a social anthropologist, has examined carefully the relationship between the development of the state and shamanic behavior in northern Asia, and she agrees that the formation of the Mongol empire relied on the Asiatic heartland's tradition of ecstatic shamanism. Her analysis shows how the shamanic tradition evolved in response to the consolidation of an established dynastic government out of the power accumulated by a charismatic, talented, and astute war leader. Rather than restricting her inquiry with a narrow definition of shamanism, she sees two significant approaches to the application of supernatural power. On the one hand, she identifies ecstatics whose spiritual quests primarily involved celestial spirits and the various expressions of the sky's power. The trance in this case allowed the shaman to summon the spirits to receive the offerings intended to elicit their assistance. A different kind of trance experience allowed the shaman to become some other being—perhaps an animal—and move through the world with that creature's power. In both cases, the shamans had to compete with other specialists in different kinds of magic capable of addressing some of the same problems. In Mongolia, Buddhist lamas in time displaced most traditional shamans but at the same time they incorporated ideas and images from shamanic belief into the symbolic system of Lamaism.

In the world of Mongol political shamanism, prophecy and ritual were the primary commodities. Although some scholars regard shamanism as a religious response of marginal peoples, an aspect of folk religion, under the Mongol khans the shamans were competing with themselves and with others to move in the mainstream of power. Although the empire began with the inspired vision of a nomadic horse warrior, it became, by the time of Khubilai Khan, a vast administrative system centered and anchored in a kingdom rooted in agriculture. The informal sky lore of the herder on horseback was transformed into a system of national observatories. Imperial investment in astronomy is evident in the *Yuan shih,* the annal of the Yuan dynasty. This official history was prepared in the fourteenth century, shortly after Mongol rule was displaced by the Ming dynasty in 1368. It contains a long section of calendrics and detailed descriptions and diagrams of astronomical instruments along with commentary on calculations and observational techniques. Guo Shoujing, one of China's most celebrated and talented astronomers, worked for Khubilai Khan. He reorganized the state observatory in Beijing and contrived a new and improved calendar to replace the *Wan nien li,* or "Ten Thousand Year Calendar," the Khan had imported from the Persian court of Hulagu Khan with the help of the astronomer Jamal al-Din.

Erdenezuu Hiid is a Mongol monastery just a mile and a quarter from the thirteenth-century capital at Karakorum. The Lamaist monastery, enclosed by the monumental, stupa-punctuated wall seen here, was built in 1586. By then, the Ming dynasty was in charge of China, and the Mongol empire was in decline. (photograph Dr. Peter H. Keller, Bowers Museum of Cultural Art)

Astronomy was internationalized by Khubilai Khan, and at the same time he also continued to develop the celestial basis for imperial authority. His new capital slightly overlapped the old city, the Jin dynasty capital, on the northeast. He called this new urban expression of royal power Dadu, or "Great Capital," and it eventually became Beijing, the present capital and center of power of the People's Republic of China. Although the cosmological symbolism designed into the layout of Beijing was not all introduced by Khubilai Khan, he did locate the imperial palace and inner city at the very center of Dadu and on its primary north-south axis. His observatory was built into the city wall at the southeast corner. The entire city was cardinally oriented, and altars for the spirits of heaven and earth were constructed on the main axis, south of the palace.

Marco Polo specifically mentioned how Khubilai Khan was stationed on the north side of the audience hall in public court. In doing so, he faced south and imitated the hub of the sky, the north celestial pole that held the cosmos in place. He was the pivot at the center of the world and was still allied with Eternal Blue Heaven, whose power seemed to radiate from the pole of the sky. Khubilai Khan, in official assembly, was the terrestrial terminal of the world polar axis and therefore in direct contact with heaven.

Khubilai Khan, the founding emperor of China's Mongol dynasty understood astronomy's value as a tool of the state. Systematic observation of the heavens had been official business of China's rulers since the Shang dynasty and the Bronze Age. Khubilai Khan continued this royal policy by subsidizing an innovative and active program of instrument development and observational research. The aggressive campaign organized by his "project manager," the astronomer Guo Shoujing, included the invention of this instrument, the *jian i*, or "simplified instrument." Alternately called the "equatorial torquetum" or the "abridged armilla," it was essentially a dismantled armillary sphere that made room for the observer. In a conventional armillary sphere, the tangle of rings used to measure the positions of celestial objects sometimes interfered with the astronomer's head and line of sight. (photograph Robin Rector Krupp)

Later, Khubilai Khan approved the construction of an altar that was symbolically color coded to the five directions (four cardinal directions and the center) and dedicated to the spirits of land. It was paired with an altar for spirits of the grain, a remarkable ideological step for an emperor fully conscious and proud of his nomadic origin. Subsequent Yuan rulers built altars for the spirits of thunder, wind, and rain.

Khubilai Khan continued to rely on astrologers for assessments of the political meaning of events in the sky. Linked to traditional shamanism through a belief in celestial spirits and through magical knowledge via divination, the Khan's astrologers were repeatedly mentioned by Marco Polo, who recognized their importance in the structure of Yuan government. They are listed along with other essential personnel, including physicians, falconers, and the rest of the Khan's officers, and they merit a section that details some of their instruments and duties. Estimating the total number of astrologers and soothsayers in Beijing at 5,000, Marco Polo reports that they have their astronomical instruments and from the planets predict the weather. For

Detailed astronomical knowledge provides the basis for the world system, and so these data are not only recorded in observing logs but are also put into the service of religion. This star map occupies part of a wall in the Temple of Five Pagodas (Wuta si), a Buddhist relic in Hohhot, Inner Mongolia. The stars are charted by their measured coordinates, and components of the astronomical reference systems—the 24 solar terms, or *qi,* and the 28 constellations (lunar mansions), or *xiu*—are indicated. A text on the map confirms the chart was based on the royal astronomical map of the Qing dynasty. Only four inscribed stone star maps are known in China, and this is the only one that identifies astronomical objects in a language other than Chinese. (photograph E. C. Krupp)

instance, they predict that "there will be thunder and storms in one month, earthquakes in another, lightning and violent rains in another, and pestilence, war, discord, and conspiracies in still another."

Polo adds, "They write their predictions for the year on small squares in booklets." Those with the best record for accuracy were elevated in worth. When the Yuan ruled China, even the mystics were accountable. Anyone planning to travel or undertake some other business would consult an astrologer to learn "the disposition of the heaven" toward the proposed enterprise.

The power of the stars and the axial organization of the cosmos never really disappeared with the close of the Mongol dynasty, but we reencounter them with a Buddhist overlay in the centuries that follow. Although shamanic techniques are certainly part of a religious response and involve a supernatural interpretation of the natural world, they do not really reflect an abstract faith that is clarified with theology, accompanied by a moral code, and tempered by a concern for ultimate destiny of the soul. Shamanism is, instead, intended to produce results in this world at this time through effective interaction with the spirits that hold the cards of power. Formal religions, however, have traditionally incorporated concepts and spiritual tools employed by shamans into more conventional, less ecstatic expressions of belief. For example, five hundred years after the era of Genghis Khan, Qing dynasty Buddhists in Hohhot (Huhehaote), or "Blue City," the capital of Inner Mongolia, embellished the Wuta si, or Temple of Five Pagodas, with a star map carved in stone relief. The stars' names are inscribed in Mongolian, and nearby, another relief illustrates the traditional world axis of both shamanic and Buddhist tradition. This Mongolian Buddhist temple is no monument to the celestial power that authenticated the empire of Genghis Khan, but it is a demonstration that power in the sky continues to trickle down to sanctify the earth. Wherever God may actually reside, Eternal Blue Heaven always holds court overhead.

THE RISING SUN

Because most modern studies of shamanic activity involve marginal or frontier peoples—usually hunters and gatherers—we are used to thinking of shamanism as primarily the business of less complex societies, people who are perhaps more comfortable with circulating through the land than with settling in one spot. Uranian power and shamanic access to it also belong to the world of the nomadic herders of inner Asia. For that reason, shamans and skywatchers don't necessarily seem out of place in the framework of Mongol social order even after those same wandering horsemen turn into empire builders. Although shamanic knowledge of the sky and shamanic concepts of cosmic order were pervasive and explicit in Mongol applications of imperial power, we might just interpret the ideology of their expansionist state as an exaggerated and inflated variant of the normal behavior of people who are not moored to the land through agriculture. While accepting components of the shaman's cosmography in the ideology of chiefs, in divine kings, and even in an empire crafted by nomadic warriors, we may doubt the likelihood of an imperial agrarian state that recognizes the sources of its power in a cosmos confronted through shamanic techniques. By the middle of the third century B.C., however, rice technology transformed the economy of Japan with intensive agriculture, yet Shinto, the religious basis of Japan's imperial tradition, still adopts the shaman's world perspective and interacts with the spirit realm according to shamanic procedures. It also threads the emperor to the sky with a celestial lineage, for Amaterasu-omikami, the Sun Goddess, is the living divine ancestor of Japanese royalty.

Probably the ancient Chinese first called Japan the "Land of the Rising Sun" in recognition of the obvious—the Japanese islands are east of northern China and the Korean peninsula, across the Sea of Japan. In Chinese, *Jih pen*, or "Japan," is translated "source of the sun." Of course, for the Japanese, the sun rises east of Japan, but they adopted the name themselves late in the sixth century A.D., during the reign of Empress Suiko. Her son, Shotoku-taishi, was the crown prince. Although active in introducing Buddhism to Japan, he still belonged to the family line of the Shinto Sun Goddess. In a letter of introduction for a cultural and diplomatic mission to the Tang dynasty emperor of China, Shotoku identified himself as the "Emperor of the Rising Sun" and greeted the Chinese ruler as the "Emperor of the Setting Sun," probably with an astro-political metaphor in mind.

Japan has continued to link its national identity with the sun, not only through its name but also through the flag it adopted in 1870. As early as the beginning of the eighth century, Japanese royalty have announced their special relationship to heaven with banners that carried symbols of the sun and moon. Today's flag has a red disk—the symbol of the rising sun—centered on a white field. This disk is called *hinomaru,* which just means "circle of the sun," and it has been used as a royal emblem since the fourteenth century. Offset from the center and surrounded by 16 rays, the red sun disk also appears on the ensign flown by Japan's Ground Self-Defense Force. It was the military flag of Imperial Japan until defeat in World War II, when it was temporarily outlawed for nine years. The flags of Japan, then, tell a story of celestial endorsement of Japan's royal line, of Japan's military power, and Japanese nationalism.

Imperial regalia of Japan include symbols of office and investiture that are relics of a tradition of celestial sovereignty. The bronze mirror is the emblem of the sun. A curved jewel, reminiscent of a shining crescent, stands for the moon. Lightning's ability to strike a deadly blow assigns to it an iron sword. In keeping with the animist, nature-oriented character of Shinto, each of the three sacred objects is associated with one of the great celestial expressions of nature's power. Because Shinto is inextricably tied to sovereignty and political power, the religious responsibilities of the ruler are performed in a setting that emphasizes natural forces and facilitates communication with the *kami,* the myriad higher spirits that populate the cosmos.

Although the Sun Goddess is the most powerful of the kami, she is not omnipotent. Shinto recognizes no supreme god but acknowledges living spirits in nearly everything. All of the dimensions of the sacred life of Japan benefit from Amaterasu's patronage and protection. Japan's well-being and destiny are her fundamental priorities. She performs her duties in partnership with the emperor, whose religious responsibility makes him Shinto's highest priest.

Because the emperor's genealogy begins with the Sun Goddess, the greatest Shinto rites are family business and a privilege of bloodline. This celestial ancestry was first promulgated by the Yamato aristocracy in the middle of the fourth century A.D. It was they, as well, who elevated the status of the Sun Goddess above all other kami. Rival tribes begged to differ with them, but the Yamato succeeded in establishing the nuclear state that would in time become imperial Japan. They legitimized their sovereignty with the claim of direct solar descent. They demonstrated their sovereignty by monopolizing the performance of national rites to the Sun Goddess. As the sole proprietors of solar worship, they accepted the ritual burden for the entire population and enjoyed the preeminent status that accompanies high public responsibility. Their domination of Japan early in the seventh century A.D. coincides with the use of the term *tenno* for their high chieftain or emperor.

Yamato territory coincided with the Nara Basin, and what is now Nara Prefecture, on Japan's largest island, Honshu. This plain is located on the thumb of land that continues southwest when the rest of Honshu bends to the west and flanks the north side of the Inland Sea. The area has many ancient monuments, historic shrines, and monumental tombs, including the great "keyhole" mounded tombs to the west and northwest of Nara. There are others near Furuichi, in Osaka Prefecture, and elsewhere. At present, archaeologists still assign these tombs to the protohistoric period and the immediate Bronze Age predecessors of the Yamato chieftain-priests. Inklings of the nature spirit tradition that evolved into Shinto seem to be buried with the high-status dead of the tomb builders.

Still older whispers of Shinto are suggested by the incomplete and ambiguous tales of Yamatai, a monarchy in the legendary land of Wa. These stories were told by Chinese travelers and traders who visited Japan late in the third century A.D. According to them, Yamatai was governed by a queen. If her land was the same territory where a couple of centuries later the Yamato chieftains turned homage for the Sun Goddess into a national charter for kingship, Yamatai may belong to a period and culture the archaeologists call Late Yayoi. The Yayoi period begins about 300 B.C., and in this era the tradition of megalithic tombs, bronze and iron tools, and rice domestication were introduced in Japan. Yayoi burials often include bronze mirrors along with many other objects. The Yayoi are also known for their granaries. These substantial raised buildings apparently became the model, in later times, for elite housing and important shrines. In fact, the structures in the innermost sanctuary of the Sun Goddess at Ise, the greatest shrine of Shinto, resemble the traditional granary.

Chinese accounts tell us that Yamatai society was ranked, and its ruling queen was a priestess known as the "Sun Daughter." The supernatural powers attributed to her suggest shamanic behavior. Archaeology and legend, then, seem to converge and lodge the antecedent of Yamato Sun Goddess rites in the first few centuries A.D.

Like the Yamatai, the tomb builders of the Late Yayoi era had a rigid social hierarchy, probably a lot like the structure of Yamato society. Slaves comprised the lowest class. The middle class included peasant farmers, craftspersons, and soldiers. The elite ran the show from the top of social ladder. Powers of the leader of the highest elite, the *uji-no-kami,* or clan chief, included community management, judicial decisions, military command, and performance of religious ceremonies.

The Yamato dynasty provided its own written confirmation of the divinely ordained events that led to their supremacy. Early in the eighth century A.D., the Yamato rulers commissioned two histories, the *Kojiki* (Records of Ancient Matters) and the *Nihongi* (Chronicles of Japan). Both rationalize Yamato dominance in the development of the Japanese state by reaching back into myth to establish the history of the rise of the Sun Goddess. Her roles as the kami of the sun and kami of the imperial line validate the Yamato claims. Through her and the emperor, the vivifying power of heaven is bonded with governing power on earth. It is a reciprocal relationship that keeps the universe on track and the Japanese imperial line in the driver's seat.

The sky is the stage of Creation according to the *Kojiki.* Everything begins in heaven, and the *Kojiki*'s recitation begins by naming the first gods spontaneously born "in the Plain of High Heaven." The first to be mentioned is Master of the August Center of Heaven, and many more subsequently appear. Finally, the two deities critical to the formation of Japan—Izanagi and Izanami—materialize. Their names mean "Male Who Invites" and "Female Who Invites." Although Izanami is Izanagi's younger sister, they are destined to marry. But first they need solid ground for the ceremony, or at least the consummation. The soon-to-be-conjugal pair glides down the rainbow, the Floating Bridge of Heaven, to get closer to the earth-to-come. There is still no land, only a formless oceanic fluid. With a "heavenly jeweled spear" supplied by the other celestial gods, Izanagi stirred the waters of the world with the point of the spear. As he withdrew it from the sea, brine that curdled at the tip dripped back onto the water's surface. The deposit of floating glutinous ooze gradually solidified and became the world's first land and the first island of the Japanese archipelago. The sexual connotations of this story—the primordial chaotic waters, the penetrating spear and its dripping point, and the birth of the first land—are reinforced by the intimacy that follows. Izanagi and Izanami learn to mate by watching the birds, and the numerous children of their embraces include the rest of Japan and all of the essential components of the living earth. The world was invited to follow the same divine pattern of marriage and procreation to ensure the continuity of life, although mortality also eventually came to the world through Izanami's death after giving birth to the last kami, the fire god.

More erotic metaphor links the world's new land with the Plain of High Heaven. Before the starter's pistol is actually fired for the Divine Propagation Invitational, Izanagi and Izanami also erect a "heavenly august pillar." Recognizing that Izanami's body "is completely formed except one part which is incomplete" and that Izanagi's body is completely formed but has "one part that is superfluous," the two agree to "supplement that which is incomplete . . . with that which is superfluous . . . and thereby procreate lands." Then they circle around the heavenly august pillar in opposite directions, as if it were a maypole, and meet on the other side "to become united in wedlock." They build a nuptial house to consummate their marriage. This honeymoon apartment is centered on their heavenly august pillar, and the *Nihongi* calls it "the pillar of the center of the land." It certainly sounds like the world's polar axis. It apparently also has phallic meaning, for fertility fountains out of the center of the world. That is also where, in time, the emperor, the sire of the imperial line, presides.

John O'Neill, an English author interested in the cosmological connotations of mythology, examined the themes of the world axis and the Pole Star in 1893. In his dense, two-volume study, *The Night of the Gods,* he quoted a late eighteenth century scholarly Japanese commentary on the *Kojiki.* That source demonstrated that the Master of the August Center of Heaven, the first god to appear in heaven, resides in the Pole Star. Returning for a moment to the *Kojiki's* narration of the origin of the gods, we learn that the fifth kami of Creation was Standing Eternally in Heaven. He condensed out of the void "like a reed shoot," also a symbol of the vertical axis, and was transformed into a god who is, presumably, the north celestial pole itself. The "pillar of the center of the land," like a central cosmic mountain, very probably then linked the earth with the eternal, stable center of heaven. All of these axial images are completely consistent with the traditional shamanic layered cosmos. Shinto, too, organizes the universe into an upper world—the sky, a middle world—the earth, and an underworld—the subterranean land of the dead.

So far, Shinto tells a standard story of Creation. The fundamental structure of the sky, the center of the world, order out of the formless fluid of primeval potential, the first pairing of male and female, and divine procreation are all part of a picture that also is intended to explain and endorse the ruler's status and the continuity of the royal line. The Sun Goddess has not, however, put in an appearance yet in the creation myth. After Izanami withdrew to the underworld, Izanagi continued to create more kami, including the Sun Goddess, Amaterasu-omikami, the "Great and August Heaven Shining Spirit." Alternatively, her name is translated "Illustrious Great Divinity Radiant in the Sky." Either way, she is the brilliant light-shedding sun that dominates the day. She was born when Izanagi, in purification rituals, washed his left eye. The kami of the moon, Tsukiyomi-nokami, "August Moon Night Possessor," was rinsed from Izanagi's right eye. To Tsukiyomi, Izanagi gave authority to "rule the dominion of the night." Amaterasu was assigned to govern the Plain of High Heaven. Scrubbing his nose, Izanagi brought to life Susano'o, or Swift Impetuous Male, and his realm was the typhoon and the sea that spawns it.

Susano'o personified the violence of seasonal thunderstorms, and as the adversary of the Sun Goddess, he embarrassed her and intimidated her through sexual aggression and outrageous behavior that violated sacred precincts. She retreated to the Heavenly Rock Dwelling, a celestial cave. It might as well have been her tomb, for her seclusion from all of the living kami was like the exile of the dead within the earth.

Amaterasu's transcendental storm cellar has a terrestrial counterpart in Shinto belief, a cave near Mount Takachiho, in Kyushu. The cave's location in southwest Japan may reflect a connection with winter solstice sunset, which occurs on the southwest horizon. Both the season of winter and the sun's daily departure are times when the sun's back is against the ropes. Renewal, however, accompanies the New Year.

Although the New Year in Japan is now celebrated on January 1, its traditional date was associated with the full moon near the winter solstice. Shinto celebrates the end of winter's darkness, the return of the sun's vitality, and Amaterasu's emergence from the cave with a New Year festival in which a ceremonial straw rope, or *shimenawa*, is strung across portals to call to mind the straw rope that the eight million kami used to block Amaterasu's return to the Heavenly Rock Dwelling after they had lured her out.

It took divine laughter, the giggles of all of the kami of cosmos, gathered together upon the Milky Way, the "Tranquil River of Heaven," to persuade the solar coquette to bring light and life back to the world. All of those spirits split their sides with shrieks of celestial laughter as they watched a spirited ecdysiast dance out of her clothes for their amusement. Her name was August Heavenly Alarming Female—*Ameno-uzume-no-mikoto*—and she must have been a match for Madonna. She wore metal headgear, a bonnet of flowers of silver and gold from the celestial spindle tree. The sash for her dance of the celestial veils was heavenly moss from Mount Kagu. Bamboo grass from the same mountain was plaited into a posy she could wiggle and wave from the overturned tub that served as her stage. Her dance owed something to shamanic trance and to the inner focus of a stripper. The *Kojiki* says she performed "as if possessed by a spirit." Of course, she was a spirit. And "pulling out the nipples of her breasts, pushing down her skirt string" to reveal her most private crowd-pleasing wonders, she headed to the top of heaven's hit parade, a celestial centerfold and calendar girl for the beginning of time.

Amaterasu heard the uproar of the crowd, and wondering how the host of heaven could be happy in the darkness, she peeked out the door of her rocky retreat. Novelty had the edge over repose. Catching a glimpse of her own luminous face in a bronze mirror brought there for just this purpose, she was attracted enough by the appealing

When Japan's Sun Goddess asserted her divine right to privacy and retreated to a cave, the cosmos turned dark and cold. She was humored out of her hideout, however, by the laughter of 8,000,000 spirits unwilling to remain in the dark. The mirror that lured her to the door hangs on the sacred sakaki tree, on the right, along with knotted tassels. The *kami* with a rope in his hand will tie a *shimenawa,* or sacred knot, across the entrance to keep Amaterasu-omikami from changing her mind. The rooster that also called forth the sun is on the scene, and on the far left we see Ameno-uzume-no-mikoto—the headliner showgirl of celestial strip—standing by the footlights of her washtub stage. (from *Shinto* (*The Way of the Gods*) by John Aston, Longmans, Green and Co.)

image to step outside. The straw rope was slipped across the entrance to the cave, and the Sun Goddess once more took her place in heaven on behalf of the cultivated fields, the weaving halls, and every other place where the dignity of the sun brought order to the world.

In Shinto, there is no reigning queen of the kami, but the myth of the Sun Goddess emphasizes her special status among the divine spirits. She is the centerpiece of the story. The future of the universe rides on her behavior. All of the kami are mobilized to modify her decision. Troublemakers moved in to exploit the gloom of Amaterasu's seclusion, and unchallenged by the light of day, those evil spirits and spooks could have shut down the cosmos. When the Sun Goddess once more sheds her light into the universe, the natural order is restored. The demons are dispelled. Tempted again by the exuberant sensuality of life, the Sun Goddess returns to reawaken the world with spring.

Long ago, when the universe was young, violence against the kami of the sun almost foreclosed our mortgage of time. Each year, the changing seasons reecho that doom, but events on the Plain of High Heaven, on the bed of the Tranquil River of Heaven, in front of the Heavenly Rock Dwelling, installed the pattern of cyclical renewal that would ever after sustain the earth. The original power and purpose of the Sun Goddess—and of all of the kami who brought her back on-line—are restated and revitalized each year in rituals performed by the emperor, the first deputy and direct descendant of the sun.

Amaterasu's family tree first sank its roots into the earth when Ninigi-no-mikoto, the Sun Goddess's grandson, descended from the sky and touched down upon the summit of Takachiho, the highest mountain on the island of Kyushu and the same neighborhood said to host the earthbound incarnation of the cave in heaven. Ninigi's full name contains an ambitious combination of divine, celestial, and agrarian imagery: August Heaven Plenty Earth Plenty Heaven's Sun Height Prince Rice-ear Ruddy Plenty. Dispatched to earth by his grandmother to establish the ruling dynasty of the sun, he learned from an advance scout that the way was barred by a giant spirit at the eight crossroads of heaven, the celestial junction of world directions. Amenouzume, still famous for her showgirl talents, was enlisted to clear the route. Her divine allies figured her looks had the best chance of turning the goliath's head. Encountering him at the Great Interchange in the Sky, she lost no time in exposing her intentions. After alarming the Spirit of the Celestial Crossroads with a view of her bare breasts, August Heavenly Alarming Female dropped her skirt below her navel to reveal a little more of the heavenly landscape. Naturally, she got his attention. The giant spirit was converted by her commitment to the mission to bring the solar guidance of heaven down to earth. He offered to lead the retinue himself. So with the three insignia of sovereignty—the bronze mirror and curved jewel that both once helped cajole the Sun Goddess out of her cave and the iron sword of lightning—the grandson of the sun marched down to earth and inaugurated a line of kings.

Myth incompletely dissolves into legend in the story of Japan's first mortal emperor, Jimmu Tenno. His divine grandfather, Ninigi, had begun to move eastward from Kyushu in a campaign of territorial expansion. With each generation, the clan of the Sun Goddess grew more powerful. After gaining control of the western half of the island of Honshu, Jimmu Tenno conquered the tribes of the Nara Plain and built the Yamato palace at Kashiwara, according to tradition in 660 B.C. More likely any historical warrior king on whom the legend is based would have been a Yayoi chieftain of the first century A.D.

Pacifying Japan as the heir of the Sun Line wasn't always straightforward. When Jimmu Tenno was battled to a standstill near Osaka, he decided there was something metaphysically wrong with trying to advance eastward. It was the prime direction of Amaterasu-omikami herself. The *Nihongi* chronicles Jimmu Tenno's way of working around the problem. "I am the descendant of the Sun Goddess," he said, "and if I proceed against the sun to attack the enemy, I shall act contrary to the way of heaven. Better to retreat and make a show of weakness." He then pulled back, sacrificed to the gods of heaven and earth, slipped out of the area, and later came back to victory from another direction. Bringing, as he said, on their backs the might of the Sun Goddess, the army of Japan's first emperor followed her rays and trampled its enemies down.

In time, Japan's emperors built a shrine for the Sun Goddess at Ise. Amaterasu's bronze mirror and the other two Sacred Treasures of sovereignty were kept there on the Shima peninsula on the south side of Ise Bay. On one of the oldest shrine sites in Japan, the Neiko, or inner sanctuary at Ise, has been the traditional earthly home of the Sun Goddess since the fourth century A.D. Access to the Inner Shrine is restricted to the imperial family and the highest Shinto priests.

The Outer Shrine, or Geku, about three miles to the northwest, has been the center for rites to Toyouke-omikami, the goddess of grain, harvest, and silkworm cultivation for almost as long. Her first shrine was built here in 478 A.D. Although the shrines at Ise have been in business for centuries, the buildings themselves are relatively new. They are rebuilt every 20 years inside a walled enclosure right next to the precincts with the shrines constructed the last time around. The most recent set was completed in 1993. This process emphasizes the theme of renewal at the heart of Amaterasu's imperial cult. Her rites, modeled on the seasonal rebirth, are ceremonies of social replication and cultural continuity.

Another sacred Shinto celestial landmark lies just off the coast of Ise, near Futamiga-ura, where two great boulders known as the Wedded Rocks, or Meoto-iwa, emerge from the waves. They are said to be natural effigies of Izanagi and Izanami, the two *kami* who came down from the sky and created Japan. Yoked together by a *shimenawa* that looks like it could moor a ship, the pair annually reties the conjugal knot on the fifth day of the first month, right after the New Year, when the rope is ceremonially replaced. A small *torii,* or ritual gateway, in the simple Ise style stands on top of the bigger rock. The word *torii* means "bird perch." It also stands for the roost out-

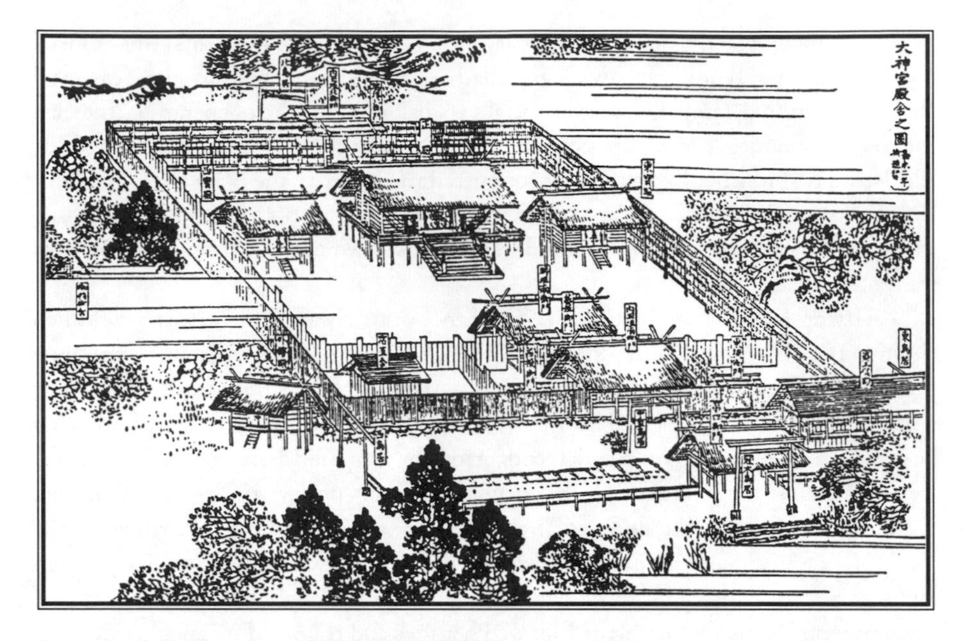

The structures that comprise the Inner Shrine of Shinto's Sun Goddess at Ise were patterned on traditional architecture that extracts religious meaning from rice cultivation, a foundation of Japanese life. (from *Weltgeschichte,* by Ullstein, reprinted in *The Stars Above Us,* by Ernst Zinner, Charles Scribner's Sons)

side Amaterasu's cave where the cock crowed to coax the Sun Goddess out of the darkness. As the formal entry to a shrine, the *torii* identified the sacred character of the zone ahead. Passing through the *torii* conducts you into an embrace with the divine. In this case, the zone ahead is the whole world. The sea stretches to the horizon where the earth and the sky are wedded, too. Pilgrims journey to this spot in reverence to the Sun Goddess and enter her presence as they watch the sun rise in the company of Mount Fuji, the great dormant volcano to the northeast and the highest mountain in all of Japan. This divine mountain also has meaning in the Imperial genealogy. It is the traditional home of Kono-hana-sakuyahime-no-mikoto, or the Princess Who Makes the Flowers of the Trees Blossom. She was an earth girl who became the wife of Ninigi, the grandson of the Sun Goddess.

The agricultural complexion of Shinto is evident in the dedication of the Outer Shrine to the spirit of the grain. The imprint of the Sun Goddess at the Inner Shrine injects more seasonal significance. The buildings themselves resemble archaic rice granaries. With timbers of cedar and cypress, their saddle roofs are thatched with reed. Steep, open gables, cylindrical ridge beams, elevated floors, squared planks, and cardinal orientation, confer upon the Ise shrines a plain elegance rooted in ancient peasant tradition. With a salute to traditional rustic architecture, Shinto erects its greatest monuments, in an evocative evergreen forest setting, to the fundamental principle of life—food that comes as the grace of heaven. Grain first arrived on earth as a gift from

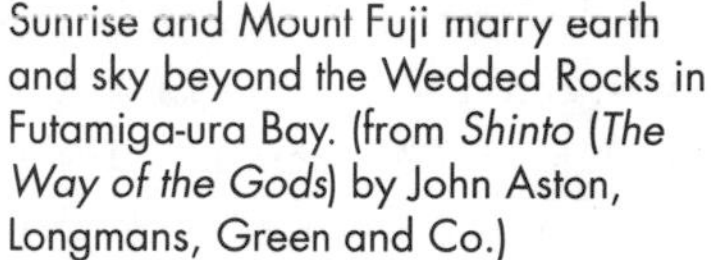

Sunrise and Mount Fuji marry earth and sky beyond the Wedded Rocks in Futamiga-ura Bay. (from *Shinto (The Way of the Gods)* by John Aston, Longmans, Green and Co.)

the Gods of Heaven carried by the emperor. With it, the world was nourished, and seasonal offerings of grain returned the favor. It takes land to cultivate the grain, and the land, too, was a divine gift bestowed in four quarters on the emperor by Amaterasu.

Japan's shrines at Ise are fused with the national identity. Despite the growth and complexity of Japanese religious tradition, largely a combination of Shinto and Buddhism, the Great Shrines of Ise transcend the personal priorities of belief and involve the destiny of the entire nation. During the last millennium and a half, the real authority of the emperor has risen and fallen with the fortunes of history, but in the Meiji era (1852–1912) of the nineteenth century, the Sun Goddess myth was repoliticized on behalf of revived imperial aspirations. Shinto evolved into a state cult and provided an ideology for extremist nationalism and militaristic expansion. The political impact of Shinto and its official status remained effective until Japan was defeated in World War II. The emperor renounced his divinity in 1946, and Japan was reordered as a constitutional monarchy. Nevertheless, the advice, assistance, and approval of Amaterasu were still sought. National affairs and imperial concerns were still brought to her attention at significant moments in the life of the country. The emperor's personal link with Amaterasu reaches from myth into this century and will apparently continue into the next. At Ise the emperor still fulfills the responsibilities of Amaterasu's highest priest. The New Year and new elections usually bring the emperor's secular counterpart, the prime minister, to Ise with his cabinet to uphold the honor of the nation through a demonstration of regard for the Goddess of the Sun.

On November 23, 1990, the *Los Angeles Times* carried a story that illustrated the continuity of the imperial bond with Sun Goddess. Japan's Emperor Akihito, the son of former Emperor Hirohito, wore the white silk robes of the Shinto high priest and entered an unvarnished shrine for Amaterasu located on the restricted grounds of the Imperial Palace on the night of November 22nd. A torchlight ritual there authorized him to ascend the Chrysanthemum Throne. His father had died two years earlier, and this ceremony was the final step in completing the royal succession. During the night, Akihito performed the *Daijo-sai,* an offering ceremony to the Sun Goddess on behalf of her people. With the help of court maidens and palace chamberlains, he thanked Amaterasu with freshly harvested millet and rice and many other comestibles. The ritual continued until dawn and joined the kami of the sun in mystical communion with the emperor in a relationship likened by some to the rapport of souls unified in sexual intercourse.

About three years later, in early June 1993, the same newspaper carried a sequential pair of reports on another aspect of the relationship between the Sun Goddess and the royal family of Japan. Crown Prince Naruhito married Masako Owada, a former diplomat, in a Shinto ceremony attended only by the Chief Ritualist. Like the investiture rites of Emperor Akihito, the wedding took place in the emperor's own sanctuary for Amaterasu on the grounds of the Imperial Palace. After the doors to the shrine were opened, silk and food were tendered to the kami. Then, attired in colorful, traditional robes of the ancient court, the couple knelt in homage to the Sun Goddess as Naruhito read the marriage pledge and petitioned Amaterasu for her divine protection while his consort remained silent. His flaming orange kimono seemed reminiscent of the sun in the formal wedding portrait and complemented the viridian that dominated her 12-layer, 30-pound, multicolor brocade. Her costume adhered to the style of the ninth century and the Heian period, when the capital was moved to Kyoto. The entire service lasted only 15 minutes. At the end they sipped sanctified sake, bowed, and returned to the secular world to be greeted outside by 800 intimate guests and to motorcade to their new residence with the help of more than 30,000 security officers and an audience of a half-million.

Celestial pipelines for imperial power in Japan are as plain as the national flag. This tradition begins with the ancient notion that Japan's very creation was the work of high gods from the sky. The theme is upheld by at least 1,500 years of formal ceremonies, explicit symbols, and official myth. From the earliest era, the emperor was affiliated with the sky. He was called "the Heavenly Grandchild." The crown prince was sometimes known as "August Child of the Sun." Other imperial titles sometimes included "Son of Heaven." Members of the royal court were styled "men above the clouds." Prayers for rain were dispatched to deceased rulers, ancestral souls who retained influence in heaven. By the end of the seventh century A.D., the cardinally gridded capital and cardinally oriented palace at Kashiwara amplified the power of the sky in the architecture of sovereignty. Territory far from the Imperial Palace was said to be "distant from Heaven."

All of this imagery of celestial power is rooted in the shamanic elements of Shinto. Himiko, a legendary woman shaman in Yamatai, illustrates the importance of shamanic practices in the earliest traditions. Shinto's acknowledgment of immanent divinity throughout nature and its recognition of the spirits that live within (and are the vital essence of) every feature of the landscape—mountains, the sea, waterfalls, trees, boulders, streams, celestial objects, and the growing rice stalk—are essentially an animist response. It may be fair to say Shinto is polytheistic as well, for mythological figures and even the emperor also may be kami. Anything, in fact, that induces a sense of the mystery, beauty, power, and life in the cosmos hosts a kami. Like the spirits of shamanic belief, the kami have power. They make the universe what it is. People can interact with them in pursuit of a result that requires supernatural assistance.

William George Aston, the scholar of Japanese language, literature, and tradition whose English translation of the *Nihongi* was published in 1896, analyzed the principles of Japan's oldest religious tradition in *Shinto (The Way of the Gods)*. He spotlighted the absence of a supreme god, disinterest in the details of a moral code, Shinto's few concrete representations of divine power, its unconscious neglect of the future destiny of the soul, and its minimal exertion of faith. He described Shinto as "rudimentary in its character" but insisted that does not mean it is primitive or no more than a cult. In fact, the depth of Shinto is invested in weaving human endeavor into the fabric of the world and harmonizing behavior with the great forces of nature.

Because the focus of Shinto was always congruence with the natural world, it integrated the spirit of a structured universe and the power of seasonal and cyclical renewal with the priorities of an agrarian state. Perhaps these shamanic features are echoes of an earlier time, before the cultivation of grain planted the foundation for imperial Japan. We find them in Shinto from the very beginning, however, and they are not out of place in an imperial setting.

Interacting with the great forces of nature is the emperor's primary task. Shinto, like shamanism, is result-oriented, but it is institutionalized with a priesthood and with highly developed ritual. It contains room for shamans, priests, emperors, and individuals in a social framework that permits all of the mediation with the spirits that is necessary.

It is, after all, the responsibility of an emperor to manage his world. His ability to do so is enhanced by a command of the power of the sky. Without a line of authority from heaven to use his resources, the ruler is at risk. Celestial ideology has, then, favored the survival of the community by stabilizing the structure of sovereignty. In Darwinian terms, success is measured in the persistence of the community's identity. It depends on reactions to internal and external stresses and on the solidity—the ongoing value—of the belief system. It all works as long as heaven remains in the king's corner.

CHAPTER EIGHT

Enlightened Self-Interest and Ulterior Motives

We don't always have enough power to make the world conform to our wishes, but we have found that knowledge is a useful alternative, especially predictive knowledge that informs us ahead of time what opportunities or catastrophes may be coming our way. Francis Bacon, the Elizabethan statesman and philosopher, had this figured out in the sixteenth century. "Knowledge," he wrote, "is power." Another philosopher, Friedrich Nietzsche, asked later, however, "Do you believe then that the sciences would ever have arisen and become great if there had not beforehand been magicians, alchemists, astrologers, and wizards, who thirsted and hungered after abscondite and forbidden powers?" Those two commentaries distill our ambivalent attitude toward specialist knowledge. We know it is valuable, but we don't trust it.

Neurobiologist William H. Calvin launched the first chapter of one of his books, *How the Shaman Stole the Moon,* with the remark from Nietzsche quoted above. Nietzsche tells us that our understanding of nature is really a by-product of the quest for power. Although Calvin's book is essentially an account of his own informal flirtation with practical techniques of prehistoric astronomy, it correctly emphasizes the importance of approximations and informed guesses in the development of systematic, scientific knowledge. Calvin, however, doesn't transport the implied theme of the title—the deceptive exploitation of knowledge—past the first chapter. Instead, he hands us a familiar cliché. With the total lunar eclipse scheduled to occur on February 29, 1504,

Christopher Columbus convinced the otherwise uncooperative Indians of Jamaica to continue delivering groceries while his crew was stranded and his ship was under repair. The Indians' reluctance to assist him ended when his prediction came true. He told them his god would pull the moon from the sky. They weren't expecting the eclipse, but he was. That knowledge put him one-up on the Indians. He was negotiating from strength.

Today, democratic instincts and multicultural bromides prompt sympathy for the Jamaicans. We say we prefer sincerity and good intentions to the bottom line and the final score. Columbus, we judge, used his knowledge cynically. In an era that celebrates the democratization of knowledge, his fraudulent manipulation offends our sensibilities. In adventure literature and popular culture, we see the past much as Calvin portrays it. It is overpopulated with shamans, magicians, seers, soothsayers, high priests, and grand viziers who collect and then guard esoteric knowledge. They release it to their neighbors or to their kings according to their own timetables and their own priorities. Their priorities always include the improvement of their own circumstances and the advancement of their own status.

Is this really an accurate portrait of the deportment of these ancient knowledge brokers? Everyday examples from our own time don't really convince us we need to revise our skepticism. Celebrity trials certainly raise warning flags about the credibility of experts. Of course, the justice system doesn't act alone to violate our virgin trust in the information elite. Entertainment and popular culture also conspire against our innocence. When a hero of knowledge like Charles Van Doren holds television audiences spellbound with answers to tough quiz show questions and then later confesses the show was rigged, our belief in learning, intelligence, and earned rewards is sabotaged. That is what happened in the 1950s when national sensations like *Twenty-one* and *The $64,000 Dollar Question* were exposed. We weren't just harmlessly bamboozled by a game-show superstar. The integrity of knowledge was compromised. Our belief system was breached.

It is hard to be generous toward the skywatching shamans and astronomer-priests of antiquity when it is so easy to run into guile today. The bald contempt for clients shared by two successful astrologers I encountered a few years ago in Hollywood probably would have surprised even their most forgiving customers. I arrived at a local studio for a television appearance as a critic of astrology's pseudoscientific claims, and while waiting to go on camera I was seated behind the pair of astrologers, who were also scheduled for the program. Most of their conversation concerned their respective rates and fees for their services, especially for high-profile entertainers. They discussed at length the personal—and presumably confidential—advice they gave to selected clients. The basis of their knowledge, the intricacies of their methods, and the reliability of their advice were not what they talked about. Rather, they tried to impress each other with the influence they exercised over the careers and decisions of celebrities.

The national profile of a tabloid prophet like Jeane Dixon doesn't do much for the reputations of the ancient oracles either. Her accuracy has been assessed with concise examples in *Visions and Prophecies,* one of the volumes in Time-Life's Mysteries of the Unknown series, and by geology professor Charles J. Cazeau in an article on prophecy in *The Skeptical Inquirer.* Despite enigmatic references to "Russian submarines near the Bolivian coastline" and explicit past predictions that the Soviets would be the first to land a man on the moon, that the Vietnam War would be over in 90 days, and that a comet would collide with earth in the 1980s, she continues to attract public attention. Despite four decades of far more misses than hits, her New Year prognostications are welcomed each year without any recall of claims for the previous 12 months. That, however, is not what makes us uneasy about her serious intent. It is the guide she wrote on dog horoscopes and her astrology cookbook that induce a pause.

Anyone who attempts to explain the lapses of the knowledge elite in our time finds little comfort in traditional cultures romantically believed to be more innocent than our own. Eighteenth-century travelers to Siberia were annoyed by the stupid shamans' tricks and amateurish deception they witnessed in the course of shamanic trances and cures. To European eyes, the wounds the shaman sustained in his wrestling match with the spirits had all of the high magic we would recognize in the arrow-through-the-head device popularized by comedian Steve Martin.

Stephan Krascheninnikow, a botanist who participated in the expeditions of Vitus Bering, described what he saw when Carimlacha, a renowned shaman in the Kamchatka Peninsula, put on a performance. Carimlacha was famous for knifing himself in the stomach. As blood copiously flowed from his gut, he collected it and drank it, to the astonishment of his audience. Krascheninnikow, however, was not entertained. He wrote the stunt was "performed in such an awkward manner, that any one, who was not blinded by superstition, might easily discover the trick." The skeptical botanist reported that he "could not help laughing at the simplicity . . ." and added ". . . the poorest player of legerdemain would have been ashamed." He continued,

> One might see him slip the knife down below his fur, and that he squeezed the blood out of a bladder which he had in his bosom. After all this conjuration he thought still to surprise us more by shewing us his belly all bloody, pretending to have cured the wound which he had not made.

Whether or not Carimlacha believed in the evil spirits that he said almost tormented him out of his senses, even a generous witness would agree that the shaman knew he was faking his injuries. Other accounts of shamanic theatrics—even sympathetic treatments—confirm the same fact. Shamans deliberately fool their clientele with parlor tricks. In *The Spirit of Shamanism,* Roger N. Walsh, a professor of psychiatry and philosophy, devotes a chapter to the hypocrisy of false shamanic magic and concludes that shamans, like the rest of us, can talk out of both sides of the mouth at the same time. This is not necessarily a talent cultivated by those seeking a profit at the expense of the naive and of the spiritual duffer. In fact, genuine shamans also buy into the belief system in which they are operating and see the tricks of their trade as legitimate resources. They are motivated by result, not self-consistency. If the trick enhances the client's belief in the shaman's power, the psychological or psychosomatic dimension of the client's problem is more likely to be conquered. This is Norman Vincent Peale's power of positive thinking in shamanic regalia.

In an essay in *Hallucinogens and Shamanism,* anthropologist Michael J. Harner clarified the ideological foundation of this approach in his discussion of shamanic curing among the Jivaro Indians of Ecuador's Amazon. The Jivaro shaman knows perfectly well that the foreign object he sucks from his patient's body and displays to the family is fraudulent evidence of the successful extraction of a disease. He doesn't see this as a deception, however, for he believes he has engaged the true cause of the malady, the invisible, supernatural power that has invaded the victim and infected him with illness. Because visible evidence of effective treatment promotes recovery, the shaman practically has a social obligation to deal in fraud. What is subjected to scrutiny is his results, not his methods.

A shaman's belief in his own power doesn't, however, keep him from criticizing his competitors. But it is usually competence, and not sincerity, that is called into question. Most, if not all, shamans resort to sleight-of-hand and deception, but they share the world view of their communities. They may be frauds, but they are "pious frauds," as Walsh puts it. They may exasperate the outsider, but within their own communities everyone agrees to be amazed.

CHEYENNE SHAMANS AND CELESTIAL WOLVES

The true renegades of esoteric knowledge are those who operate outside the belief system that the knowledge serves. Unpious frauds promoting merchandise of questionable value can appear, of course, at any time in any culture, but possession of esoteric knowledge doesn't always mean abuse of privilege and power, despite what we might conclude from watching all-night infomercials. When shamans mobilize their communities through ritual and myth, their special status legitimizes and reinforces the beliefs and behaviors that sustain the community's corporate and ethnic identity. Although shamans may benefit, through status and income, from unique personal powers and knowledge, they mostly work as shareholders, not as robber barons. Their familiarity with the cosmic powers of nature, including astronomical knowledge, benefits them through their service to their neighbors.

Shamanic knowledge of the sky and shamanic concepts of world order were put to work on behalf of tribal unity in the seasonal ceremonies of the historic Cheyenne of the Dakotas. Karl Schlesier, an anthropologist and an authority on the Cheyenne, examined the function of this Cheyenne ceremonialism in the context of a prehistoric arrival of the proto-Cheyenne in their northern plains homeland. His book, *The Wolves of Heaven—Cheyenne Shamanism, Ceremonies, and Prehistoric Origins,* traces Cheyenne traditions—especially an earth-giving ceremony in which the world's creation was symbolically restaged—to the emergence of the Cheyenne cultural identity in about 500 B.C.

Schlesier's reconstruction of the early development of the Cheyenne, or *Tsistsistas*—the name the Cheyenne call themselves—departs from the conventional story of the Great Plains told by archaeologists and historians. Usually said to be latecomers to the northern plains, the Cheyenne lived east of the Missouri River until the eighteenth century. They regard the Dakota prairies and Montana grasslands, however, as their original homeland. Their primordial ancestors, they say, inhabited lands far to the north of the territory where they became the Tsistsistas. Schlesier argues that these earlier peoples, who had not yet developed into the Tsistsistas, were various bands of Northern Algonquian taiga hunters in prehistoric Canada. He believes they preserved many of the religious traditions and cosmographical concepts their distant ancestors had brought from Siberia around 8000 B.C. When these Algonquian groups moved south from Saskatchewan into the Dakotas, they gradually coalesced into a distinct tribe with ceremonial traditions that owed a debt to northern Siberian shamanism and that set them apart, as Cheyenne, from others in the Great Plains of North America. At some unknown time between the middle of the first millennium B.C. and the 1700s, they migrated to west-central Minnesota, but Schlesier believes their earlier presence on the plains can be detected archaeologically.

The structure of the Cheyenne universe conforms to the cosmos we have already encountered in traditional cultures throughout the world. We, and the Cheyenne, reside in the middle world, the earth's surface and the junction between the sky and the underworld. The underworld begins below the grassy plains, in "deep earth," where tree roots end. Maheo, the Creator, occupies the highest realm of the sky, "blue sky lodge." The power that runs the world radiates from Maheo past the territory of the sun, moon, and stars in blue sky lodge, through "near sky space," where clouds, thunder, and rain preside and migratory birds fly, to the solid ground where plants, animals, and the rest of us all live. Mountains reach into the highest zone of the lower sky, the kingdom of wind, air, and the breath of life. Because they lift deep earth into contact with the lower sky, mountains are particularly sacred stations on the cosmic power line. There are special caves in deep earth that operate as animal-spirit hostels and other caves where shamans make contact with spirits that guide and protect the world's living creatures. The terminology for all this spiritual geography is part of a hidden language known only to, and used only by, Cheyenne shamans and ritual specialists. This vocabulary is part of the special shamanic knowledge that activated the ceremonial and magical dimension of bison hunting and of rituals of cosmic solidarity.

According to Schlesier, Cheyenne origins are most closely linked with *Massaum,* a shamanic tribal ritual that reenacted the Creation and the establishment of the Tsistsistas as a distinct and sovereign people. The ceremony was, in fact, the charter for Cheyenne institutions. As a symbolic recitation of tribal articles of incorporation, its performance ensured and celebrated Cheyenne survival.

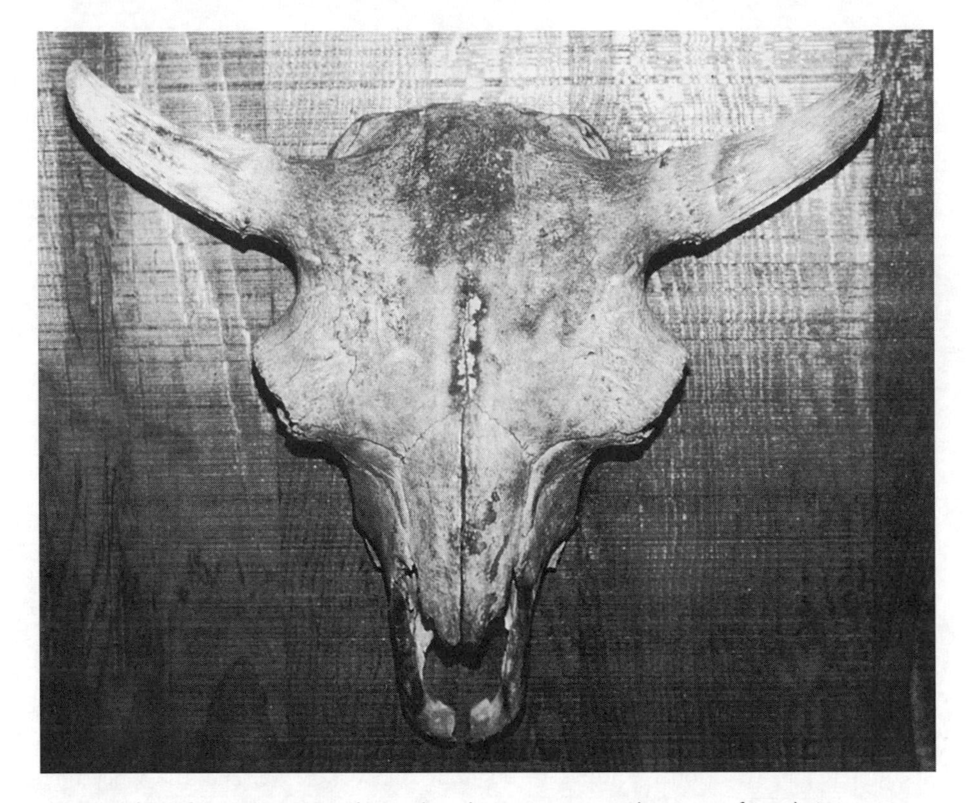

When Maheo, the Creator, gave the earth to the Tsistsistas, or Cheyenne, of North America's northern plains, his gift included the game animals, who agreed to allow themselves to be taken for food. The bison, one of the most important animals of the Cheyenne world, is a powerful spirit and a key symbol in Cheyenne shamanic ritual. A bison skull placed in the Massaum ceremony lodge represents deep earth, which is the great reservoir that makes life possible. This Northern Cheyenne painted bison skull, displayed in the Colter Bay Indian Arts Museum in Grand Teton National Park, was used in another ceremony, the Oxheheom, or New Life Lodge ceremony, also known as the Sun Dance. (photograph E. C. Krupp)

Details of the Massaum ceremony are reported and reconstructed in Schlesier's book, but here we confine our attention to its astronomical components and the responsibilities of the shamans. From encampments scattered through their hunting territory, the Cheyenne bands were called together for the five-day ritual. First, a cottonwood tree was cut and reerected at the center of the Massaum camp. Although it was stripped of its boughs, seven branches were left near the top. Acknowledged as the world axis, the Massaum pole stood at the symbolic center of the Cheyenne world and linked the seven levels of the universe—deep earth, four zones of the middle world, and the two layers of sky. Schlesier adds that the number seven also references the Seven Brothers of Cheyenne celestial myth, but he equates them with the Pleiades. This deserves another look. Other Algonquian peoples of the Great Plains tell stories of seven celestial brothers who are really the seven conspicuous stars of the Big Dipper. Because the Big Dipper is a constellation of the northern sky, it is linked with the world axis and the north celestial pole, around which it demonstrably turns.

After the central pole was raised, four posts were installed around it at the intercardinal directions and tipped toward the center like the frame of a tipi. Then 24 more lodge poles were leaned against the vertex, and the entire structure was covered. The central world tree-pole extended through the smoke hole at the summit. The sod inside was cut deep enough, in a ring around the center pole, to expose the interior to the upper surface of deep earth. Just east of the center pole, a small fire was lit.

Schlesier confirms that the Massaum tipi is the wolf lodge. The wolf is closely affiliated with the presiding spirit of deep earth and also with the spirit of near sky space. Like the Cheyenne hunters, it is a predator. The meaning of the wolf's role in the Massaum is confirmed by the myth of Cheyenne origins. In this tale, Wolf Man and Old Woman saved the lives of a shaman named Motseyoef (Sweet Medicine) and his traveling companion. Undertaking a sacred search on behalf of their people, they had journeyed far outside their home territory. Because the animals on this unfamiliar ground would not let themselves be taken in hunt, the two young men were starving. As they headed to a mountaintop to die, one of the pair was seized by a huge water snake with horns. With the help of Wolf Man and Old Woman, the shaman retrieved his companion from the serpent's waters and killed the monster. Together, they returned to Wolf Man's lodge and cured the victim of the great snake in a sweat lodge ritual. The two custodial spirits fed both of the men, let them see the animals in the four sacred directions, and sent them home with Wolf Man's daughter, an attractive young woman who became the wife of the man who was snatched by the snake. She is the yellow-haired human incarnation of the bison spirit and perhaps the prototype for White Buffalo Cow Woman, whom we met in Chapter One on our way to the Center of the World.

As the gift of the two spirit keepers of the game of the northern plains, Yellow-haired Woman brought the Tsistsistas/Cheyenne the right to live on the prairies and to hunt the bison. Through Wolf Man, who was also the spirit of thunder, and Old Woman, who was also the spirit of the earth, Yellow-haired Woman gave the Cheyenne their cultural identity. Once she joined them, they became relatives of the spirits. That kinship intimately bonded them with the animals and with the land. In the myth, however, Yellow-haired Woman failed to shield her heart from pity for any bison calf brought to the Cheyenne camp, and so after eight years she was obligated by the contract with her supernatural parents to return to Wolf Man and Old Woman. Her span of time with the Cheyenne is interesting, for with a simple moon-count adjustment, eight years comprise a cycle of reasonable congruence between the phase of the moon and the seasonal progress of the solar year. There is also an eight-year cycle of the movements of Venus. We have no evidence of Cheyenne interest in this kind of celestial compatibility, but the specification of eight years is suggestive.

Yellow-haired Woman's departure put the Cheyenne's compact with the spirits at risk. For that reason, the shaman who had helped to bring her to them decided to return to Bear Butte, at the northeast edge of South Dakota's Black Hills. In sacred Cheyenne geography, it is the place where Wolf Man and Old Woman have their lodge and where Yellow-haired Woman was first given to her human husband. Accompanied by some of his people, the shaman performed the first Massaum to commemorate the first time the world was given to them, in a ceremonial re-creation. Cultural unity and a distinct identity were required for partnership in any contract with the spirits. The ceremony was the signature that closed the deal. Every time they performed it, the Cheyenne reestablished their credentials and their claims. Because the performance was itself evidence of cultural identity, it fortified social bonds and contributed to cultural survival. This, in fact, is a primary function of sacred ritual. Reflecting a shared world view, it enhances social cohesion.

By the end of the first day of the Massaum ceremony, the wolf lodge embodies the structure of world order, but it is a cosmos-in-waiting. It is still empty, and it must be activated by Maheo the Creator. On the second day, the shaman enters the lodge alone, sits down to the west of the world-axis tree, and through symbolic activity impersonates Maheo. First, he makes a small imprint in deep earth with his right thumb, and by his own hand begins the universe with a mark that represents the center. Next, he makes four more marks—one at each intercardinal direction—roughly a hand's width from the center. Earth dug from the center point was then placed at each of the intercardinal depressions, and the shaman drew, in white gypsum, the arms of a cross, cardinally oriented and radiating from the center hole. The small mounds of earth at the southeast and southwest were sprinkled with red pigment. Those at the northeast and northwest were powdered with black, and together, all four of the little piles represented the sacred mountains of the four spirits that protect the four corners of the world. On the outside of the lodge cover, the shaman, still acting as the Creator, painted a red disk for the spirit of sun and the moon's spirit in a blue-black crescent. More paintings were executed. Bundles of white sage were attached in key places. Through these and other steps in the ritual, the shaman symbolically re-created the entire universe. A bison skull, eyes and nose cavity plugged with tied bundles of grass, was painted with lines and colors that stood for day, night, the earth, the sun, the moon, and the "blue star."

The Massaum shaman followed the lead of Motseyoef, the mythical first shaman who inaugurated the custom, and performed as Maheo the Creator, but the real organizational responsibility for the ceremony belonged to the Cheyenne woman who acted as Yellow-haired Woman. She helped select and train participants for the ritual. She consulted with ceremonial leaders and chose the location for the event. And she brought the painted bison skull into the Massaum tipi on the second day and seated it in the earth on the west side of the lodge.

Later in the ceremony, three spirit impersonators enter the cosmic lodge. One is costumed in the skin of a male, red wolf. The second puts on the hide of a female, white wolf, embellished with the tips of bison horns and a strip of bison hair along the back. The third wears the skin—partly painted yellow—of a kit fox. Each canine is a celestial gamekeeper spirit, for each one is a star. The red wolf is both Wolf Man and the red star Aldebaran. Old Woman, who appears as the white wolf, is also the white star Sirius. Yellow-haired Woman, returned for the ceremony as the kit fox, is also present in the sky as Rigel, which the Cheyenne know as a blue star. Their astronomically informed shamans schedule the Massaum ceremony with an eye on Rigel. They look for its reappearance in the predawn sky on the morning of the fifth and last day. The star-spirit impersonators then hunt the game, represented by more costumed actors, and guide the animals down a ritual drive lane and into a sacred enclosure, where they were symbolically killed.

The world and its animals, then, are given back to the Cheyenne in every performance of the Massaum. To be given at all, the world must be created, and that is accomplished with emblems of cosmic architecture—the world axis, the center of origin, and the intercardinal and cardinal directions. Then the universe comes alive and accommodates the presence of human beings with a seasonally sanctioned hunt, timed and guided by the stars.

Schlesier believes the ancient Tsistsistas/Cheyenne ceremonial season was governed by the same three stars that dominate the myth of cultural origin and the ceremony that recapitulates it. Beginning in June, close to summer solstice, with the first predawn reappearance of Aldebaran, it ended in mid-August, about 56 days later, when Sirius rose in the same way. Twenty-eight days after Aldebaran's return and twenty-eight days before the arrival of Sirius, Rigel performed its heliacal rising in the third week of July to close out the Massaum. Blue star symbolism actually pervades the ceremony in the painted bison skull, on the faces of nine of the participants, and in the blue lines drawn on the face of the kit-fox performer.

Now the significance of an astronomical analysis of boulder configurations—usually known as medicine wheels—undertaken by John A. Eddy in the 1970s, was not lost on Karl Schlesier. Eddy had first discovered a summer solstice sunrise alignment

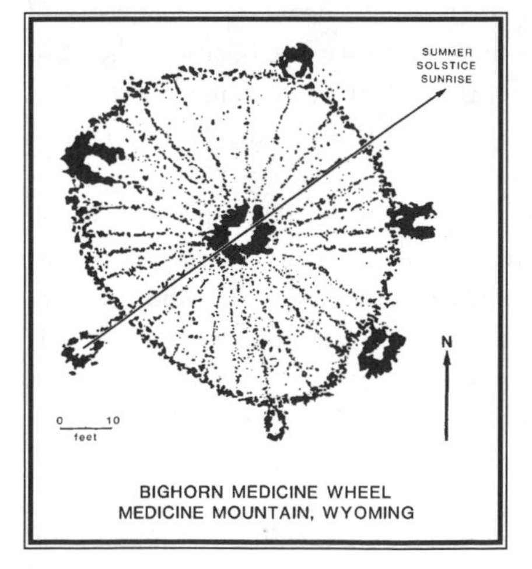

The axis of symmetry of the Bighorn Medicine Wheel, on Medicine Mountain, Wyoming, coincides with the direction of summer solstice sunrise and is emphasized by a spoke that extends southwest beyond the wheel's rim to an outlying cairn. Stellar alignments to Aldebaran, Rigel, and Sirius are formed by the cairn on the west side of the wheel and the cairns to the north, the east, and the southeast, respectively. (Griffith Observatory drawing, Joseph Bieniasz)

in the Bighorn Medicine Wheel, high above the tree line on Medicine Mountain, near Sheridan, Wyoming. Subsequently, he identified alignments with Aldebaran, Rigel, and Sirius between the cairn on the west rim of the wheel and cairns on the north, east, and southeast sides, respectively. A similar set of stellar alignments, along with a summer solstice sunrise line, was also found in southern Saskatchewan's Moose Mountain Medicine Wheel. The age of the Bighorn Medicine Wheel is unclear, and it is possible it is no older than a few centuries. The Moose Mountain Medicine Wheel, on the other hand, is dated to the fifth century B.C. The presence of these and more than 130 other medicine wheels, most on the Great Plains, led Eddy and others to examine the possibility of a connection with surviving Plains Indian ceremonial tradition. A variety of explanations have been offered, and frequently the Bighorn Medicine Wheel, with its summer solstice sunrise dimension and its 28 radial spokes (originally they may have numbered 27), has been interpreted as a symbolic representation of the lodge many of the Plains tribes construct for the Sun Dance, which is held in conjunction with full moon near summer solstice. The Cheyenne know this as the *Oxheheom,* or New Life ceremony, which is concerned with world renewal.

By 1990, Canadian astronomer David Vogt had completed a study of all 135 known medicine wheels and concluded they comprise a disparate set of monuments with no overall pattern that would permit us to reach a comprehensive conclusion. Individually persuasive arrangements such as the Bighorn and Moose Mountain wheels may reflect the astronomical intentions of the builders, but it is impossible for us to know. Nevertheless, Vogt judged that most wheels were intentionally oriented commemorative monuments directed to symbolic or ceremonial use. Calendric observation at functioning observatories just doesn't seem to be part of the scheme.

Without taking on the challenge of the entire medicine wheel catalog, Schlesier appropriates the Bighorn and Medicine Mountain wheels into Massaum ceremonialism. They are, he says, what the Tsistsistas/Cheyenne call *oxzemeo*—spirit wheels—made of stone and, as mountaintop monuments, aimed toward the sky. Shamans may have drawn protective celestial spirits to earth through them. As fingerprints of human presence, they may have marked rights to the territory as explicitly as the Massaum ceremony, and Schlesier associates them, through the stellar alignments, with the Massaum and the appearance of the Tsistsistas in these lands. Accordingly, he rejects the summer solstice sunrise line. It belongs instead, he argues, to the Oxheheom ceremony and has nothing to do with the formation of Tsistsistas culture.

Alignments in prehistoric monuments are inevitably circumstantial and usually problematic, and a thoughtful objection to solstitial alignment in the Moose Mountain Medicine Wheel has been raised. Even though the line between its southwest cairn (the backsite for the summer solstice line) and the central cairn (the foresite) is in the right direction, the elevation of the central cairn and the slope of the hilltop prevent a skywatcher from seeing the sun's first light. Instead it first shows up to the right of the central cairn and off the line. If all the celestial lines are symbolic, however, and not observational, this may not mean much. But whether the summer sol-

At an elevation of 9,642 feet, the Bighorn Medicine Wheel is free of snow for only about two months during the summer, the period for which the first predawn risings of Aldebaran, Rigel, and Sirius mark the beginning, the middle, and the end. The same three stars play significant symbolic roles in the Cheyenne Massaum ceremony, and some researchers think the Bighorn Medicine Wheel allowed shamans to call down celestial spirits. In this picture we are looking northeast along the spoke from the outlying southwest cairn. It continues through the central cairn and points to the place on the mountain ridge where the summer solstice sun comes up. (photograph E. C. Krupp)

stice sunrise put in an aligned performance on Moose Mountain or not, the Bighorn Medicine Wheel is another story.

Summer solstice sunrise is emphasized by the design of the Bighorn wheel. The backsite cairn is the only one on a spoke that extends beyond the wheel's rim, and the spoke line itself draws the eye to the northeast horizon. In addition, the shape of the Bighorn Medicine Wheel is actually more like a flattened circle, and the summer solstice sunrise line coincides with the ring's axis of symmetry.

Certainly we don't know that either medicine wheel was built by the ancestors of historic Cheyenne. If, however, we join Schlesier and recognize in them—perhaps more plausibly in the Moose Mountain Medicine Wheel, which is much closer to prehistoric Algonquian territory—evidence of early Tsistsistas occupation of the grasslands, we don't have to reject the solstitial alignments. More than one sign of celestial power might be incorporated into a monument meant to protect tribal interests.

The Tsistsistas/Cheyenne may have brought celestial power to earth in ancient medicine wheels, and they may regard Bear Butte as the spiritual center of the cosmos where the Tsistsistas originated as a people. They were, however, nomadic hunters who followed the bison and the rest of the game. They traveled with the center of the universe, and they unpacked it every time they performed the Massaum. It was unfolded for them by their shamans, who, from the very beginning, put their specialized training and their esoteric knowledge of the cosmos and the stars at the service of their own people.

EVERYBODY WANTS TO GET INTO THE ACT

The prehistoric Cheyenne bison hunters of the northern plains may have taken the ceremonial re-creation of the universe on the road, but fully committed cultivators like the ancient Egyptians weren't about to uproot the place of Creation from their own neighborhoods. The claims were only as good as the power of the priesthoods that promoted them, and so the specialized knowledge of Egyptian priests served the consolidation of power. While this may seem like little more than a story of local squabbles, obscure doctrine, and the elevation of the learned elite, Egypt's cultural and social evolution was a response to the challenge of natural selection in the arena of power politics. Political power flowed to the place of mythic Creation, for that is where the god at the top of the pecking order got his big break.

Of course, as more people are persuaded or coerced to accept the legitimacy of a particular site for the beginning of life, the universe, and everything, the more influential and effective become those who are in charge of that place. In a complex society, this has adaptive value, for when power is centralized, it is easier to apply. Centralization of power may also diminish local autonomy—and nobody local ever wants that—but it provides a more effective defense against predators. It encourages, as well, that celebrated best of all defenses—the good offense. First, internally, and later, internationally, there were plenty of predators eyeing the breadbasket of the Nile. The rivalries of the priesthoods that administered the country are reflected in their competing claims on Creation and in the esoteric ideology of the gods.

Prior to Egypt's first dynasty and roughly 2900 B.C., the Valley of the Nile was split into two kingdoms—Lower Egypt (the Nile delta) and Upper Egypt (the narrow river valley south of the Nile delta). They, in turn, had emerged from neolithic regional centers. Legend credits Menes, the ruler of Upper Egypt, with defeating the north and uniting the two lands into one. Historians identify Menes with one of the known kings of Egypt's first dynasty, usually Narmer or Aha, but the actual circumstances were probably more complex and probably involved more than one king. Prior to unification, competing regional states in Upper Egypt, centered on the one hand at Hierakonpolis (Kom el-Ahmar, between Esna and Edfu) and Nekheb (el-Kab, across the river from Hierakonpolis) and at the market center of Nubt (near present-day Tukh north of Luxor) on the other, were merged by Narmer's victory. Something similar must have occurred in the delta, where Buto (Tell el-Fara'in, in the northwest part of the delta) and Iwnw (Heliopolis) were among the most influential cities. Menes, whoever he was, gets the credit for bringing the whole thing together and establishing a capital at Memphis, near the apex of the Nile delta.

We know the priests of Memphis localized the Creation in their own hometown because we have a text describing how Ptah, the divine patron of Memphis, managed to make a world out of the primordial fluid darkness. The story is inscribed on a stone now displayed in the British Museum. The inscription was only committed to this mass of black basalt in the eighth century B.C. and the Twenty-fifth Dynasty. Originally thought to document an archaic narrative, recent scholarship now argues that it was deliberately written in artificial, antique style. Nevertheless, the tradition behind it could belong to a much earlier time.

Ptah was not only the first dweller in darkness, he was the primordial fluid darkness—the unformed, undifferentiated state that preceded Creation and that Egyptians knew as Nu. Through his heart (that is, mind) and his tongue (that is, commanding power), Ptah brought Naunet, a dark and fluid female counterpart to Nu, and Atum, who would become his executive officer in the creative enterprise, into being. Because this account of the creation of the gods is modeled on human reproduction, Naunet gave birth to Atum. Expectorating Shu, possibly the divine personification of air or part of the atmosphere, and vomiting Tefnut, or moisture, Atum brought the atmosphere on stage. While Ptah remained hidden in darkness, Atum became the Heart of Ptah, and Thoth, the ibis-headed god of wisdom, writing, and calendrics, appeared as

Only the lowest course of stone wall and the column bases of the Temple of Ptah survive at Memphis, Egypt, where Ptah was given credit for creating the world. (photograph E. C. Krupp)

the Tongue of Ptah. These gods were part of the Great Mouth of Ptah. When Atum thought and Thoth issued the order, Ptah's mouth spoke, and his words brought everything into being, including the king, or pharaoh, as the god Horus, who was another aspect of Ptah.

This creation myth not only proclaimed the preeminence of Ptah—and therefore the preeminence of his city, Memphis—it also made a god out of the king. We can see in these assertions a reaction to other Egyptian power centers whose claims on Creation were coopted by a myth in Memphis.

The Memphite story of the origin of the cosmos assigns titles to Ptah that figure in the tale they told at Iwnw, or Heliopolis, as the Greeks later called it. Heliopolis means "Sun City," and it was the primary home of Egypt's sun god, Re. The town's name in hieroglyphics includes a symbol that some have read as "pillar" and that looks like the support column for the pyramid-shaped cap of an obelisk. The obelisk was, in fact, the primary emblem of Egypt's sun cult. Through the minipyramid at the summit, the obelisk supported the creation myth of Heliopolis.

Pyramidal stones symbolize the *benben,* the first mound of solid ground to rise out of the dark and formless waters of Nu. In the beginning, Atum, the divine patron of Heliopolis, had nowhere to stand, and so he had to create the primordial hill. It

The Temple of Re, the sun god, has entirely disappeared from Heliopolis, but one solar emblem still stands—a pink granite obelisk, a little more than 65 feet high and erected by the pharaoh Senwosret I. Its inscription tells that Senwosret is the "son of the sun" and "beloved" by the spirits of Heliopolis. Two obelisks later erected by Thutmose III at Heliopolis were subsequently removed. One is now in London, the other in New York. (photograph E. C. Krupp)

emerged from the water like the inundated fields that returned when the Nile's annual flood began to recede, and in some versions of the story, Atum himself was the world's first earth.

Downriver from Memphis, Heliopolis was located in what is now Matariya, a Cairo suburb northeast of the city. The temple of Re was regarded as the original piece of good earth willed out of chaos by Atum. The world began in earnest there when Sirius, the stellar signal for the Nile flood, in its first return to the predawn sky, alighted as the *benu,* the bird of creation, upon the benben and then took wing as the sun followed it into heaven to bring light, life, and order to the cosmos.

Fields and apartments now occupy the spot where the benu bird first spread its wings. Although the temple at the point of Creation is long gone, along with most of the rest of old Heliopolis, an obelisk erected in the Twelfth Dynasty by Senwosret I (ruled 1971–1926 B.C.) still stands to mark the sacred precinct of the sun. Fragments of sculpture and inscribed panels of stone at least as old as the Third Dynasty (2649–2575 B.C.) are scattered around the area.

The sun temple of Heliopolis commemorated the world's first sunrise and housed Atum and four cosmic couples. Conflated with Re, Atum made a creator god out of the sun and then, according to some of the ritual texts, masturbated the other gods into existence. It may seem like a questionable way to install a pantheon, but in the beginning Atum-Re was alone in the universe. He had no female companion, but the Egyptians understood the procreative power of semen. With special handling, then, Shu and Tefnut were created. After them came Geb, the earth, and Nut, the sky. They were followed by Isis and Osiris, multidimensional deities whose stellar identities were Sirius and the stars of Orion. They were the parents of Horus, the divine personification of the king. Finally, Nephthys and Set completed the group, and Set, in time, became the adversary of Osiris and his heir, Horus. The plotline of the story of Osiris is really the structure of the cycle of cosmic order that renews time, the moon, the seasons, the land, all living things, and the institution of kingship.

Starting with Atum-Re, the Heliopolis myth of creation culminates in nine major gods. Most of what we know about this story and its genealogy of divinity comes from the *Pyramid Texts,* spells and incantations carved into the chamber walls of the pyramids of pharaohs from the Fifth Dynasty like Wenis (or Unas). These inscriptions are attributed to the priests of Heliopolis, and they indicate the supremacy of the sun god Re in Heliopolitan tradition.

Meanwhile, in Memphis, Ptah was creating Atum and all the other gods through his Heart and Tongue and taking credit for Atum's efforts on behalf of dry land. The Shabaka Stone, the inscribed slab in the British Museum that details the creative effort of Ptah, gives him names that equate him not only with the vast, still, and watery darkness before the beginning of the world but also with the world's first mound, the hill Heliopolis said belonged to Atum. And both Memphis and Heliopolis engineered their accounts of Creation to elevate the status of the pharaoh. In doing so, they announced their own allegiance to the divine ruler.

Related notions about Creation were voiced by the theologians of Hermopolis, but they cultivated them in their own backyard, a regional center the Egyptians knew as Khnum. The ruins of this site are in Middle Egypt, at the village of el-Ashmunein, five miles from the town of Mallawi. Mallawi itself is about 170 miles south (and upriver) of Memphis. Hermopolis is a Greek name that refers to Hermes Trismegistos, or "thrice-great Hermes," and by the Ptolemaic era, Hermes—the messenger, magician, patron of commerce, and psychopomp of Greek mythology—was equated with Thoth.

Although the same topography of Creation advocated at Heliopolis was embraced by the priests in Hermopolis, their primordial mound exploded out of the primeval watery darkness with a burst of light and the birth of the sun. The original chaos was a kind of cosmogonic bouillabaisse. The primeval fluid murk concealed eight supernatural creatures of darkness, dampness, and strange, hidden power. They were paired as four primordial couples of cosmic potential. Those symbolically linked with the male aspect of nature were metaphorically treated as frogs. Their "female" counterparts carried the associations of snakes. Amphibious and reptilian characteristics imply affinity with water and land. These creatures had what it takes to engineer the universe from one state into another. All of them represented abstract talents of a cosmos on the verge of prime-time broadcast. Somehow the equilibrium of unexpressed elemental power in the unformed inky universe was fractured, and the first land was whipped and blended out of the cosmic soup. The procreative energy of the eight divinities of Creation brought forth the mound the Hermopolitan priests called Isle of Flame.

Emphasis on the number eight is no accident. Its factors reflect the principle of male/female partnership and the four cardinal directions. The Egyptian name for the city means "eight town," and the elite thinkers of Khnum believed their eight creatures of Creation possessed a claim on the universe prior to the charter in the hands of the nine gods of Heliopolis myth. Accordingly, a ritual version of the ancient waters was constructed at Hermopolis in the form of a sacred lake to accompany the temple. An "island" in the middle of that pool symbolized the Isle of Flame, where the sun first ignited. Later, the priests of Hermopolis brought the god Thoth into their cosmological system. Thoth embodied perception, measurement, and organization of knowledge, skills all linked with the installation of cosmic order. Keeping track of the changing moon and counting out the days of the year, he transformed celestial cycles into the regulators of time. At the heart of all Egyptian accounts of Creation we can discern the magical power of word, and Thoth, as the inventor of hieroglyphics and divine scribe, put words into print. We can guess that his mastery of the magic of language persuaded the priests of Hermopolis to make him the true agent of Creation. Thoth was given credit for bringing the first great eight beings of Creation into existence by invoking their names.

In the campaign to add luster to Thoth's reputation, a new metaphor was added to the story of the first sunrise. According to those in the know at Hermopolis, the sun hatched from an egg, an apt symbol of creation. In one version, this egg was laid by

Hermopolis also claimed to be the place of Creation, and its divine patron, ibis-headed Thoth, was believed to be the Creator. As a god of the lunar calendar, Thoth was associated with the waning crescent moon. It was seen in the dawn, which was greeted by the raucous howls of baboons, and so baboons also came to symbolize Thoth. This huge statue of a baboon is one of a pair at Hermopolis. (photograph E. C. Krupp)

an ibis, the talismanic bird of Thoth. Thoth's emblems also included the baboon that greets the rising sun. Both it and the ibis are connected with waning crescent moon, which rises before dawn and is seen in the east as a herald of the sun. Thousands of mummified baboons and ibises were interred at Tuna el-Gebel, the necropolis of Hermopolis, in apparent homage to Thoth.

During the New Kingdom of Egypt, 1550–1070 B.C., the seat of central power shifted to Waset, or Thebes, as it was known to the Greeks. Its greatest temples still stand in the town of Luxor, about 459 miles south of Cairo, and at the nearby village of Karnak. Across the river, on the Nile's west bank, New Kingdom pharaohs buried their dead in the Theban necropolis, which includes the famous Valley of the Kings and monumental royal mortuary temples like Queen Hatshepsut's stately shrine at Deir el-Bahri.

At Thebes, another god was the captain of Creation—Amun, the invisible animating power of life. In other theological venues, he was affiliated with the air. Although air can't be seen, as wind it makes its power known, and as breath it is an essential feature of human life. By the time Amun was combined with the sun as Amun-Re, he was the supreme god of Egypt. The Great Temple of Karnak, enlarged by one pharaoh after another, was his home and, naturally, also was the place where the first land rose out of the uniform, motionless waters of Nu.

Amun's status and the story of his creative instincts are documented in a variety of ancient texts, including *Papyrus Leiden I 350,* a long hymn composed near the start of the Nineteenth Dynasty (1307–1196 B.C.) and the reign of Ramesses II (1290–1224 B.C.), probably in Thebes. The prayer tells Amun his name is "high and mighty and powerful" and confirms that he fashioned himself when nothing else existed. Like the sun in Hermopolis, he had his own egg. He made it himself and then incubated himself out of it. He was the first of the eight creator gods of Hermopolis. He took shape from the lightless, watery Nu as the hill of Creation and drew the other seven out of himself. The sun was "united with his body," and that made Amun "the great one who is in Heliopolis." "All gods came into being, after he came to be," including the nine gods of Creation in Heliopolis. His noise and his light inspirit everything. The trees "stir at his presence." The fish "leap in the water." The birds "dance with their wings," and the cattle all "frolic before his countenance." Everything lives because everything sees him every day. And yet, although he is formed, his form is "unknown." The papyrus also informs us that Amun remains hidden because he withdrew to heaven when he concealed himself. He is too mysterious and too powerful to know in detail. His hymns understandably link him with Thebes, which is "the Mistress of Cities" and "stronger than any city." What's more, as the center of Creation, Thebes is the navel of cosmic order for the rest of Egypt's cities. All cities are "under the supervision of Thebes." "The water and the land were in her in the beginning of time," and it was on her ground that the "land came into being."

Early references to Amun appear in the Old Kingdom *Pyramid Texts,* which already place him in the universe before it takes form out of Nu. Later, his connection with kingship is highlighted by titles of rank: "chief of the Two Lands," and "chief of the gods." He is first called "king of the gods" in an inscription carved on his earliest monument at Karnak, the elegant little kiosk of Senwosret I. There he is also shown with an erection as the "bull of his mother." The name refers to his fertile energy, and the "mother" he impregnates is the sky, sometimes represented as a cosmic cow. Her four legs are the world's cardinal directions. The stars travel along her belly, along with the sun and the moon in their celestial boats. Shu, the air, supports the cow's underside with upraised arms, and the eight primordial gods of Creation buttress her legs. Through sexual union with this bovine heaven, Amun, the transcendental bull, refertilizes the sky. In perpetual renewal, the lights of heaven, time, and the world are ever reborn.

In *Genesis in Egypt,* a carefully reasoned analysis of cosmology and Creation in ancient Egyptian thought, Egyptologist James P. Allen cautions against reducing the independent Creation myths to a competitive exercise in smoke and mirrors that hides the quest for power behind the apron strings of theology. The Egyptians were serious about their philosophy, and not particularly inconsistent. Throughout these traditions runs the same theme: a cosmos renewed through celestial cycles that return to and restate the circumstances that activated a habitable world out of the liquid, unlit, and inert chaos. Somehow the Many must come out of the One. A concrete world with distinct properties and filled with diversity emerges from the quiet, homogeneous chaos. In one way or another, something stirs on the face of the deep, and by the time it is out in the open, it is land—the first hill and the place of Creation. Simply claiming to be that place doesn't turn an enterprising village into a capital. Power

Amun, the Creator god of ancient Thebes, can be recognized by the two high plumes of his crown, which, as feathers, symbolize the sky and his celestial dimension. He is portrayed on the walls of the inner sanctuary at the Temple of Amun-Re at Karnak. This chamber was built in 320 B.C., on the location of an older sanctuary, by Philip Arrhidaeus, the half-brother of Alexander the Great and the king of Egypt in the era of Alexander. In the image of Amun on the left, an erection worthy of the gods announces his fertilizing power. Enthroned in the portrait to the right, he holds the *was* scepter, a symbol of power and dominion. (photograph E. C. Krupp)

gravitates to a city in response to the political and economic influence it has already achieved through a variety of forces. Any city that has the potential to centralize the power of the state, however, has to insist that it is the spot where everything began. The city that actually becomes the center of state power must be granted recognition as the place of Creation to make it clear why the gods have smiled on it and given it their blessing to act on their behalf.

ADVISING THE KINGS

Ashurbanipal, the king of Assyria (669–627 B.C.), must have been convinced knowledge is power, for he assembled a massive library of tablets inscribed with cuneiform, the writing developed in Mesopotamia by 2400 B.C. It was a Sumerian invention that evolved from the pictographic script in use as early as 3100 B.C. and was adopted by the speakers of Semitic languages when they superseded the Sumerians in the Tigris-Euphrates river valley. Ashurbanipal inherited some of his "books" from his father Esarhaddon (ruled 680–669 B.C.) but supplemented his library at Nineveh with more literature, history, mythological texts, legal codes, medical handbooks, magical incantations, mathematical treatises, astronomical records, and astrological omens relating to the welfare of the state. Of the 2,000 or so tablets comprising the library, a respectable fraction was devoted to reading the future.

These rock-hard documents confirm that the kings of Mesopotamia subsidized specialists in esoteric knowledge of the sky not only to establish and maintain the calendar but to read the will of the gods in "the writing of heaven." The behavior of celestial objects was pertinent in this respect because they were recognized as gods or as the emblems of gods.

Astronomy was important because what happened overhead meant something down below. Arrivals and departures of familiar planets from familiar stations among the stars offered insight into divine intentions. R. Campbell Thompson, an expert on Egyptian and Assyrian antiquities in the British Museum, collected and published many of these astronomical omens in 1900 in *The Reports of the Magicians and Astrologers of Nineveh and Babylon.* One of these seventh century records illustrates how the Assyrians read the sky.

Mesopotamian kings were cognizant of the power of the sky and subsidized astronomy to read the will of celestial gods. Allying themselves with the gods in heaven, they portrayed themselves in the company of their emblems. This portrait stela of Assurnasirpal II, the king of Assyria from 883 to 850 B.C., includes five symbols of divinity. The helmet-shaped object nearest the king's head is a horned cap. Although assigned to different gods at different times, it was probably at first the insignia of An, the high celestial god of the sky and Creation. To the left is the elaborate feathery rayed disk of the sun, and it is followed by a crescent for the moon. The "tuning fork" next to it is stylized lightning, the sign of Adad, the storm and weather god of heaven. Finally, at the far left, the eight-pointed star of Ishtar symbolizes the planet Venus. This monument, now on display in the British Museum, comes from the Temple of Ninurta (an Assyrian war god) at Kalhu, the ancient city of Nimrud, which is in northern Iraq, on the Tigris River south of Nineveh. (photograph E. C. Krupp)

> When Jupiter stands in front of Mars, there will be corn and men will be slain, or a great army will be slain. . . . When Mars approaches Jupiter, there will be a great devastation upon the land. When Jupiter and a planet, their stars face, evil will befall the land. When Mars and Jupiter approach, there will be a slaughter of cattle. Mars has approached Jupiter. When Mars approaches Jupiter, in that year the king of Akkad will die and the crops of that land will be prosperous.

What may be good for the fields is bad news in the palace, and almost any conjunction of Jupiter and Mars means some kind of trouble is on the way. Their meetings in heaven were inevitable, however, and you could see them coming. Eclipses, on the other hand, appeared suddenly and intensified anxiety with dramatic departures from the customary order.

> Of the eclipse, its evil up to the very month, day, watch, point of light where it began and where the moon pulled and drew off its eclipse—these concern its evil . . . it is evil for the fourteenth day [full moon]. . . . It is evil for Elam and Aharrû [two cities]. . . . From the east and north, when bright, it is lucky for Subarti and Akkad [two more cities]; it says they will have favor [uncertain text]. The omen of all lands:—the right of the moon is Akkad, the left Elam, the top Aharrû, the bottom Subartu. . . .
>
> The omen is favorable and the king, my lord, may be content. Jupiter stood in the eclipse; it portends peace for the king, his name will be honourable, unique. . . .

The Mesopotamians, however, had learned to predict eclipses—or were at least able to anticipate when they might be seen, and the meaning of this lunar eclipse varied from one city to another. The king's fortunes also were tied to eclipses:

> On the fourteenth an eclipse will take place; it is evil for Elam and Aharrû, lucky for the king, my lord; let the king, my lord, rest happy. It will be seen without Venus; to the king, my lord, I say "there will be an eclipse."

More recently, in the 1970s and 1980s, Assyrian records have been studied by Professor Simo Parpola of the Academy of Finland, and some of his work has been summarized by Michael Baigent in *From the Omens of Babylon.* Of 370 divination texts, almost 25 percent are astrological. About half of those were written during a short, four-year period of Esarhaddon's reign. One astrologer, whose name is recorded, sent 11 oracular messages to the king in a two-month span, and Parpola judged the surviving tablets represented the typical level of astrological and astronomical communication. We also know these star diviners were part of the priesthood. They held the official title "court astrologer." There is also evidence that they interpreted astrolog-

According to Diodorus Siculus, Chaldaean astronomers observed the stars from the summits of the high temples that towered from their cities. Diodorus traveled widely during the first century B.C. and synthesized his own knowledge and what he read in other authorities in the volumes of history he wrote. He may have correctly identified the old Mesopotamian observatories, but we have no independent evidence to confirm this. From this romantic nineteenth-century illustration from *Astronomical Myths* by John F. Blake, we would also conclude that the Babylonian skywatchers also anticipated the present fashion of large research teams.

ical dreams for the Chaldaean king in the sixth century B.C. The chief astronomer was called "chief of the month," a reflection of his calendrical responsibility. Other court credentials documented in tablets carry the official character of Mesopotamian astronomy to at least as early as the Akkadian period late in the second millennium. It probably relies on even earlier Sumerian precedent.

Remarks in Classical sources tell us the Greeks believed the Egyptians and Babylonians invented astrology. In the *Epinomis,* the Greek philosopher Plato specifically credits Syria, Assyria, and Egypt and confirms it belongs to an old tradition. Manilius, the Roman author of *Astronomica,* an astrological poem written in the age of Augustus, reported that astronomy's origins could be found in the lands of the Euphrates and the Nile, where the study of the stars was sponsored by kings. Macrobius, another Latin writer, also roots astronomy in Egypt in the *Commentary* he composed in the fifth century A.D. on *Scipio's Dream* by Cicero, the famous orator, lawyer, and politician. According to Diodorus Siculus, the Roman historian of the first century B.C., the observational foundation of Mesopotamian astrology was acquired at the top of ziggurats, high, pyramidal temples of brick that rose out of the heart of the city. In Chaldaean Babylon, the observatory was the temple of Marduk, the divine patron of Babylon and the king of the gods. The great tower was 300 feet high, and to the Babylonians, it was known as the "foundation of heaven and earth." Its seven stories have been attributed planetary symbolism, and its height put whoever climbed to the summit—astronomers, priests, or kings—in touch with celestial gods.

Real records and the opinions of the ancients produce a picture of ancient astronomy as the prerogative of trained experts, specialists who acquired knowledge of the sky to help the king organize the affairs of the realm and to understand the will of the gods.

The same situation prevailed in ancient China where astronomy was officially chartered by the emperor. Its bureaucratic character was clear by the Han dynasty (202 B.C.–220 A.D.), when the Five Classics were written and compiled. These texts form the foundation of Confucian thought about the universe, government, and behavior, and in one of them, the *Shu ching*, or "Book of Documents," the commission granted by Emperor Yao is detailed. Although Yao's reign belongs to the mythical era of Chinese history and is assigned, without confirming evidence, to the second half of the third millennium B.C., Yao's astronomical goals certainly reflect the priorities of the Han era. The account of them provided by the *Shu ching* is probably, in fact, as old as the seventh or eighth century B.C.

Emperor Yao appointed two brothers, Hsi and Ho, to observe the "august heavens" and, in particular, to "compute and delineate the sun, moon, and stars, and the celestial markers" in order to "deliver respectfully the seasons to be observed by the people." Their first responsibility, then, was calendric, an assignment completely consistent with imperial prerogative to establish and regulate the calendar. The calendar, of course, governed the agricultural cycle and helped manage the distribution of goods and services. Calendric control demonstrated imperial authority. It was a planning tool, and Chinese emperors used it in long-range thinking. Believing that shortages and surpluses traveled through time in congruence with longer celestial cycles, the royal calendar makers kept track of the sky in support of the emperor's effort to maintain political, social, and economic equilibrium. We get a glimpse of how this worked, for example, in the *Chi Ni Tzu*, which documents a belief current in the Zhou dynasty and the fourth century B.C. Agricultural success and reversal were linked in this ancient reference with the 12-year cycle of Jupiter through the stars of the zodiac. After describing how abundance and catastrophe alternate every three years, with a bumper harvest every six years and a famine every 12, the decision-maker's self-help book advises, "Thus, the sage, predicting the recurrences of Nature, prepares for future adversity."

A claim to the throne by a new ruler was naturally reinforced by the issuance of a new calendar, and it took qualified astronomers to work out the details. The organization administered by the royal astronomer was known by various designations, but all of the titles conveyed the nature of the office. Names given to it during the Tang dynasty (618–907 A.D.), for example, include the "Platform for Scrutinizing the Sky," the "Inspectorate of the Spherical Sky," "Inspectorate of Spherical Gauge [armillary sphere]," "Inspectorate of the Grand Notary," "Board of the Gallery of Secret Writing [the sky]," and more. The staff comprised five senior "star officers"—one for each

The *Shu ching,* the official Classic of history, tells that state-sponsored astronomy in China began officially when the legendary astronomers Hsi and Ho received their commission from Emperor Yao, who was said to have brought the arts of civilization to China. This illustration of the astronomical appointment was published during the Qing dynasty (1644–1912 A.D.) and was reprinted in Joseph Needham's *Science & Civilisation in China, Volume III.*

planet, observers, astrologers, record keepers, clock monitors, ritualists, omen compilers, and others needed to collect information from the sky and integrate it with terrestrial affairs.

A variety of celestial phenomena conveyed messages of heaven's intent. Comets, conjunctions, occultations, "guest stars," odd appearances of the moon, solar and lunar halos, and almost any deviation from celestial business-as-usual were greeted with interest and concern. In general, old Chinese reports of "guest stars" document the temporary visibility of a stellar explosion, a nova or supernova, which ignites a faint and distant star, usually far below the threshold of the unaided eye, into the known and mapped company of stars. Probably the most famous of these reports are the Chinese records of the Crab supernova in 1054 A.D., and one of the accounts also tackled the meaning of the event. Although twice as bright as Venus, the third brightest object in the sky, the Crab explosion guest star "did not infringe upon Aldebaran," which is the brightest star in Taurus the Bull. The Song dynasty astronomer then added, "this shows that Plentiful One is Lord, and that the country has a Great Worthy." His remark was a compliment for the emperor, who, he reassured his imperial audience, faced no challenge in the appearance of the celestial interloper.

The night sky didn't always tell a story of sweetness and starlight, however. Sometimes the news for the emperor was gloomy. A Tang dynasty omen offers a warning: "When a Visitor, or Sweeper, or Exploder enters in the Dipper, the barons will contend for authority and oppress the Son of Heaven." A comet, resembling a broom with its head and long tail, is the "Sweeper" in this caution, and the "Son of Heaven," of course, is the emperor.

Official Chinese astronomy was observational, mathematical, and systematic. Its complexities and high responsibilities required an organized bureaucracy. Its core, however, was simple. It delivered a calendar to govern the flow of ordinary affairs, and it monitored changes in the sky—and interpreted what they meant—to enable the ruler to make good decisions that conformed to divine intent. Although the astronomical and astrological aspects of Chinese skywatching were fused, distinctions were made according to function. Dynastic histories compiled and digested the era's celestial lore, but different kinds of information were incorporated into different chapters. Omens and prodigies were listed in a section about "Patterns of the Sky." On the other hand, unencumbered observations and predicted movements of celestial objects based on past performance were consigned to the discussion of the nuts and bolts of imperial calendars.

The calendar, however, also defined the ritual cycle, and through imperial ceremony, the ruler interacted magically with the supernatural powers of Nature on behalf of his people and his land. This cosmo-magical dimension of Chinese astronomy was also applied to the layout of cities—especially the capital—and to the design and placement of altars, halls, and shrines to model the universe in monuments and architecture. By establishing a sacred topography, based on the concepts of the center, the world axis, and astronomical directionality, the emperor tried to harmonize the will of heaven with human enterprise.

Heaven's judgment made a difference because the Chinese, like everyone else, invested the sky and the phenomena that occur there with supernatural power. The king was part of an analogy of power, for he ruled with heaven's blessing. Divine power was possessed by the sky, and the emperor held power on earth. He was the intermediary between gods and mortals, and although as the Son of Heaven, he was regarded as divine, heaven could withdraw its blessing and confer it on another. Successors are the ones who get to write the history, and the replacement of one dynasty with another was often explained by a subsequent regime as heaven's dissatisfaction with the previous administration. Grounds for dismissal usually had something to do with corrupting the natural order and acting unjustly against divine purpose.

Recognition of the sky's power over the state is illustrated in the story of Geng xun, an official of the Sui dynasty (581–618 A.D.) in the twilight of its imperial tenure. Geng xun had advised his emperor, Yang di, that a military campaign would fail, and dissatisfied with the assessment, the ruler ordered Gen xun's execution. Before he was beheaded, however, news of the failed campaign reached the court, and Yang di instead appointed Geng xun as Assistant in the Bureau of Astronomy. In 618 A.D., Yang di was assassinated, and Geng xun, now an astronomer, could see the "secret writing" of

The first "ancient" Beijing Observatory was completed during the Yuan dynasty, in the thirteenth century, by Guo Shoujing, the chief astronomer of Emperor Khubilai Khan, but the foundation for the walls of the present "Terrace for Observing the Stars" dates to the fifteenth century and the Ming dynasty. In this view, we are looking south from Changan jie Avenue, and some of the instruments—including the quadrant (on the far right), the new armilla, and the top of the azimuth theodolite—are visible on the roof. All of these instruments are Jesuit replacements of earlier apparatus. (photograph Robin Rector Krupp)

heaven on the wall. He told his wife, "I have studied the affairs of men below and investigated the patterns of heaven above." He concluded that he was affiliated with the wrong side in the ensuing struggle for power and tried to escape to the contender the stars had told him would win. He was right about the founder of the next dynasty, but he didn't get out of the old court alive.

Rulers in both ancient Mesopotamia and China, then, consolidated the power of the state with the help of an astronomical bureaucracy committed to understanding the will of the gods through careful observation of the sky. Notices of divine displeasure appeared upon the celestial bulletin board as bad omens, but it took experts to recognize and read them. The Mesopotamian king was "elected" to rule by the gods as the gods elected their own chief, and the king served at the pleasure of the gods. The Chinese emperor was the divine heir of the gods and ruled with a divine mandate. Both trafficked in heavenly power and governed with transcendental authority. Neither, however, could engage the gods without interpreters of their celestial communiqués. Although both exercised extraordinary power, neither was strictly despotic, for the signs in heaven offered opportunities for criticism and correction. These "institutional checks," as Wolfram Eberhard characterizes them in a paper on "The Political Func-

We may not be certain that the ancient skywatchers of Babylon climbed to the heights of the ziggurat to get a better look at the stars, but the astronomers of old China ascended this stairway on the west side of the Beijing Observatory for access to their bronze instruments and an unobstructed view of the heavenly vault. (photograph Robin Rector Krupp)

tion of Astronomy and Astronomers in Han China," help avoid excesses and abuses by limiting royal power with the belief system. Control over important elements of the belief system is relegated to advisory specialists—politician-astronomer-priests—who are not part of the chain of command. Ostensibly objective and independent witnesses of nature, the astronomers might not be able to manipulate economic and political resources directly, but their information influenced the stability and legitimacy of the sovereign.

As part of the educated elite, the royal skywatchers provided testimony in the court of public opinion. This "public" was not the democratized electorate we embrace today but the uppermost levels of the social pyramid. It included elevated officials, regional rulers, and high-status families. Their favors, fortunes, and failures all depended on the effectiveness and the fortitude of the king. As long as the ruler fulfilled his role in accordance with rhythms of nature and the order of the cosmos, he was "virtuous" and "worthy" for office. When he departed from good judgment and indulged personal whim, celestial gods could put a ruler on probation. If he tolerated improprieties in his subordinates, he could be called on the carpet of heaven. A Chinese imperial proclamation in 178 B.C., quoted by Eberhard, makes the process clear: "If the ruler of men has no spiritual power, if the exercise of government is not balanced, Heaven shows portents in order to warn against bad government." As we already learned from the Amazing Spider-man in Chapter Two, with great power there must come great responsibility. Society cannot endure, and heaven will not abide, bad government indefinitely. In time, the skies become a celestial lighthouse warning of a dangerous approach to the reef of error or a supernal press release publicizing divine admonition for mistakes already made.

ROYAL CONFIDENTIAL

Power to modify the behavior of the king, no matter how well it may be contained, retains the risk of exploitation. A cynical or ambitious member of the astronomical elite—or even an advisor who has only the best interests of the country at heart—may be tempted to distort the celestial record and tilt the direction of decisions. Eberhard's analysis of the portents registered by China's royal interpreters of celestial signs verifies the presence of some fictional data entered into official annals on behalf of a political agenda. These were usually contrived by historians after the fact. We also know that celestial portents played a part in real time in some factional disputes.

Deliberate deception also occurred in Mesopotamia and is documented from the eighth century B.C. An astrological auditor on the payroll of Sennacherib, the Assyrian king, exposed the royal astrologer of Sargon, Sennacherib's father. Operating on the principle that equivocal news is better than bad news, Sargon's astrologer conspired with the other diviners. He persuaded them to conceal any sinister omens from the king, and instead, he lobbied, "If an untoward sign appears, let's tell the king: 'An obscure sign has appeared.' " According to the whistle-blower, "they systematically censored all unfavorable predictions." They seem to have been a crew that already knew the story about the wicked messenger.

The Bible's Book of Isaiah also had some choice commentary (47:13) about the celestial soothsayers. In his judgment of Babylon, Isaiah warns,

> Thou are wearied in the multitude of thy counsels: let now the astrologers, the stargazers, the monthly prognosticators, stand up, and save thee from the things that shall come upon thee.

Doom was on the march, and it had Babylon's name written on it. In Isaiah's mind, God's judgment made astrology obsolete.

The Book of Daniel also offers little solace to the king's skywatchers. Nebuchadnezzar II, the Chaldaean king of Babylon (605–562 B.C.), with measured royal skepticism, tested his magicians, enchanters, sorcerers, and Chaldaeans (astrologers) to tell him the contents and meaning of a troubling prophetic dream he said he had forgotten. Believing anyone with the powers they claimed to possess ought to be supernaturally informed of his private dreams, he threatened to cut them in pieces if they didn't deliver. All of his prophets protested. The astrologer argued that no one but the gods could know what the king wanted them to reveal. Outraged by their response and their failure, he decreed that all of Babylon's wise men should be killed. Daniel, who had been brought to Babylon with the capture of Jerusalem, was trained by the Chaldaeans in astrological lore. Although he was able to save himself and the lives of his three companions from the king's fury by correctly revealing the dream, he relied on God's inspiration, and not on astrological skill, to do it. His story, then, is another biblical rejection of the value and validity of Mesopotamian astrology.

Astrology in ancient Rome also received short shrift from Cicero, the celebrated orator and author of *On Divination.* Although he was first converted to believe in the power of the stars by the philosopher Posidonius, Cicero's conviction steadily eroded, and he eventually insisted astrology is an "incredible mad folly which is daily refuted by experience."

There have also always been skeptics of prophetic powers. Wang Chong, a Han dynasty philosopher, also challenged the validity of omens in the sky. Disputing the assessment of a more ancient authority's gloomy commentary on comets and certain appearances of Mars, Wang Chong concluded there "can be no possible connection between celestial events and moral judgments."

Our personal experience with human nature persuades us that the special knowledge and special status of the palace skywatchers might encourage hypocrisy in any omen reader who understood the king's character better than the behavior of the sky. In *Pacing the Void,* an elegant and artfully crafted examination of astronomical imagery in the Tang dynasty, Edward H. Schafer mentions a poem in which Po Ju i, a ninth-century magistrate and internationally known author, criticizes the imperial star-watchers because "they approach their sovereign only with flattery and misleading talk about felicitous clouds and longevity stars, then, serenity guaranteed, back to their precious instruments." Po condemned the astronomers' superficiality and preoccupation with their tools, but not because he doubted the value of reading the sky. It was their lack of serious prophetic intent that troubled him.

Of course, true belief in celestial revelation does not extinguish personal priorities and professional conflict, and we find both in the astrological tablets of Mesopotamia. Michael Baigent repeats the vitriolic, but concise, criticism of a competitor by Akkulanu, an astrologer with high-performance standards: "This omen is nonsense: the king should disregard it." Simo Parpola's studies turned up another colorful example of the astrological pecking order. King Esarhaddon wrote to ask the chief astrologer Ishtar-shumu-eresh if a report he had received from another seer were correct. The king had been told that Mercury would rise from conjunction with the sun in the coming month, and he wanted to know if the prediction were accurate. Ishtar-shumu-eresh knew that Mercury was already visible and replied with the facts and with an editorial about professional incompetence: "An ignorant one frustrates the judge, an uneducated one makes the mighty worry." There is no substitute for good information, but additional comments about Venus in the critic's report confused Esarhaddon, who concluded that Venus had already appeared. Always inclined toward independent corroboration, the king called in another astrologer and asked him about Venus. Venus, of course, was not yet visible, and the third skywatcher, without accurate knowledge of what Ishtar-shumu-eresh had really said, regressed to backbiting. After bringing Esarhaddon up to speed on Venus, he wrote that Ishtar-shumu-eresh was "a vile man, a dullard and a cheat" and asked, "Who is this person that so deceitfully sends such reports to the king. . . . Why does someone tell lies and boast about these matters. If he does not know, he should keep his mouth shut." One more astrologer joined the dispute, which culminated in an open argument in court that finally resolved the misunderstanding.

The court astrologers had to deal with ordinary obstacles and professional frustrations like everyone else. One of the same court astrologers who participated in Esarhaddon's Venus debacle also complained about the unnecessary inconveniences that subordinates have to bear thanks to policy at the top. Although he knew in one season no solar eclipse would occur, he had been directed to remain alert for it day after day. The useless waiting made him irritable and turned his official reports testy. Household headaches prompted other astrologers to petition the king for food, for a donkey to relieve aching astrological feet, for a doctor, and for help in retrieving a runaway handmaiden.

Evidence from the Mediterranean, Mesopotamia, and China shows that the political astronomers of antiquity didn't always command respect. Their knowledge and skill generally were valued, but they were only as infallible as their last accurate prediction. No doubt, there was astrological intrigue from time to time, but the soothsayers served the king, not vice versa. We have, then, manipulations of fact and of meaning, but neither of these manipulations dominate the discussion. Most of what we see reflects a more or less honest attempt to understand just exactly what bee was under nature's bonnet. Power and politics were the primary concern, but most of the time most of the participants were investors in the belief system.

Astronomical sincerity aside, the king had an interest in keeping some kind of lid on the flow of celestial information. Palace astronomy was a loaded situation for a king who could be compromised by the official astronomers and their information. Naturally, the ruler acted defensively.

At times, astronomical data were state secrets. Imperial China restricted access to the astronomical bureaux. Joseph Needham's monumental, multivolume *Science & Civilisation in China* quotes written evidence of the government's policy of astronomical discretion. In 840 A.D.,

> an imperial edict was issued ordering that the observers in the imperial observatory should keep their business secret. "If we hear," it said, "of an intercourse between the astronomical officials or their subordinates and officials of other government departments or miscellaneous common people, it will be regarded as a violation of security regulations which should be strictly adhered to. From now onwards, therefore, the astro-

> nomical officials are on no account to mix with civil servants and common people in general. . . ."

Omens and calendrical intricacies were imperial property dispensed according to need-to-know, and policy violations were taken seriously.

Although the rank of astronomers who headed the office of celestial observations was always relatively low in the imperial bureaucratic ladder—below, for example, the office that prepared official histories—the political sensitivity of astronomical data is evident in the evolving jurisdiction over astronomical endeavors. During the Han dynasty, the office of astronomy was part of the Ministry of Imperial and State Sacrifices, preserving the protocol of the preceding Zhou dynasty (1122–221 B.C.). Astronomy's place in the Han organizational chart emphasizes the role of calendrics, which governed the administration and performance of elite ritual. By the Tang dynasty, roughly four centuries later, clocks, calendars, and celestial observation all belonged to the Secret Secretariat, which treated the information these offices supplied as a guarded imperial resource.

Chinese emperors could express imperial power in a variety of ways, including astronomical research. Although Guo Shoujing's system of observatories was expansionist, it was not militaristic. Rather, it demonstrated that the reach of imperial astronomy extended throughout the country. This truncated brick pyramid is Guo Shoujing's state-of-the-art facility at Gao cheng (Yangcheng), in Honan province, in north-central China. A times, it was the primary center for celestial observations. Guo Shoujing's network totaled 27 field stations. Some were located in what is now Siberia, Mongolia, Korea, and an island in the South China Sea. (photograph E. C. Krupp)

As Matteo Ricci, the Jesuit astronomer, was on the way to Beijing in 1583, the officials confiscated his mathematics books. Mathematics was correctly understood to be an integral component of astronomy, and according to Ricci, "In China, it is forbidden under pain of death to study mathematics without the king's authorization." Actually, in Ricci's time, the prohibition was relaxed, but the spirit of state-sponsored astronomy was alive and well. There were neither schools that taught astronomy in sixteenth-century China nor manuscripts for study outside the imperial umbrella. Edward Schafer affirms that Tang dynasty astronomy and calendrics were a "monopoly of the court." Private ownership of certain almanacs and diagrams was prohibited, and violations were punished with two years at state labor. Unsanctioned investigations into the behavior of heaven were regarded as treason. Imperial religious ritual, which put the emperor in contact with divine celestial power, was governed by astronomical cycles. Independent astronomical intimacy therefore compromised the emperor's exclusive contract with heaven.

The political use of astronomical portents and the political value of the promulgation of the calendar ensured an interest in astronomy in the imperial court of China. Although the observed data were vulnerable to exploitation, it was challengers to the throne, rival factions in the court, critics of the emperor's performance or of the behavior of his entourage, historians, and the emperor himself who were mostly likely to manipulate the celestial signs for political advantage. The Chinese astronomers were not the real movers and shakers in the government. Rather, they fulfilled a high duty. In Babylonia and Assyria, the situation was similar. Observations could be checked independently. Interpretations, on the other hand, carried the liabilities and rewards of real power. Skywatching astrologers were consulted by kings, but healthy skepticism often courted second opinions. Mesopotamian reports also confirm the uncertain status of the court astronomers. They earned their living with the grace of the king.

CELESTIAL SECURITY CLEARANCE IN ROME

In 331 B.C., the Hellenistic empire of Alexander the Great absorbed Mesopotamia, which had been ruled by Persian kings since the sixth century B.C., and the reputation of Chaldaean astronomy continued to grow. Data, observational techniques, and computational methods were funneled to the Greek world and eventually became the mathematical astronomy in the foundation of modern science. At the same time, according to ancient authority, Alexander's retinue in Babylon "swarmed with astrologers, soothsayers, and prognosticators." Alexander encouraged them and consulted them.

Roman emperors and imperial *wanna-bes* also consulted astrologers in affairs of state, but they didn't monopolize the art. By the Roman era, the democratization of the horoscope had put the power of the stars into anyone's personal destiny. Scholarly speculation attributes to the Roman astrologer Theogenes an auspicious astrological omen for Octavian, who later was granted the title *Augustus* and became Rome's first emperor. No doubt Augustus Caesar thought highly of Theogenes and the astrological craft. According to the Roman historian Suetonius, Augustus was enamored enough of astrology to mint the sign in ascendance at his birth on the coin of the realm. Presumably mindful, however, of the power of independent prognostication, he outlawed astrology in Rome.

Other expulsions of astrologers are documented, and Tamsyn Barton, a Classics scholar, explains in a paper on "Astrology and the State in Imperial Rome" that all of them are connected with political instability. When Tiberius followed Augustus on the throne of Rome, he, too, banned the astrologers, and records of trials from his reign inform us that seeking astrological advice could attract the charge of treason. Tiberius is said to have sent astrologers he interviewed to their deaths if he were unsatisfied with their skills or wanted their predictions kept quiet. In this story, the astrologer Thrasyllus successfully avoids death by recognizing the intentions of Tiberius and correctly predicting his immediate danger. Tiberius, in fact, promoted Thrasyllus after the incident.

Nero, Rome's fifth emperor, relied on his own private astrologer, who calculated for him the most propitious time to take command of the empire, and his second—and scheming—wife, Poppaea, subsidized her own team of astrologers. The twenty-first emperor, Septimus Severus, relied on horoscopes to find a suitable wife. Ambitious for leadership, he kept looking until he found a woman with a chart that predicted marriage to a monarch.

Official response to astrology in ancient Rome really followed an older pattern. Powerful leaders attempted to restrict access to influential information. The Romans already had lots of practice at limiting "access to the knowledge of destiny," as Barton puts it. Long after Tarquin, one of the legendary Etruscan kings of Rome, purchased the three books of destiny from the Sibyl of Cuma, they eventually passed into the control of the Roman Republic. The Cumaean oracle actually had first offered Tarquin nine books for 300 pieces of gold. After he turned down the expensive opportunity, the Sibyl burned three of the books and tendered the remaining six for the same price. Tarquin apparently still couldn't see where the law of supply and demand was headed and turned her down again. Finally, on her third try, Tarquin decided he better save the books, and much later, the Roman Senate commissioned a rewrite of the missing volumes. At first, two aristocratic citizens were assigned to the task. Gradually their number increased to a maximum of 15, and no one but these 15 men were allowed access to the first three books of prophetic verse. Consulted for guidance in times of crisis, the texts were still important sources of inspiration for the government in the era of Augustus.

We can get an idea about the kind of crisis that would prompt consultation by reviewing an earlier emergency that called out the *Sibylline Books.* In 217 B.C., at the same time Hannibal was upping the ante in his challenge to Rome, celestial prodigies

Oracles are believed to be vessels of inspired, privileged information, and rulers, in general, like to have exclusive—or at least first—claim on any data relevant to affairs of state. It took Tarquin, the Etruscan king of Rome, however, three offers to convince him he should purchase the *Sibylline Books*, prophetic texts allegedly written by the Oracle of Cuma. Once they were in the hands of the authorities, however, access to them was a lot tighter than the Vatican Library. The Cumaean Oracle, or Sibyl, also offered advice and knowledge of destiny to clients who came to visit her sanctuary at Cuma, an ancient town founded by the Greeks on the Italian coast, north and west of the Bay of Naples and overlooking the Bay of Gaeta. This gallery, monumentally cut 430 feet into the rocky hillside, is by tradition the Cave of the Sibyl and the place where she held her interviews. In the *Aeneid*, the Roman poet Virgil wrote that it was "dug out like a cave" from the flank of the Euboean cliff. He added, ". . . a hundred broad passages lead to it, a hundred doors, from which the answers of the Sibyl sound forth in as many voices." This tunnel is vaulted with a ceiling more than 16 feet high, and the Sibyl's audience chamber is at its dark and distant end. (photograph E. C. Krupp)

prompted a call for prophetic counsel. First, the size of the sun miraculously diminished, and then it seemed to collide with the moon. Two moons where there should have been one, a luminous rip in the sky, celestial conflagration, and the uninvited impact of one of the extra moons with the earth all followed. It was the kind of day that would compel authorities to seek advice.

We don't really know what was in the books, for they were all lost in fire. We may get an inkling of what the real *Sibylline Books* were all about, however, from an alternate version of the prophecies composed about two thousand years ago. It is known as *The Sibylline Oracle.* Perhaps contrived by a group of insubordinate Christians and Jews as competition for the official prophecies, it details celestial events at least as troubling as those that reputedly plagued Rome in 217 B.C. The sun was surrounded by stars in the daytime sky. The moon fired lightning. The heavens themselves were at war, and the Morning Star directed the battle from the back of Leo the Lion. The rest of the passage describes apocalyptic reversals of the places of the stars and ruptures of cosmic order. Libra displaced Orion. The Pleiades went dark. Aquarius the Water Bearer set fire to one of the planets. The entire sky was thrown into disarray, and finally the heavens themselves, in the form of Uranus, rose and flung all of the starry warriors to the earth. They sank in the oceans and ignited the land, and the night sky, now vacant of stars, went dark. By the time we encounter this kind of prophecy, the sky no longer signals just the fortunes of kings and personal destiny. It spells out in heaven the end of the world.

Until Christianity condemned the Sibyl's books as relics of the pagan past, they were kept under guard in an underground compartment in the foundation of the Temple of Jupiter on Capitoline Hill. A death sentence was the price of unauthorized use. Like astrology and other forms of prophetic revelation, they represented knowledge that could destablilize the structure of power. The curbs imposed on general access—in China, Mesopotamia, and Rome—all add up to a considered understanding of the relationship between power and the natural world. Nature, of course, is as visible as the sky, but knowing what it means is a route to power. For that reason, the knowledge of power is restricted. Those who acquire it through observation also possess power, but as the word gets out, they also risk conflict and competition. Hypocrisy, cynicism, and exploitation all may blossom in the garden of privileged information. The record shows that the officially appointed skywatchers were susceptible to such lapses, but careful review of the relationship between astronomy and power also tells us the real manipulators of celestial knowledge in organized states were just who we would expect them to be—the emperors, the nobles, and kings.

CHAPTER NINE

It Pays to Advertise

Comic book superheroes usually advertise their special attributes with costumes that reflect their unique and extraordinary powers. It really all began with Superman, in 1938. His first appearance, in *Action Comics #1,* costumed him in blue tights with red trunks, red boots, and a red cape. The bright colors made him a standout in any crisis, and the yellow *S* ensconced on his chest in a red shield confirmed he was super. In time his adventures informed us he had powers far beyond mortal men: super strength, super speed, invulnerability, enhanced senses (including X-ray vision), and the ability to fly. Superman operated under a secret identity and went into action when he changed his wardrobe, usually in a phone booth. His futuristic skintight outfit emphasized his muscles and suggested he came from some other place or time. Joe Shuster, the illustrator who created Superman with writer Jerry Siegel, said he added the cape "to help give the effect of motion," but the cape also conveys the impression of elevated status. It's not your basic street wear. Clothes make the Superman. You can tell he is something special just by looking at him.

Thematic costumes, emblems, and paraphernalia conferred distinct identities on other newsstand superheroes. Batman, who started swooping down on villains in 1939, possessed no superhuman powers, but his arsenal of bat-inspired weapons and his relentless attitude meant big trouble for the agents of evil. Declaring war on all criminals to avenge the death of his parents in a holdup, he adopted the bat costume. His cape looked like a bat's leathery wings, and his cowl masked him with tall, pointed ears. He had an eerie look, and the bat silhouette on his chest made the symbolism transparent. "Criminals," he said, in a nocturnal epiphany, "are a superstitious cowardly lot. So my disguise must be able to strike terror into their hearts. I must be a creature of the night, black, terrible . . . a . . . a . . . a bat!" The rest is comic book–television–motion picture history.

Talismanic animals took another turn in the superhero renaissance of the 1960s. Peter Parker, a high school science student, was bitten by a spider accidently irradiated at a museum exhibit on atomic science. The teenager acquired the proportionate strength of a spider plus the ability to scale walls and to sense danger. He designed himself a blue and red costume, complete with armpit webbing, spider insignia, and webshooters. As the Amazing Spider-man, he battled supervillains, some of them also inspired to adopt specialized animal identities, including the Chameleon, the Vulture, Doctor Octopus, the Lizard, the Scorpion, and the Rhino.

Captain Marvel was Superman's primary comic book competitor in the 1940s, and his emblem of power was celestial—a yellow lightning bolt that struck across the breast of his red leotard. Whenever newsboy Billy Batson said the word "Shazam," he was transformed into the superpowered Captain, a metamorphosis accompanied by a bolt from the blue. With the talents of ancient heroes and gods, funneled to him through the magic of an ancient wizard, the paperboy became the World's Mightiest Mortal.

The symbolism of time and space invested the uniforms of many other superheroes with badges of cosmic power. The original Flash was a high-speed hero who wore the winged helmet and boots of Mercury, the swift messenger of the gods, and also had lightning on his chest. Green Lantern's power emerged from a ring that had to be

Comic book superheroes intimidate their adversaries with costumes rich in symbols of power. These die-cast toy figures include, from left to right, Superman, Batman, Green Lantern, Captain Marvel, and Spider-man. (photograph E. C. Krupp)

recharged every 24 hours by a lantern forged from mysterious meteoritic metal. Explaining his green pants, red shirt and boots, high-collared cape, and black mask, Green Lantern said, "I must have a costume that is so bizarre that once I am seen I will never be forgotten." The green lantern on the front of his shirt told his adversaries who he was and where he got his power. Hourman ingested Miraclo, a powerful substance today we would call a dangerous drug, and acquired 60 minutes of miraculous strength to battle crime. His hourglass necklace told the bad guys who they were facing and told him when his time was just about up. Dr. Mid-Nite's damaged eyes restricted his activity to night. The crescent moon on his forehead, his crescent-moon buttons, and the owl perched on his shoulder confirmed his affiliation with the dark. There was a yellow star on the front of Starman's red uniform. His Cosmic Rod plugged him into the power of "infra rays" from outer space, and manipulating the energy of the stars, he could transform materials, move objects, and fly.

All of these heroes have unusual talents, and their costumes tell the rest of us we aren't dealing with ordinary people. In some cases, the clothes they wear are advertisements for the power they possess. In other instances, the costume is actually part of the hero's power. Either way, the costume and other emblems influence the perceptions of the audience and so have their own cultural power.

Now, popular culture is one venue of the mythology of our time. Comic books may only be vehicles of popular culture, but they put the mythology of our time on display and help reveal how our own symbolic system works.

Each culture's emblems of power reflect its social complexity and reveal some of the functions of its symbolic system. Symbols of power may focus prestige and influence into portable, exotic, and rare objects. Through possession and display, they tell who has power, how much, and where they got it. Large, permanent monuments also advertise power, and public evidence of power reinforces belief in the validity and permanence of the social order. Symbols of power, then, are not just indulgences of the elite. They are tools that craft cultural character. They are a protective additive for the social engine. When the system fails, the symbols respond. Through them, we can track the creation, maintenance, and collapse of power in a community.

Symbols of power, like a superhero's costume, persuade us that power is at work. For traditional peoples, the universe is saturated with power, and proper acquisition of some of it was considered useful. Because the celestially tempered power at the command of rulers, and at the disposal of those who talked with the spirits, belonged to this great reservoir of power, the authority of kings and the credibility of shamans were legitimized by their affiliation with the sky. The celestial and seasonal sources of power were apparent, but those who claimed to call upon them and said they had procured some of that power also had to publicize their success. They documented the power at their command and affirmed its value to their communities through celestial symbols, equipment, and rituals that functioned like Captain Marvel's thunderbolt and magic word, Dr. Mid-Nite's crescent moons, and Starman's stellar trademark and Cosmic Rod.

THE WARDROBE OF COSMIC POWER

Siberian shamans often brokered cosmic power in costumes that incorporated celestial symbolism. These costumes were as specialized, personalized, and as infused with power as anything a superhero might don for action. Clients knew their shamans were open for transcendental business when they were attired for trance, and in some communities, possession of a costume was the confirmation of shamanic maturity. The shaman was not believed to be a confidant of spirits without one, for it was the spirits who revealed to the shaman how the costume should be designed. Each shaman cultivated his or her own network of spirit allies, and the idiosyncracies of each costume were the natural product of an individual relationship with the spirit world. Despite these eccentricities, the shaman had to operate within a shared symbolic system. In Siberia's far east, for example, shamans of the reindeer-hunting Tungus (or Evenks), a Mongoloid people whose language is part of the Altaic family, were often outfitted with deer antlers. Apart from any other connotation carried by the deer headgear, the antlers were infused with the power of seasonal renewal and so provided an analogy from the natural world for the shaman's own transformations. Anyone in the community could recognize it.

This caftan worn in performances by a nineteenth-century Altaian shaman has celestial pendants fastened on the back: a "lunate" half-disk for the moon, a complete disk for the sun, and a ring for the rainbow. (photograph E. C. Krupp, Peter the Great's Museum of Anthropology and Ethnography, St. Petersburg)

Shamans' costumes also mirrored other universal shamanic themes, including the shaman's travel to the skyworld and to the underworld. The Nganasan, for example, are reindeer herders of the far north. They live in the Taymyr peninsula and speak a Samoyedic language, and their shamans use one costume to negotiate the underworld and another to reach the upperworld. Because the shaman must ascend to the upperworld, the costume he wears for that journey is lighter in weight. The design of its parka incorporates the principle of complementary opposition. One half is colored red, and the other is black. The dark component alludes to winter and to night while the red refers to spring and to day. As the Nganasan shaman rises to the sky, he is garbed in cyclical renewal. Beneath this coat, the shaman wears a breast cover on which is attached an iron mask that represents Moudjali, a spirit whose name means "daylight." Moudjali controls the plants, the birds, the deer, and human beings, all of which are affiliated with daytime and light. Another metal pendant fastened below the mask stands for Moudjali's nine celestial maidservants.

The Tungus also saw their shamans slip on different outfits for different missions. The reindeer costume, complete with iron "reindeer bones" and many more metal ornaments could weigh nearly 90 pounds. Unsuitably heavy for traffic in the sky, it was worn on journeys to the underworld, but it was reinforced with celestial power. Among the metal pendants sewn to it are a disk that stands for the sun, a half-disk that symbolizes the moon, a rainbow ring, star and thunder emblems, and a square perforated with a square hole to invoke the polar entrance to the skyworld. The Yakuts language is Turkic, but their shamans use the same kind of celestial ornaments and say the sun and moon light the way as their spirits travel through the darkness of the world beyond. Turkic-speaking Altay shamans of southern Siberia also pin the sun and moon to their coats, and at least one of them has a set of small dots sewn to the back of the collar. They are the daughters of Ülgen, the high celestial god, and they help lift the shaman into heaven.

The Tungus duck or bird costume facilitated travel to heaven, for birds, understandably, were believed to have access to the sky. Fringe on the sleeves simulated wings. Crosses sewn to the jacket of a Yukaghir shaman turned his suit into a bird skin for shamanic flight. The Yukaghir speak a Paleo-Asiatic language but have assimilated many Altaic traditions.

The whole universe was painted on the costume of a Goldi shaman. The Goldi, also known as the Orochi, occupy a small part of the Khabarovsk region, in far-eastern Siberia, across from Sakhalin Island on the Tatar Strait. Their language is related to Manchurian, part of the Tungus-Manchu branch of the Altaic family. They were maritime and forest hunters with an elaborate concept of the cosmic tree. According to the Goldi, there are really three world-axis trees—one in our world, one in the underworld, and one in the sky. Three different trees are illustrated on the Goldi shaman's attire now displayed in the American Museum of Natural History in New York. This costume was collected from the lower Amur River area in the late nineteenth century by Berthold Laufer, and although he did not obtain any information about the meaning of its images, we know enough about Goldi shamanism to recognize the cosmological content of the outfit. A leopard and a tiger are posted by the tree in the underworld, on the back of the tunic, along with snakes, lizards, and a pair of black bears. Above the tree, a set of concentric rings may symbolize the center of the earth and the portal between worlds, for there is a similar set of rings on the front of the shirt, and below it, on the front of the shaman's skirt is the tree in the middle of the earth. It too is guarded by animals, including a leopard and a tiger, swans, and snakes. The tree in the sky shows up on the back of the skirt. Also protected by felines and reptiles, it is haloed by hovering birds. These are human souls waiting their turn to fly down to earth.

The Goldi shaman is believed to have worn this type of costume when escorting souls to the underworld. He first traveled all over the cosmos to retrieve the wandering soul, and once the psychopomp found his client, he climbed a notched tree—perhaps the Milky Way—to locate the highway to the land of the dead. Wherever he roamed, his costume kept him at the center of the world.

Siberian shamans wandered through the spirit worlds with cosmic symbols on their clothes, and the universe through which they traveled was often mapped on their drums. Drums also were vehicles that put the shamans on the transcendental highway, for they transported themselves into altered states by beating those drums. Symbols on those drums disclosed celestial forces and other supernatural agents they might encounter in their trance journeys.

The Goldi are upriver Nanays of eastern Siberia's Amur basin. Like other Tungus peoples, they symbolize the world axis as a tree. This portion of a Goldi shaman's costume—the front of the skirt—is illustrated with the cosmic tree in the middleworld that connects earth with sky. (photograph E. C. Krupp, American Museum of Natural History, New York)

Each group conjured its own image of the drum. Some Altaic peoples saw the drum as the shaman's horse. To some Mongols, it was a stag. So the drum was not just a guide to the spirit realms, it was the magic steed that carried the shaman to the center of the world, up and down the cosmic staircase, and through the kingdoms of gods and demons, spirits and souls. For that reason, some shamans treated the drum as if it were world's center or the axis between worlds.

Altaic drums were usually ornamented with circles that stood for the stars. The field of the drum was the Upper World, and figures painted there represented the daughters of Ülgen, who facilitated the shaman's ascent. In one survey of 50 Altaic drums, the sun, moon, and Venus were documented on 38 of them. Shors drums also depicted the celestial realm and populated it with the sun, as a red rayed disk, and with spirits of the sky. The Teleuts divided their drums into two zones. The upper territory naturally illustrated the sky and included rayed disks that symbolized the sun, moon, and Venus. These celestial lights were accompanied by geese, migratory waterfowl that carried seasonal meaning and in some cases alluded to the Milky Way. The stripe dividing the upper and lower sections of the Teleut drum was the earth. Baraba Turks included a rayed sun, the full moon, and a star on their drums.

The Khakass are a Turkic-speaking people of central Siberia, and their shamans split their drums in half with a zigzag framed inside parallel lines. These horizontal strips are "the remote way of the shaman," the route by which the adept enters the supernatural world. The upper zone of the drumhead is the skyworld. (photograph E. C. Krupp, exhibited in "Nomads: Masters of the Eurasian Steppe")

Lapp drums usually illustrated three cosmic realms, and the sun was often placed at the center of the drumhead. The moon and astral deities appeared on these Lapp drums, too. Some Lapps and other Siberian peoples regarded heaven as the face of the high sky god. As the sky, the drum was also the face of celestial divinity, and the sun and moon were the eyes of the sky.

The specific role of Venus on shamans' drums is not known, but the Buryats—and probably other central Asian peoples—associated the planet with fertility. Certainly this is consistent with the shaman's efforts on behalf of the health and continuity of the community, and Venus may also have been linked with the idea of transformation and cyclical renewal.

Other components of the sky—the polar axis, Orion, the Great Bear, the Pleiades, rainbows, and lunar calendar tallies—made less frequent appearances but still are known from some shamanic drums. Like the sun and moon ornaments on the Yakuts shaman's costume, the sun and moon on the drum were believed to have the power to illuminate the route as the shaman glided through the shadowed corners of the cosmos. The polar world axis oriented the shaman in flight. With constellations such as Orion, painted on the drum, the shaman found the way to heaven and back with the help of transcendental celestial navigation.

Alive with supernatural power, the shaman's drum was a coy charmer that invited the spirits to alight within the shaman's reach. The symbols on it didn't just advertise celestial power. They activated it.

BILLBOARDS OF CELESTIAL POWER

Over the last two decades, accelerating study of prehistoric, and more recent, rock art throughout the world has led researchers to conclude that much of it was painted or carved in conjunction with shamanic activity. The specific function of the art varies with context and character, but petroglyphs and pictographs are believed, in many cases, to comprise sacred art that reflects some kind of contact with the spirits. One aged San has explained, for example, that the surface of the rock was a veil between our world and the spirit realms. When properly activated, this membrane permits the shaman to cross the frontier to other worlds and allows the spirits to emerge in ours.

Anthropologist Thor Conway has reported what an Ojibwa elder had to say about a pictograph site on Lake Huron known as Rooster Rock. Dreams were obtained there, and spirits inhabit it. Shamans might experience trances at such places or record visions from their dreams there. Their art endorsed, contributed to, and released supernatural power at these spots.

Although many rock-art sites seem to be places of hidden power, the Comanche Gap petroglyphs, in the Galisteo Basin of northern New Mexico, offer a public message on a natural volcanic dike and on the boulders scattered below this vertical wall of extruded stone. Most face south. Some face east and west, but none face north.

The dike itself is one of the most conspicuous features in the terrain, and the break that splits the ridge into two parts is the gap for which it is named. Traveling either north or south on the highway that passes just east of the site, you see the distinctive dike on the open mesa long before you reach it. At its highest, it towers about 230 feet above ground level. Although widened by the talus at its base, the wall is very narrow at the top, and its face offers many flat surfaces for carved images. The petroglyphs are more concentrated in some places than others, however, and appear to congregate at locations where the rampart drops a bit or opens. Many of the figures depict shield-bearing warriors, which suggests a connection with warfare. Celestial imagery is also abundant, particularly in the form of personified—and aggressive—stars.

The symbols at Comanche Gap and elsewhere in the Galisteo Basin have been studied by Polly Schaafsma, an authority on rock art in the American Southwest. Detailed style analysis prompted Schaafsma to conclude most were produced between 1350 and 1525. The area, she notes, was occupied by the Southern Tewa, an historic Pueblo people, between the thirteenth and sixteenth centuries. Not long after 1500, the nearby pueblos, which survive today as ruins, were abandoned, possibly in response to increasing attacks by Plains intruders or war among the Pueblos themselves.

Along with the abundant war imagery at Comanche Gap, we encounter curious variations in celestial elements, most intertwined with the warfare theme. Many of the

Protected by a celestial shield, a petroglyph warrior defends Comanche Gap with the magical power of the sky. The whole shield is the rayed disk of the sun, and the figures inside include diamond-shaped stars with faces that, in part, probably represent Venus as the Morning Star. (photograph E. C. Krupp)

shields are decorated with four-pointed stars and with zigzag borders reminiscent of the sun's rayed disk. Others have eagle feathers and claws, as do some of the star symbols. Capable of flight, the eagle is also emblematic of the sky. Ethnography tells us the designs on real shields were not just decorative additions but powerful symbols, and it is reasonable to think they operated the same way in the rock art. Associated with war and defense, to the Southern Tewa they may have had magical power or the ability to invoke supernatural help. The stars with belligerent faces also seem to have something to do with war and magic.

Through the eagle attributes of the star petroglyphs, Schaafsma has convincingly linked the rock art with Knifewing, an important figure in Pueblo mythology. To the Zuni, another Pueblo group, the high-flying eagle is the Beast God of the Zenith, and he commands the direction that belongs to the highest point of the sky. The eagle was also equated with the sun and with Achiyalataba, or "Knifewing." Affiliated with fertilizing power of the rain, he fired lightning bolt arrows with a rainbow bow. As his Zuni name implies, Knifewing had flint knives for feathers in his tail and wings and was the supernatural sponsor of scalping. Certainly scalping is indicative of war, but it was really sacrificial violence connected with rainmaking and agricultural fertility magic. The flint knife, used for scalping, was a tool for sacrifice, for the scalp was taken to induce rain. Because sparks can be struck from flint, the knife mirrors the power of lightning. Lightning, of course, accompanies the rain and strikes from the sky like an arrow.

Sotuqnang-u, a Hopi star and lightning war spirit of the zenith, also scalps, kills, and fertilizes. The analogy between war and fertility in these Pueblo traditions appears to be rooted in the concepts of male predators, aggression, and penetration. Weapons are the agents of the aggressive power that renews life. The rain enters the earth, and semen enters the womb. The delivery in either case has the penetrating power of an arrow, knife, or spear.

Knifewing, like the eagle, is a predator, and his abduction of a woman to the skyworld eventually results in his own death and celestial transformation. The woman's husband pursues Knifewing into heaven, kills him, and flings his head into the east, where it is turned into the Morning Star. Knifewing's heart is hurled into the west and becomes the Evening Star. Other parts of his corpse take their places in the night as key patterns of stars, such as Pleiades and the Belt of Orion. At first a talismanic directional ruler of the sky, Knifewing acquires astronomical meaning through conflict and death. Through his transformation into Morning and Evening Star, Zuni myth merges his patronage of violence, fertility, and rain with appearances of Venus, and Schaafsma believes the celestial references at Comanche Gap reflect a warfare tradition centered on Venus as the Morning Star.

At the Tewa village of Hano, Morning Star was associated with war, and they referred to scalps as "Morning Star." Hano is actually on First Mesa next to the Hopi village of Walpi, in north central Arizona. The presence of Southern Tewa there, in the middle of Hopi territory, was itself a product of war. They were invited by the Walpi chiefs to help defend the Hopi against the Utes and Paiutes.

The stars with eagle claws, wings, and feathered tails at Comanche Gap appear to depict Knifewing, the celestial war patron and Morning Star. The mean-looking stars there may also be portraits of the Morning Star, or perhaps they symbolize other sources of stellar power. The armed figures on the rocks are either different incarnations of the star spirit of warfare or pictures of Southern Tewa warriors fortified for war with celestial magic. But why was there so much war-oriented imagery at Comanche Gap?

Pueblo warfare was a possible cause of abandonment of so many sixteenth-century pueblos in northern New Mexico. Non-Pueblo raiders also might have pressured the pueblos. A regular pattern of warfare prevailed before 1540—probably both types. The tradition of warfare persisted strongly among the historic Pueblos, who had a well-developed system of war chiefs, war societies, and related ceremonialism. Large, fortified pueblos took advantage of high ground and incorporated permanent barricades for defense on the eastern edge of Pueblo lands, and it is known there was historic conflict with the Teyas (Caddo-related plains nomads) in the 1520s.

Schaafsma concludes that Comanche Gap "may have functioned as a Southern Tewa war shrine, rendered more powerful by the presence of these images." The Comanche Gap dike was essentially a billboard that advertised Pueblo power and Morning Star war magic to all comers.

With the claws and tail feathers of an eagle, the four-pointed star from Comanche Gap is a hybrid representation of Knifewing, a Southern Tewa war spirit associated with the zenith, the sun, and the Morning Star. (photograph E. C. Krupp)

CELESTIAL DECLARATIONS OF WAR

The Venus warfare symbolism we encounter in the American Southwest probably owes something to a similar, but more developed, tradition in Mesoamerica that has been traced back to Teotihuacán, in Mexico's central highland plateau. Teotihuacán was the first real city in Mesoamerica and the largest pre-Columbian urban center in the New World. Its tentacles of trade, tribute, and conquest extended as far north as the tropic of Cancer, where Alta Vista, near what is now the town of Chalchihuites in the state of Zacatecas, was an outpost of Teotihuacán influence. Evidence of the impact of Teotihuacán's military-mercantile power is also apparent as far south as Tikal, the great Maya ceremonial center in the northern Petén lowland forest of Guatemala. Although the roots of Venus-regulated warfare seem to be as old as the third century A.D. and Teotihuacán's Pyramid of the Feathered Serpent, the set of symbols that links rain, fertility, sacrifice, and war with Venus transcends the centuries and the cultural and linguistic territories of Mexico and Central America.

Teotihuacán's enduring prominence and expansionism and the persistence of its legendary meaning after its fall in the middle of the eighth century A.D. all probably explain the wide dispersion of Venus-regulated warfare throughout Mesoamerica. Specialists now usually call this militaristic celestial symbolism the Tlaloc-Venus war-

Heads of feathered serpents once monitored the traffic on the stairway of the Pyramid of the Feathered Serpent at Teotihuacán. (photograph E. C. Krupp)

fare cult. Tlaloc was the fanged and goggle-eyed central Mexican god of the storm, rain, and lightning. Prosperity was also part of his portfolio, for his meteorological power controlled agricultural fortunes.

Tlaloc's emblems and the insignia of divinities connected with the planet Venus converged in the ceremonies, regalia, and weapons of ritual warfare. It was a theatric kind of war, more directed toward acquiring elite captives for sacrifice and establishing the hierarchy of power and control of trade than toward physical acquisition of land. Combining the agricultural abundance implicit in the rains with the status and wealth that accompany tribute won through war, the imagery of Tlaloc-Venus aggression operated as propaganda. Through its symbols, expressed in costume, architecture, and sacrificial ritual, winners advertised their imperial aspirations and their power.

Venus, the second divine patron of ritual warfare, was the celestial avatar of the god the Aztecs called Quetzalcóatl. The name means "Feathered Serpent" in Nahuatl, the language spoken by the Aztecs. We don't know what language was spoken earlier by the people of Teotihuacán, but the feathered serpents all over Teotihuacán's Pyramid of the Feathered Serpent are obviously a reference to their version of Quetzalcóatl. In Yucatán, far to the south, the Maya called the same god Kukulcán. That name, too, means "feathered serpent," and there are plenty of plumed snakes in the ceremonial centers of the Postclassic Maya. The earliest known rendition of this deity is a feather-crested rattlesnake on an Olmec relief from La Venta, in Mexico's tropical Gulf Coast zone. The Olmec are credited with originating many fundamental Mesoamerican traditions, including the calendar, writing, and monumental sculpture, and the La Venta feathered serpent belongs to the first half of the first millennium B.C.

Quetzalcóatl was a multifaceted god, too complex to be pigeonholed into a single role. He was regarded as a divine benefactor who brought the arts of civilization to human society. In his role as Éhecatl, the Aztec god of rain-bringing wind, he embodied the notion of life-sustaining air—and by extension, the "breath" of life. In the era of the world's beginning, he was one of the four creator gods. With the feathers of the iridescent quetzal bird and the body of a snake, Quetzalcóatl combined the elemental principles of earth and air. Water and the renewal of life were implied by his serpentine character. As Tlahuizcalpantecuhtli, he was "Lord of the House of the Dawn," the Morning Star. An Evening Star skeletal alter-ego that descends like Venus back into the western underworld sometimes appears back-to-back with Éhecatl/Quetzalcóatl. Both apparitions of Venus were apparently greeted with fear and apprehension, and Tlahuizcalpantecuhtli is usually shown equipped with an atlatl, darts, and a shield.

In addition, Quetzalcóatl shared his name—and perhaps some of his mythology—with a legendary Toltec ruler. Banished from Tula, a Postclassic site in central Mexico, this Quetzalcóatl traveled east to Chichén Itzá, a center of Postclassic Maya power in northern Yucatán. In one crucial episode of Quetzalcóatl's career, he immolated himself to pay through sacrifice for his transgressions on earth. His heart, however, rose from the ashes into the sky to become the Morning Star.

To both the Aztecs and the Maya, Venus was a dangerous heavenly power. The Aztecs nourished him with the blood of captives slain when he reappeared as the Morning Star, and both the Aztecs and the Maya kept track of his movements through systematic observation. The Venus almanac in the Dresden Codex, one of only four pre-Columbian Maya manuscripts known to have survived, coordinates the behavior of Venus with eclipses, with the moon, and with the 365-day solar calendar via the Maya 260-day ritual calendar.

Venus is shown five times as a Morning Star god in the Venus almanac of the Dresden Codex, and in each appearance he is attacking a victim in a panel below him, launching deadly darts from his celestial perch. He also shows up armed in five Morning Star appearances in pictorial hieroglyphic documents of the Borgia group, which were prepared prior to European contact and reflect central Mexican tradition.

Although the capacity for violence Mesoamerica attributed to Venus has been evident since scholars first really began to study these sources, about a hundred years ago, the political significance of this material was not understood until 1982. By then, Maya monuments were known to carry dated, hieroglyphic texts that referred to specific

Putting in an appearance as the god Kukulcán, probably in the ninth century A.D., Venus rides a feathered serpent on a carved jadeite plaque recovered from the Sacred Cenote at Chichén Itzá. He is poised to fire a dart with an atlatl. We also see him in this posture in the Venus table of the Dresden Codex. He wears a star skirt around his waist. (photograph E. C. Krupp, Peabody Museum, Harvard University)

events—battles and sacrifice—and that included the Maya glyph for Venus. In 1982, however, anthropologist Floyd Lounsbury demonstrated that those dates coincided with significant stations in the 584-day interval through which Venus cycles from morning star to evening star and back to morning star again. In particular, he argued that the best interpretation of a date painted in Maya numerals on a mural at Bonampak—illustrating battle, sacrifice, and the subjugation of prisoners taken in war—is August 2, 792 A.D. (Julian calendar date). On that day, he added, the sun crossed directly through Bonampak's zenith (an event that happens only twice each year), and Venus was in inferior conjunction, a short period of invisibility before its reappearance as a morning star. Bonampak, a Maya ceremonial center in the jungles of eastern Chiapas, was already famous for its three rooms of elaborate murals, discovered in 1946, but their Venus-modulated conflict focused attention on what had been an unappreciated dimension of Maya power politics and on what Maya experts started calling "Star Wars."

The Bonampak paintings document several politically significant events, including heir-designation rites organized by Chaan-Muan, the Bonampak king. Those ceremonies began on December 14, 790 A.D. (Julian date), and concluded with a Venus-battle a little more than a year and a half later. All of these activities, including the capture and sacrifice of prisoners of Venus-war, are thought to be part of the magical preparation that transformed a child of the king into a divine being authorized for future rule.

As Venus-timed warfare in the eighth century ends in victory for the Maya city-state of Bonampak, its ruling lord grips one of the defeated warriors by the hair. (photograph E. C. Krupp, replica of Bonampak's Room 2 murals, National Museum of Anthropology, Mexico City)

In Bonampak's first painted room, Chaan-Muan, enthroned on a bench, is accompanied by other members of the royal family, while a lord or retainer holds the young heir in his arms and presents him to the audience of assembled nobles, who are dressed in white robes for the occasion. Another scene in this room records a ceremony that took place 336 days after the presentation of the child. Members of the elite, including Chaan-Muan, are elaborately costumed for dance with feathered backracks and are accompanied by a procession of nobility and musicians that Mary E. Miller, the foremost authority on the Bonampak murals, calls "the boys in the Bonampak band." They play gourd rattles, drums, turtle shells, and trumpets, and some of them are wearing grotesque masks of animals connected with water and perhaps fertility. The ceremonial dance was performed as part of official dedication of the building, and presumably to honor the announced heir. It took place on November 15, 791 A.D. (Julian date), and Venus was just reappearing as an evening star.

Room 2 contains the Bonampak battle scene, and assisted by allies from Yaxchilán, Chaan-Muan subdues his primary target as the raid ends. Nine prisoners, publicly tortured and sacrificed, are displayed on the steps of a pyramid in the mural on the north wall. Chaan-Muan, still attired in his jaguar jacket, remains in the center of action as he holds one of the captives by the hair. All of the prisoners are naked in a scene of humiliation and violence. Some are bleeding from their fingernails. Another, sprawled across the stairway, appears to be dead, and a decapitated head rests on one of the steps.

We see more references to Tlaloc-Venus warfare in other dynastic propaganda of the Maya. At Dos Pilas, in Guatemala's Petexbatún region, one of the inscribed carved monuments, stela 2, confirms the military subjugation of Ceibal, a rival center of power on the Río Pasión, by Ruler 3 of Dos Pilas with a "star over Ceibal" glyph that accompanies a date when Venus first reappeared as an evening star. Ruler 3, portrayed on this stela in Tlaloc-Venus war regalia, is standing on top of his captive, Jaguar Paw, the king of Ceibal. The same inscription also references Jaguar Paw's coerced public bloodletting six days after becoming a war trophy and his sacrificial death a dozen years later at the time when Venus was hidden by the sun during inferior conjunction and about to reappear as a morning star.

Ruler 3's war costume on stela 2 crowns him with a "balloon" headdress that looks like an inflated deer pelt. His owl pectoral is also part of the visual vocabulary of ritual warfare. He carries a flexible rectangular shield and a spear, and the goggle-eyed face of Tlaloc stares from his apron. The skull on the bar he wears as a pectoral around his neck also belongs to this set of war symbols. Other elements that sometimes appear in a ruler's wardrobe for war are butterfly nose ornaments, Tlaloc shields, spear-throwers, obsidian-bladed war clubs, Tlaloc headdresses, and jawless, fanged, and goggled war-serpent bonnets like the masks paired with the faces of Quetzalcóatl on Teotihuacán's Pyramid of the Feathered Serpent.

Ruler 3's headgear is also ornamented with one of diagnostic symbols of the Tlaloc-Venus battle fatigues—a trapeze-and-ray emblem that is often called the Mex-

Although we don't know the real name of Ruler 3, a Maya king at Dos Pilas in Guatemala, his portrait is preserved on stela 2. The hieroglyphic inscription above his "balloon" headgear documents his victory over Ceibal, a formidable adversary. Fully costumed for Venus-Tlaloc warfare, he wears a ceremonial bar over his massive, jade-bead collar. Beneath the face of the skull on the bar, an owl hovers with outspread wings. Tlaloc stares goggle-eyed from his apron below the owl. His rectangular shield and a spear are on his left arm, near the bottom of the picture, and there is another spear on his right. (photograph E. C. Krupp)

ican year sign. It is a combination of interlaced geometric figures, usually a rectangle and a triangle, but sometimes more complex, and is known from its appearance in pre-Columbian manuscripts to have some calendrical meaning. John B. Carlson, an astronomer who has undergone Maya metamorphosis and become a leading interpreter of Mesoamerican Venus iconography, has argued convincingly that this trapeze-and-ray symbol refers to Venus. It is a stylized element of the bundled sticks that were strapped together at the completion of fundamental calendrical cycles. We know the Aztecs prepared these bundles for the "Binding of the Years" ceremony at the close of a 52-year period, when the 365-day calendar based on the length of the solar year and the 260-day ritual calendar both returned to the same starting point on the same day. Carlson believes the same kind of calendrical symbolism was incorporated into the eight-year interval in which Venus completed five 584-day cycles of movement. The total number of days involved—2,920—is also naturally built into the Venus almanac in the Dresden Codex. Thirteen of these intervals are, in turn, congruent with two of the 52-year cycles. Maya war lords not only fought, then, according to the movements of Venus, they fought with Venus almanacs symbolized on their combat caps.

Prior to the discovery of the Bonampak murals, with their emphasis on kingship and warfare, the renowned Mayanist J. Eric S. Thompson constructed a very different picture of the Maya and their hieroglyphic inscriptions:

> These texts, to the best of our knowledge, contain no glorification of ruler or recital of conquest, such as are customary on the monuments of other peoples. Instead, they are an impersonal record of steps in the search for truth, as the Maya saw it, that is the whole philosophy of time with its interlocking cycles of divine influences.

Thompson envisioned the Maya as a high-minded, passive, intellectual society of astronomer-mathematician-philosopher priests. Bonampak portraits of battle and dynastic ritual, however, inspire a less romantic image. The Maya may very well have been concerned with the philosophy of time and interlocking cycles of divine influences, but it was an applied philosophy, forcefully exerted on their neighbors with the power of the sky. Their priests manipulated an elaborate system of symbolic numerology and cyclical resonances with a calendar designed to apply the sacred structure of the cosmos to the affairs of the earth. Priests juggled the cycles and calculated when several would coincide in their search for cosmic congruence. In the Maya world as elsewhere, celestial phenomena revealed the affairs of dynasties and kings.

Hieroglyphic inscriptions revealed the role Venus played in ideology of power among the Maya, but it was not so clear just how thoroughly the tradition of Tlaloc-Venus warfare permeated Mesoamerica until a whole new set of war and sacrifice murals was excavated at Cacaxtla. Cacaxtla, however, was not in Maya territory but in central Mexico, about 80 miles east of Teotihuacán in the state of Tlaxcala. Fortified and occupying a hilltop for defense, Cacaxtla's location and construction mirror the destabilization and conflict that followed the fall of Teotihuacán. Other centers of regional power, such as Xochicalco in Morelos and Cholula in Puebla, established early claims as Teotihuacán's successor and were perhaps challenged by Tula, in Hidalgo, after it was founded in the tenth century. All of them possessed major architectural monuments incorporating allusions to the Feathered Serpent. Cacaxtla, apparently, was also a contender, and it was founded by the Olmec-Xicalanca, the same group that dominated Cholula after 800 A.D. Despite the name given to them by archaeologists, they were not related to the Preclassic Olmec of the Gulf Coast. Their ethnicity may have become a mixture by the time they moved into prominence in highland Mexico, but the second half of their hyphenated name also belongs to a Postclassic commercial center on Campeche's Gulf Coast. This was territory of the Putún Maya, who were Mexicanized culturally but spoke a Maya language. They operated as warrior-merchants who consolidated their own prosperity through trade connections enforced and defended by military power. The Olmec-Xicalanca in the highlands appear to have developed a similar military-mercantile complex, and the murals at Cacaxtla proclaim its power.

One scene appears to be a battle between Cacaxtla warriors in Tlaloc-Venus gear and what are thought to be adversaries from the Maya lowlands fully feathered in bird outfits. Carlson concludes, however, that the fight is already over. Some of Cacaxtla's enemies are already painted and dressed for sacrifice, and the defeated Bird Captain, in the center of the painting, is being forced to spill his own cheek blood with the victorious war leader's dart. In a second painting, Bird Captain's arms are crossed in submission, and what originally was interpreted as a white cloak, studded with five-pointed half-stars—the Teotihuacán Venus symbol—seems to be draped over his shoulders, identifying him as a Venus sacrifice. Certainly he is a victim of Venus-regulated warfare on his way to Morning Star misery, but Carlson demonstrated that his starry white robe is not a garment for death but a stylized rendition of the Star Chamber—the room where high-status captives were sacrificed with the sanction of Venus. According to Carlson, that room has been found. It is a closet of death, a small chamber on the west side of Cacaxtla's acropolis with interior access. Each of the two columns that front the room is painted with a mural of a bizarre blue but human figure that seems to be standing—or dancing—on tiptoe above a base illustrated with aquatic animals. One is male, the other female, and both are wearing a jaguar skirt with a Oaxaca Venus glyph apron. Their arms are fringed with feathers. The woman's head and much of her upper body have fragmented away, but we can still see that the man has eye rings like Tlaloc's goggles, white feathers radiating from his head, and a scorpion's tail. Teotihuacán Venus symbols accompany the frame of the column, and the scorpion man also has one in his upraised palm. There are other explicit portrayals of the Venus god with a scorpion tail from the Maya region, and so almost all of the details in this ominous cell at Cacaxtla indicate Venus sacrifice. Agricultural connotations—fertilization through penetration, rain, and corn—are, however, also part of Maya scorpion symbolism. Cacaxtla's Venus scorpion man stands, then, at the threshold of death and acts as one more witness for the blood sacrifices that nourished the gods, who through rain reciprocated with water, fertility, renewal, prosperity, and power.

All of this began at Teotihuacán, a place that was a centuries-old ruin by the time the Aztecs took their place in Mesoamerican power parade. To them, it was "the place where the gods began," and it is likely that the people of Teotihuacán regarded their city as the center of the world and the place of origin and creation. Many aspects of its monumental architecture and its overall layout offer circumstantial evidence in favor of that interpretation. Its most massive pyramid, which the Aztecs named the Pyramid of the Sun, is built on top of a cave that was used ritually and perhaps symbolized the place where people first emerged from the underground womb of Mother Earth. Astronomical and topographical alignments of the city welded human enterprise with the will of cosmic gods and the forces of nature. These ideas are advocated in a review summary by René Millon, who supervised the Teotihuacán Mapping Project, itself a monumental project that opened Teotihuacán to modern scholarly scrutiny. Millon also believes the best explanation of the orientation of the Pyramid

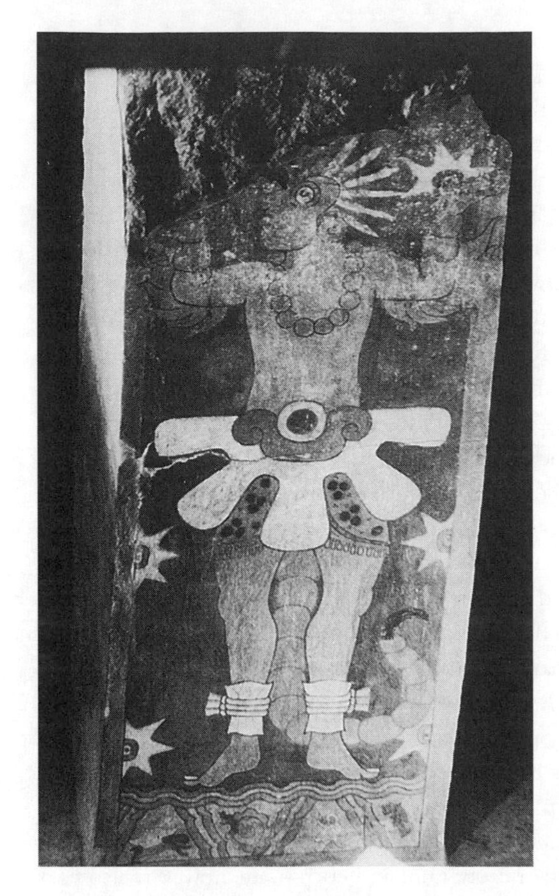

Cacaxtla's scorpion-tailed version of Venus is painted on a pillar that fronts a small chamber where prisoners of ritual warfare may have been sacrificed. His skirt is a Oaxacan Venus emblem, and the half-stars on the frame of the pillar and in his upraised hand are the symbol for Venus once used in Teotihuacán. (photograph E. C. Krupp)

of the Sun and the whole Teotihuacán city grid is the direction of sunset on the day when time, according to the traditional calendar, and the world began.

Teotihuacán was not only Mesoamerica's foremost economic engine and political powerhouse, it was a center of religious pilgrimage where residents, neighbors, and distant partners could see the power of the gods on display. It was, after all, as Millon puts it, "the custodian of the place where the universe began," the keeper of the cave of creation. Its architectural monuments demonstrated its rulers' power to mobilize labor and belief, and the mobilization of belief, through cosmic ideology, solidified the intent of an otherwise disparate and displaced urban population.

Teotihuacán's great public works tell us that the leaders who built them possessed the kind of power that is sustained by ideology. We suspect, but cannot confirm, that some of these rulers were buried in the Pyramid of the Sun and the Pyramid of the Moon, but the looted tomb below the center of the Pyramid of the Feathered Serpent contains the remains of 20 rich burials and probably included the ruler responsible for that monument and for the Tlaloc-Venus cult of war and sacrifice that accompanied it. The Teotihuacán Quetzalcóatl, represented explicitly as a plumed snake swimming in water and also as a serpent-head war bonnet, on the flanks of the Pyramid of the Feathered Serpent was allied with the rain and storm god in Teotihuacán's version of Venus warfare. Within the pyramid's foundations, at the corners and on-axis at each side, the bones of victims of ritual mass sacrifice guarded the elite burial in the center of a structure that stood for the world itself and for the mountain at the center that touched heaven and shed the water of life. Ornamented with the symbols of Tlaloc-Venus warfare, this pyramid announced, in no uncertain terms, the basis of Teotihuacán's supremacy—agricultural success and conquest-trade. The validity of this assessment is supported by a more recent discovery in the La Ventilla zone. A small chamber there, southwest of the Pyramid of the Feathered Serpent, is decorated with a band of Venus symbols, stylized blood drops, and storm-god goggles painted around the base of its walls. On seeing it for the first time, John Carlson identified the room as one of Teotihuacán's own Star Chambers.

Carlson studied and consolidated Venus warfare symbolism in the Maya region and then demonstrated its impact in central Mexico through analysis of the Cacaxtla

In 1990, Dr. David Webster uncovered a bench ornamented with a band of full-figure astronomical glyphs at Las Sepulturas, a residential suburb of Copan, the famous Classic Maya ceremonial center in Honduras. With an arm looped through a Maya Venus symbol and a scorpion's tail, his identity is certain. This bench is part of a modest platform where it is believed an elite official, of second-order rank, lived and presided. (photograph E. C. Krupp)

paintings. When he applied the same concepts to the iconography of Teotihuacán, he realized its performance as an expansionist trade and tribute state was propelled by astronomical ideology. The luxury goods collected for the city's inhabitants—jade, shell, and mica and presumably items unlikely to survive archaeologically, such as jaguar pelts, quetzal feathers, rubber, copal incense, and cacao—symbolized Teotihuacán's economic power and were privileges of commerce protected by force.

The ideology of war and sacrifice didn't end with the fall of Teotihuacán and the collapse of Classic Maya civilization nearly two centuries later. The same themes can be seen at Chichén Itzá in northern Yucatán, where the semilegendary events recorded in post-Conquest transcriptions of historical prophecy and the absence of dynastic inscriptions both imply the Postclassic Maya shared power in a council of joint rule.

Feathered rattlesnakes slither all over the architecture of Chichén Itzá. The stairways of four-sided platforms, pyramids, and other buildings are ornamented with them. They frame the doorways and support the entrance lintels of temples, and they coil through panels of carved relief that illustrate Chichén Itzá's campaigns of military conquest.

Feathered serpents and militaristic themes in the public art of Chichén Itzá once persuaded the experts that its original Maya inhabitants had been conquered and

At Chichén Itzá, a Postclassic center of Maya power in northern Yucatán, Kukulcán, the feathered serpent has the rattles lodged in his tail. They can be seen among the feathers at the level of the second step on this balustrade of the Platform of the Eagles and Jaguars. The head of the plumed rattlesnake protrudes from the top of the stairway. (photograph E. C. Krupp)

transformed by Toltec invaders from the Mexican highlands. Archaeological knowledge, however, now suggests that Chichén Itzá's "Toltec" art was not a late transformation of the site by invaders but part of an international style adopted and spread by those who regarded themselves as legitimate heirs of Teotihuacán's legacy, including Maya speakers of the Yucatán peninsula. The Itzá, who mobilized Chichén Itzá first as an inland foothold—and later as the capital of an expansionist state—were Putún Maya and merchant-warriors from Yucatán's northern coast. Economically and militarily, they successfully challenged the established centers of Maya power in east and west of northern Yucatán and dominated the northern lowlands from the ninth century until the thirteenth century, when rival cities, particularly Mayapán, advancing the same feathered serpent ideology, defeated Chichén Itzá in war.

Chichén Itzá means "at the mouth of the well of the Itzás." The well in this name is the Sacred Cenote, a large natural sinkhole north of the Temple of the Feathered Serpent—or Kukulcán, the main pyramid at Chichén Itzá. Offerings and human sacrifices were ceremonially consigned to this well, presumably to facilitate acquisition of supernatural power and put it at the disposal of Chichén's rulers. Rain for agriculture and military victory were their primary desires.

Certain natural features in the landscape like the Sacred Cenote, along with temples and shrines, were believed to be portals to the Otherworld, the spirit realm. Chichén Itzá's sacred well and imposing ceremonial architecture not only provided access to the gods for agriculture and conquest, they advertised those divine alliances to their competitors. The Itzá name, in fact, seems to contain a Yucatec Maya root referring to "sorceror," "wizard," or "shaman." The imperialists of Chichén Itzá were effectively promoting themselves as shamans in possession of a sacred well.

The Itzá inaugurated their military campaigns and commemorated their victories in a symbolic environment of monuments saturated with the imagery of their version of Tlaloc-Venus warfare. The exact meaning of all the feathered serpents, four-sided pyramids, four-staircased platforms, battle scenes and costumed warriors, predatory celestial bird mosaics, and jaguars on parade still eludes us, but we recognize in them symbols of power encountered throughout ancient Mesoamerica. Chichén Itzá seems to replicate the mythic themes of Teotihuacán and probably regarded itself as the place of creation and the center of the world. The steps in the four stairways of its main pyramid, the Temple of Kukulcán, probably totaled 365 and so symbolized the number of days in the year, while other decorative details incorporated more calendric numerology. Quartered by its sides and staircases, the pyramid mimics several crosslike symbols that respectively stood for cosmic directions, celestial intersections, the planet Venus, intervals of time, and the completion of calendrical cycles. The whole pyramid could have symbolized the mountain of creation at the center of the world, and one researcher, Clemency Coggins, argues that this structure was built as a foundation monument, involving the symbolic ignition of "new fire" on March 13, 840 A.D., at the close of one *baktun*, a 400-"year" calendric cycle, and the start of another. Although there is some dispute over this monument's era of construction, this date is possible. It also coincided with a reappearance of Venus as a morning star and is comfortably close enough to the vernal equinox, which usually falls around March 20, to support the claim that the Maya of Chichén Itzá intended their pyramid to display the equinox serpent of light-and-shadow that seems to descend the west balustrade of the north stairway in the last couple of hours before sunset. This phenomenon has now turned into a great public event, with its own modern cultural and political dimensions, and draws more than 40,000 visitors to the site each March. If this effect really were intended as an additional advertisement of a title to celestial power in the ninth century, the old lords of Chichén Itzá would probably be delighted to know their influence has not waned.

Other interpretations of Kukulcán's pyramid have been offered. Art historian Marvin Cohadas believed the northwest-southeast diagonal of the pyramid was intended to highlight the summer solstice sunset as part of a seasonally-tempered rain-bringing ritual involving the Sacred Cenote. Anthropologist Susan Milbrath emphasizes the orientation of the western face of the pyramid toward the direction of sunset on about May 25, the first day in the year the sun crosses directly overhead,

In the equinox season, an undulating, luminous profile forms from seven triangles of sunlight on the north side of the Pyramid of Kukulcán as the sun approaches the western horizon. The head of a feathered serpent, sculpted in stone at the bottom of the stairs, turns the display into a descending snake, intent perhaps on the Sacred Cenote farther to the north. We can't be certain the Maya celebrated a public ritual in conjunction with this event, but today tens of thousands of persons fill the plaza in the belief they are witnessing the seasonal return of the Plumed Serpent. (photograph E. C. Krupp, March 20, 1995)

through the zenith at noon. This event is traditionally associated with the transition from the dry season to rainy season. Milbrath also thinks that sunrise on one of the two days the sun passes through the nadir at midnight was significant, for the east side faces that direction.

Without abandoning any of these options, we should not forget the prominence of the feathered serpent. Whatever else this plumed rattlesnake means in terms of seasonal change, it is an incarnation of Venus, and just north of the Pyramid of Kukulcán, there is a smaller, four-sided platform known as the Platform of Venus. It is named for the Highland/Oaxaca-style Venus symbols on its wall reliefs, and it has four feathered-rattlesnake stairways. They are paired with bundled sticks topped by the trapeze-and-ray Mexican year sign that refers to the eight-year/five-cycle Venus almanac. A similar panel from an earlier version of the Venus platform actually frames the wrapped rods with eight dots to symbolize the number eight and ties the star to a bar, the Maya symbol for the number five.

Not far south of Chichén Itzá's Pyramid of the Feathered Serpent, there is a smaller version of the same kind of monument. It is called the Osario ("ossuary"), or sometimes the Tomb of the High Priest. It has four sides, four serpent stairways, and a temple on top, but it was built over a natural subterranean cavity that contained burials and offerings of ceremonial objects. This small cave beneath a pyramid is reminiscent of the consecrated cave beneath Teotihuacán's Pyramid of the Sun. The axes of the Osario are twisted, like the axes of the Pyramid of Kukulcán, from the true cardinal directions, but the Osario faces east-southeast instead of north-northeast. It is also associated, however, with an alignment of structures that mimics features found between the Pyramid of Kukulcán and the Sacred Cenote.

In the small plaza just east of the Osario, there is a low, round platform. If there were anything like this at the foot of Kukulcán's big pyramid, no trace of it remains. Both pyramids, however, are then fronted by a Venus platform. Beyond the Osario's Venus platform there is another, stairless platform. An elevated causeway connects each pyramid to a cenote. A natural outcrop of rock occurs on both lines, and each cenote has its own temple at the rim.

We have, then, two similar arrangements of sacred structures terminating at a portal to the watery underworld. They are, however, laid out in different directions. We don't know which of the two pyramids was constructed first, but there is a dated inscription on the Osario. One possible reading of this text provides the date June 20, 842 A.D. If correct, this is the earliest inscribed date at Chichén Itzá. It is also suggestively close to summer solstice. The calendrical content of Chichén Itzá's ceremonial architecture may, then, be more complicated than we think. No matter what the calendrics mean, the two different Venus platforms tell us Venus was part of the story of public ritual.

Certainly the peregrinations of Venus were part of the feathered-serpent war pageantry at Chichén Itzá, but how could the Itzá—and the Teotihuacános before

A relatively small relief, saturated with astronomical and calendrical symbolism, survives from an earlier version of Chichén Itzá's Venus Platform, which is located in the plaza, north of the Castillo, or Pyramid of Kukulcán. The symbol on the left, with starlike points, is a Venus symbol from Mexico's central highlands. It is looped to a bar, the Maya cipher for the number 5. What looks, on the right, like a tied bundle of bent rods is a reference to calendric closure, and it is topped by another calendric emblem, the Mexican trapeze-and-ray "year" sign. The eight dots around the bent and bundled sticks is a reference to the number 8. In eight years, Venus completes five of its 584-day Morning Star/Evening Star cycles, and the entire panel is a reference to that congruence. (photograph E. C. Krupp)

them—link the multiyear cycle of Venus with the annual cycle of agriculture? Recently, Ivan Sprajc, a Slovenian archaeologist working in Mexico, completed a comprehensive study of the connections between Venus, rain, and maize cultivation in Mesoamerica and their role in the ancient world view. Certain configurations of the planet, he confirms, recur seasonally. For example, Venus, setting as an evening star, would reach a northern limit once in each 584-day cycle, or five times in eight years, but the northernmost of those five northern settings always occurred in early May. That only happened once in eight years, but when it did happen, it was seasonally consistent. Behavior like this must have been noticed in the era of primordial agriculture, long before the establishment of great ceremonial centers, and even though Venus had no physical effect on the seasons, to the ancient farmers it looked like there was a connection. In a traditional cosmovision, everything that is congruent with an event is regarded as part of the supernatural pattern of cause and effect. Early on, people in Mesoamerica got the idea that Venus had something to do with seasonal change, the rains, and the growth of corn. Rain was regarded as a gift from the gods, a divine sacrifice that had to be repaid in human blood, and so sacrifice became part of agricultural calendar and celestial cycles. With the development of regional ceremonial centers, trade, war, and tribute complicated and enhanced the economy, and by the time political and economic power were concentrated in a state like Teotihuacán, the ritual components of agriculture and conquest were transformed into a standardized vocabulary of power that everyone understood and no one could ignore.

Proprietary Rituals of Celestial Power

Although the syntax of power is a product of several obvious influences—the natural environment, individual art and inspiration, and cultural history, the pattern of power's symbolic expression always involves controlled access. Access to these symbols, like access to power itself, is limited and reserved. Only the elect and the elite may activate them. Their effectiveness is linked, however, to their visibility, and so symbols of power—perhaps a pyramid, a palace, or a tomb—often operate on a large scale. Location itself carries meaning, and so symbols of power are linked to the symbolic landscape and help map it. Finally, the detailed imagery also always borrows from the natural order and from the sources of power that appear to stabilize the world and energize the cosmos.

It is possible, however—and also necessary—to maintain power through symbolic activity in the landscape of cosmic power. That, in fact, was the consequence of seasonal ritual in imperial China. Lawfully designated by heaven, the emperor performed ceremonies at suburban altars outside the palace and outside the walls of the nuclear city. During the Ming (1368–1644 A.D.) and Qing (1644–1911 A.D.) dynasties, the palace was in Beijing. China was the "Middle Kingdom," the land in the center of the world, and Beijing, the capital of China, was the center of the center.

In China, the north celestial pole, at the center of the turning sky, was regarded as the visible face of Shang di, the unseen power of heaven, and the emperor was the terrestrial counterpart of heaven. The Imperial Palace, or Forbidden City, corresponded on earth to the zone of circumpolar stars that negotiated the corridors of power around the north celestial pole. The Chinese called the palace the "polar forbidden city," and their name for the region around the unmoving hub of heaven was the "forbidden polar palace." Those stars that surrounded Shang di in his forbidden palace were like the court that hovered around the emperor.

China's fundamental principle of cosmic order was the daily rotation of the sky. Singling out the pole naturally emphasized the cardinal directions, and the basic layout of the Chinese capital was built upon a north-south axis that focused the city on the palace and on the emperor, who held audiences from a throne in the Hall of Supreme Harmony. It was there he presided over the ceremonial inauguration of the New Year. Although commoners were not permitted in such places—it was, after all, the Forbidden City—people knew that the emperor interacted with the gods on behalf of the entire country. Attendance at royal functions reflected the hierarchy of power, and even those privileged to attend such events had to approach the throne from the south. By occupying a place that was north as far as the audience was concerned, the emperor restated the analogy with Shang di and the north celestial pole.

At the proper times, however, the emperor would depart the palace and carry out rituals designed to maintain the dynamic balance of cosmic forces in a world of ongoing change. This change was most evident in the seasons and in the cyclical patterns of time. To be effective, the emperor had to carry out his duties at those key moments when the balance was either jeopardized or in flow. For example, at the New Year, which was timed through a complicated combination of progress of the sun and the moon, the emperor went to the south side of the city and offered sacrifices at the Hall of Prayer for Good Harvests (or Hall of Yearly Prayer), the largest and most elaborate building in a complex of temples, altars, and auxiliary buildings known as the Temple of Heaven. An open-air altar, usually called the Round Mound, in the same precinct was the site for the emperor's Great Sacrifice at winter solstice. Everything he did then was intended to enhance *yang*, the active "male" celestial character of the cosmos in the Chinese concept of complementary opposition. In winter, *yin*, the passive "female" terrestrial aspect of the universe was dominant. The sun's reserve of yang was at its ebb as it reached the southern limit of its annual excursion, and it was the emperor's duty and prerogative to replenish the cosmos with yang through magical ritual saturated with symbolism of the sky. Six months later, when yin was on the run, the emperor performed a summer solstice sacrifice at the Altar of the Earth, on the north side of the city. On that occasion, all of the symbolism—color, number, direction, and behavior—referred to the earth, or really to the watery underworld power of the realm below heaven.

The color of heaven, blue, dominated the winter solstice ceremony, while all of the symbolic materials at summer solstice, including the emperor's robes, were yellow, the color for earth. Red, the sun's color, prevailed at vernal equinox. White, for the moon, was the color at the autumn sacrifice. The emperor commuted to the Altar of the Sun, east of the Imperial Palace, on the vernal equinox, and performed the sacrifices

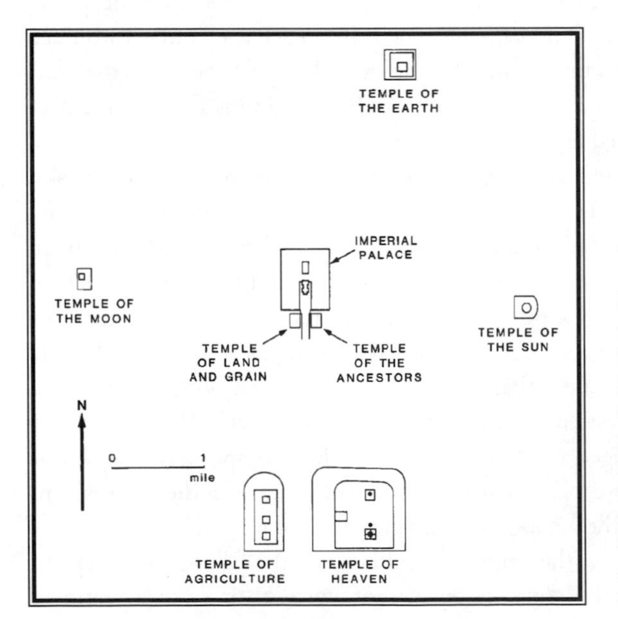

Suburban temples for imperial ritual surround the Imperial Palace of Beijing. Dedicated to heaven and to earth, to the sun and to the moon, they provide stages for seasonal sacrifice, and periodically the emperor would return the ceremony to the center at the Temple of Land and Grain and at the Temple of the Ancestors. (Griffith Observatory drawing, Joseph Bieniasz)

while facing the direction of the rising sun. The Temple of the Moon, in the western suburbs, was visited at the autumnal equinox for offerings to the autumn moon.

Color symbolism of a different sort was incorporated into another altar for imperial ceremony. The Altar of Land and Grain, just outside the Imperial Palace's Meridian Gate, and close to the center of the city, was divided into five sections—one for each quarter of the empire and a square zone in the middle for the center. Unlike the other stages for imperial ritual, this altar did not have a hard, stone surface for the emperor and his paraphernalia. Instead, the top of the altar was soil, and the soil in each section had a different color: blue-green in the east, red in the south, white in the west, black in the north, and yellow in the center. These colors belonged to a system of symbolic correspondences with directions, seasons, talismanic animals, zones in the sky, the five fundamental "forces" or "elements," geographic landmark mountains, planets, atmospheric conditions, grains, musical notes, and many other aspects of the world. Anchored in cardinality, these associations turned the cosmos into a

By 1992, the Chinese had restored the altar surfaced with soils of five different colors in the Beijing's Temple of Land and Grain (the *She ji tan*). It was originally commissioned in 1420 by Emperor Yong le. He and subsequent emperors returned the ritual calendar to the heart of the city when they performed "great sacrifices" here in the second and eighth moons of the year on behalf of the productivity of the growing season. The potted plants and the freewheeling dragons are probably not quite what the emperor used to encounter when he came here, but they have more charm than the decrepit generator that occupied the center of the neglected altar before it was renovated. (photograph E. C. Krupp)

cohesive, ordered experience, and it was part of the emperor's job to uphold that cohesion and order.

The emperor's excursions to the Altar of Land and Grain took him to a symbolic center of the living land. It was not the same kind of center as the palace itself, which was the center of imperial authority, but it was a navel where the Emperor's seasonal magic nourished the world in the second and eighth moons of the year. It periodically brought the calendar of ritual back to the city's center.

Balanced symmetrically with the Altar of Land and Grain, east of the primary axis of the Imperial Palace, the Temple of the Ancestors put the emperor in touch with the honored spirits of his own lineage. In the same sense as the spirits of the land and grain, the emperor's lineage resided in the center of the city, and he performed sacrifices on their behalf five times a year—on the first day of the first, fourth, seventh, and tenth moons and on the twenty-ninth day of the twelfth moon. Both altars provided a platform for the emperor to announce a new project or military enterprise with prayers for a successful outcome and to thank the tutelary spirits for success in war, agriculture, or some other major endeavor.

The sacred imperial shrines of Beijing reflect a hierarchy of cosmic power. Heaven and earth top the list. The ancestral spirits follow, and behind them march the spirits of land and grain. Sun and moon were not always accorded imperial respect, but once their worship was renewed with their own altars, they fell in line right behind heaven and earth. The emperor could delegate some of these sacrifices to other high-level officers of his court, but the winter and summer solstice rites belonged to him alone. No one but the emperor could approach Shang di directly. That was evidence of imperial prerogative, but these ceremonies also made it clear that the emperor's power was not absolute. He was not only the mandated delegate of heaven, he was heaven's servant. Exalted by imperial ritual, the emperor was also visibly subordinate to greater powers. His rule could be neither arbitrary nor unjust. His pilgrimages to the altars outside the protective walls of the inner city eroded the insulation of the palace and put him to work. It was a privileged vocation, but it also advertised the limits of power.

THE SHADOW OF IMPERIAL POWER

Imperial power was demonstrated in China through the emperor's announcements of the calendar, declarations of intercalations, and performances of seasonal or calendrically scheduled sacrifices to cosmic forces of nature. In Rome, on the other hand, Augustus Caesar displayed his status and the power of the empire with an Egyptian obelisk that put his control of the calendar and his integration with the fundamental order of the world in plain sight on a giant sundial inscribed into the pavement of the Campus Martius.

According to the Roman writer Pliny, Augustus retrieved his obelisk from Heliopolis, the center of ancient Egypt's sun cult, and erected it in 10 B.C. on a huge grid marked with gilt metal. This network of lines and curves included a north-south graduated meridian that showed the changing length of the sun's noon shadow and so qualified the white marble pavement as a solarium. The rest of the inlaid design was a horologium that revealed the hour. Although it has been speculated that the Egyptians may have used some obelisks as giant gnomons in this way, there is no evidence from Egypt to support the notion. Augustus must have gotten the idea somewhere, but the source of his inspiration remains unknown. Egypt's calendar, however, did travel on one of all those roads that led to Rome.

Julius Caesar had been sufficiently frustrated with Rome's awkward and disengaged lunar calendar to commission an Egyptian, Sosigenes of Alexandria, in 46 B.C. to adapt the civil, solar calendar of Egypt for use in Rome. With 365 days comprising each year, the system added an extra day every four years to keep the calendar synchronized with the seasons. It is known as the Julian calendar, and it wasn't superseded until 1582, when Pope Gregory promulgated a reform and advanced the date by 11 days.

A reliable, accurate calendar is essential for organizing resources and mobilizing activity, and no empire can afford to be without one. Simply by functioning properly,

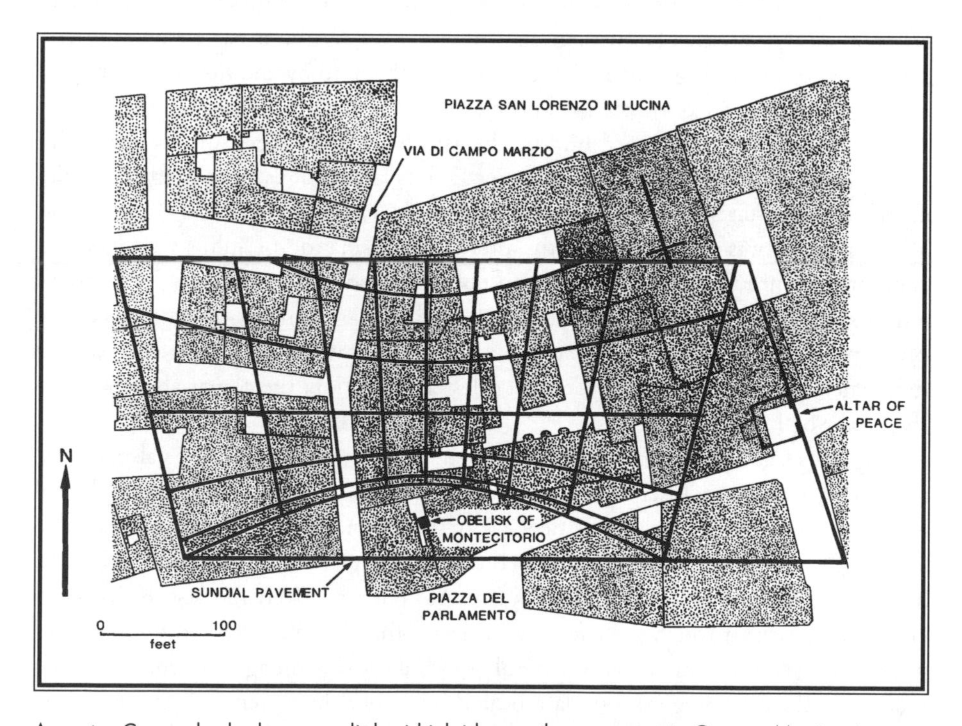

Augustus Caesar had a large sundial grid inlaid upon the pavement in Campus Martius, in Rome. Although the pavement is now beneath buildings near the Piazza del Parlamento and the Piazza San Lorenzo in Lucina, it still survives. The Egyptian obelisk Augustus enlisted to cast an imperial shadow is also still around, reerected in the Piazza di Montecitorio, not far to the south of the area mapped here. (Griffith Observatory drawing, Joseph Bieniasz)

the calendar—and uniform timekeeping—are transformed into symbols of the power and effectiveness of the state. When Augustus put a monumental sundial in the Field of Mars, he wasn't building a better chronometer, he was putting an image of empire on display.

German archaeologist Paul Zanker has analyzed the relationship between visual symbols and the Augustan program to revitalize a war-weary, pessimistic, and increasingly decadent Rome. Uneasy with new burdens of empire, transformed by the intrusion of foreign influences, and uncertain about the future, Rome was ready for an effective leader who would both symbolize what was best in Roman civilization and have the talent to carry out necessary reform. Zanker's book, *The Power of Images in the Age of Augustus,* details how Augustus Caesar's reign brokered a new mythology. It gave Rome 45 years of unprecedented prosperity, tranquility, and belief in the future. Under Augustus, family values became Roman law. He accomplished this as a participant in the belief system, not as a cynical manipulator. Once his power had been consolidated, Augustus had the good sense to project an image of restrained statesmanship and credible piety. In fact, historians of his era portray his lifestyle as dignified and simple. He was disciplined and respectful . . . and always in command.

Manipulation, of course, was also necessary, but societies always manipulate themselves in an unconscious Darwinian quest for survival. Without real belief, the symbols have no power. Augustus had already been influenced by and taken advantage of cosmic symbolism that reflected his belief. When he was still a student, the spontaneous worship directed toward him by the astrologer Theogenes, on examining his conception horoscope, left a big impression. Much later, Augustus minted coins with Capricorn, his sun sign. It and his birthday were judged bearers of good fortune. When a comet was seen a few months after the assassination of Julius Caesar, Octavian (at that time the Roman Senate had not conferred on him the title "Augustus," with its connotations of semidivinity) promoted Caesar's connection with the comet on coins and statues. Augustus was Julius Caesar's great nephew, adopted son, and designated heir. As evidence of Caesar's celestial exaltation, the comet also enhanced the honor of his heir.

Augustus Caesar's monumental timepiece in the Campus Martius was also a celestial political symbol. Closely affiliated with the *Ara Pacis Augustae,* or Augustan Altar of Peace, the sundial announced the new age of peace and social stability. The marble altar, enclosed within a rectangular structure with an open roof, was originally located just east of the sundial's pavement template. Although relatively modest in size—the enclosure measured roughly 34 feet by 38 feet—the altar and its surrounding walls were elaborately decorated with symbolic and allegorical images, including scenes relating to the mythological foundation of Rome and a divine female personification of peace flanked by Tellus, goddess of the productive power of the earth, and Venus, goddess of fertility and love. Intended by the Roman Senate to honor Augustus for the peace achieved with his military victories in Spain and Gaul, the altar was going to occupy a place in the Curia, the house where the Senate met. Perhaps judging that location to be loaded with too much presumption, Augustus insisted the monument be built in the Campus Martius, an altar of peace in the field of Mars. The altar was consecrated on July 4, 13 B.C. (a procession of the imperial family on that day is carved in relief on the outside of the south wall), and completed and dedicated on January 30, 9 B.C. It is thought the altar was sited close to the corner of Via in Lucina and Via del Corso, in the area north of the Pantheon. Over the centuries, the altar fragmented and disappeared. Scattered pieces of it were finally recognized in the nineteenth century, and in 1938 it was reconstructed in Via di Ripetta, closer to the Tiber River, where it is still displayed.

The *Solarium Augusti* also survives, but not intact. The gridded pavement, with names of the winds at appropriate places on its perimeter, the names of the zodiac signs along the date-indicating meridian, and other useful legends, is now below the buildings of today's Rome. It has been studied, and its plan reconstructed, by Edmund Buchner, a German archaeologist. The Egyptian obelisk, no longer in position, probably occupied a spot now covered by the building that fronts the north side of the Piazza del Parlamento and near its junction with Via in Lucina. Like the Altar of Peace, the solarium obelisk toppled and disappeared. Excavations relocated it in 1587. Reburied, the red granite needle was found once more in 1748, but it was not reerected until 1792, when

Augustus Caesar's time-sharing obelisk originally came from Heliopolis. After the *Solarium Augusti* ceased to function, it was permitted to go to ruin, and for centuries the fallen obelisk was lost. Now reerected in front of Bernini's seventeenth-century Palazzo di Montecitorio, the obelisk cedes its former timekeeping responsibility to the clock on the wall. (photograph E. C. Krupp)

it was placed in the Piazza di Montecitorio. It can be seen there today, just a piazza south of its original position. Now about 72 feet tall, the obelisk has probably lost some of its original height. At the time of Augustus, it may have been closer to 95 feet. An Augustan inscription on its pedestal mentions the emperor, his Egyptian triumph, and the dedication of the sundial to Sol, the deified sun. Hieroglyphics on the obelisk inform us the ancient Egyptians had something similar in mind. That inscription glorifies the pharaoh, Psamtik II (Saite, Dynasty 26, ruled 595–589 B.C.), with references to the sun. According to the text, he is "the Golden Horus," "lord of Heliopolis," and "beloved of Re-Harakhte (the rising sun)."

According to Zanker, the annual performance of the Solarium Augusti included a special alignment of the obelisk's shadow with the Ara Pacis Augustae on the emperor's birthday, September 23. Actually, this doesn't make a lot of sense. It must happen when the sun is in the southwest, sometime between noon and sunset, and the sun will actually attain the right azimuth at different times in the afternoon throughout most, if not the entire, year. Buchner, on the other hand, judged that a bronze device placed on the summit of the obelisk reflected September 23 sunlight onto the altar when the shadow was aligned and formed a suggestive luminous figure there. If the speculation is correct, Augustus may have been telling his countrymen it was the luster of his own heaven-charmed birth that brought the great era of imperial peace to Rome. When it comes to an ordered empire, it pays to say it in sunlight.

CHAPTER TEN

UPWARD MOBILITY

Roman emperors, like other rulers, were upwardly mobile seekers of celestial status. They allied themselves with the sky to boost their prestige and consolidate their power with the divinity of heaven. When they died, the status and mandate of their successors were enhanced, in turn, through verification of the dead king's divine celestial franchise. For that reason, monuments that acknowledged dynastic links with heavenly powers and rituals that celebrated the king's commerce with lofty gods were predictable priorities in any ideology concerned with the continuity of established power.

For example, the divinity attributed to Augustus Caesar during his lifetime, and expressed in the state cult of the deified emperor, was fortified with more celestial imagery at his death. At the funeral, an ascending eagle, liberated when the pyre was ignited, was understood to symbolize his soul on its way to the sky. In fact, when the Roman Senate met to confirm the deification of Augustus posthumously, official testimony under oath included one senator's eyewitness report of Augustus rising up to heaven. His migration to the sky was regarded as clear evidence of his transformation into a god.

Immortal like the unchanging stars, the triumphantly returning sun, and the rejuvenated moon, the gods, of course, had a right to reside in heaven. As a god, Augustus, too, belonged there after death. In his era, however, the destiny of most mortal souls was neither celestial nor immortal. The shade of the average person instead resided in a subterranean netherworld that made contact with the land of the living at the door of the tomb. In time, the soul of the deceased would disintegrate spiritually and cease to exist, just as the body decomposed physically.

The concepts of celestial divinity and of heaven as the final destination of the worthy soul seeped into the Mediterranean world from Egypt and Babylon. A belief in the deceased pharaoh's translation to the sky was at least as old as Egypt's Old Kingdom (2686–2181 B.C.), and in the East, Persian belief in the celestial privilege of the soul merged with the astral religion of the Chaldaeans and amplified the concept of heavenly immortality. These ideas had actually reached Greece long before there was a Roman emperor to be launched into heaven, but the sky remained off limits for the ordinary soul until religion and philosophy driven by the quest for personal salvation took root in the Roman world.

In the meantime, the institutionalized divinity of the emperor provided a vehicle for legitimizing the rule of Rome at home and in the distant, conquered territories, where it was often grafted onto indigenous beliefs. This process domesticated foreign ideology and made citizens out of adversaries. A great altar, for example, was dedicated to the worship of Augustus and Roma, the divine personification of Rome, in the city in southeastern France now known as Lyon. To the Romans, it was Lugdunum, the capital of a province of Celtic Gaul. The Latin name includes the root *Lug*, which refers to an important god of the Celts—Lugh, the "Shining One." When the Romans transformed the city of Lugh into a provincial administrative center, all of the chiefs of the area's Gaulish tribes were summoned to Lugdunum in conjunction with the altar's dedication on August 1. This date coincided with a traditional Celtic festival, the Lughnasa. Intended to acknowledge the opening of the season of harvest, the holiday survives as Lammas in the church calendar in Britain. For Augustus, the date commemorated his triumphant entry into Egypt and was, of course, the premier day of the month named in his honor. Stephen C. McCluskey, an historian of science who has studied the persistence of pagan calendrics and astronomy into the Medieval era, believes the Romans contrived their Altar of Augustus in Lyon to line up with an avenue that reached east-northeast from the old Roman amphitheater. If so, it framed the rising sun between its two monumental columns on a day that was mutually significant to the Romans and to the subjugated Celts.

HANDSHAKES IN HEAVEN

You didn't have to be a Roman emperor to stake a claim of divinity in the sky. Antiochus I, the ruler of a small but wealthy kingdom that buffered Rome's eastern frontier from its Parthian rival, transformed the top of a mountain into a tomb intended to place him in the company of the highest gods. Located in what is now southeastern Turkey, this peak is called Mount Nemrut, or Nemrut Dag. With an elevation of 7,053 feet, it is the highest summit in the upper reaches of the Tigris and Euphrates

Rivers, but Antiochus I capped it with another 164 feet of piled, loose rock. Already more or less isolated from the neighboring range, the peak is a conspicuous landmark, and its artificial apex endows it with a distinctive profile. From a distance, this tapered tumulus seems to grow naturally from the mountain's own naked slope. I have picked it out easily from the town of Andiyaman, 51 miles away.

Antiochus could have seen his final resting place from at least half of his domain, but the land he ruled was not very big. Known as the Commagene Kingdom between 69 B.C. and 72 A.D., it was part of the region sometimes recognized as "Northern Syria" or "upper Mesopotamia." At its greatest extent, during the reign of Antiochus I, it encompassed little more than the plain between the Taurus Mountains on the north and the Euphrates River on the south, a territory roughly 140 miles long and 30 miles wide. Its capital, Samosata (Sun Capital), was 34 miles southwest of Nemrut Dag, and from the Euphrates, it enjoyed a view of the sanctified mountain. Although Samosata was a significant archaeological site, it is now unfortunately flooded by the waters of the Atatürk Dam.

Intentionally exploiting the height and visibility of Nemrut Dag, Antiochus I built what he called a *hierothesion*—a tomb consecrated as a shrine through its dedication to the gods. His sepulchre is believed to be somewhere within the huge heap of stones, which is almost 500 feet in diameter, but the burial chamber has never been found. Extensive inscriptions, however, verify the mountaintop monument belongs to him. The texts, carved in Greek and Persian, establish the stratospheric bloodline of the Commagene king and also name the gods who escort him on the mountain.

Through his mother, Antiochus traced his ancestry to the Macedonian rulers of the Hellenistic Seleucid Empire and even to Alexander the Great. The roots of his father's lineage were Persian and as imperial as Darius the Great. At one time, the province had belonged to the Persian empire. After Alexander defeated Darius II, Commagene dropped into Macedonian control. When the Seleucid Empire began to disintegrate, Rome moved in. After some initial conflict, Antiochus I cut a deal with the Roman general Lucullus, who allowed him to remain in power. Commagene continued to do what it had always done best. It controlled the passes through the Taurus Mountains of eastern Anatolia and monitored a crossing of the Euphrates. Goods routed between Persia and the Ionian coast all moved through Commagene, and its prosperity was linked to this royal trade route. In addition, the land was an agricultural cornucopia. Well watered by the river, its pastures delivered the goods. Although its military power was limited, Commagene could usually maintain a respectable local autonomy thanks to imperial expediency and to its own agribusiness and international trade.

Antiochus pictured himself as a fusion of Greek and Persian civilization, and the colossal statues that accompanied him on top of Nemrut Dag represent gods in which Eastern and Western tradition are unusually and deliberately combined, in content and in style. They belong, in part, to the Zoroastrian doctrine the Iranians carried to Asia Minor. Mithraic religion developed from these ideas, and, in fact, the

Enthroned gods, beheaded by time, occupy the east terrace of the monumental tomb of Antiochus I of Commagene, on the summit of Nemrut Dag, in southeastern Turkey. The artificial mound believed to contain the king's burial chamber rises behind the giant statues, and some of their fallen heads may be seen on the ground below the platform. (photograph E. C. Krupp)

shrine on the summit of Nemrut Dag is the oldest known Mithraic monument. Its Parthian gods, enthroned on the east and west terraces that flank the tomb, comprise a colossal who's who of hybrid Greek and Persian divinity.

Zeus-Oromasdes combines the highest god of the Greek pantheon, the king of the Olympian gods, with a variation in the name for Ahura Mazda, a universal celestial creator god of ancient Persia. Both are supreme rulers in heaven and in the Commagene context are divine sponsors of kingship.

Haloed with a sunburst, the god Antiochus called Apollon-Mithras-Helios-Hermes links the deified visible sun (Helios) with the spirit of divine inspiration represented by Apollon (or Apollo), who by this time was also identified with the sun. These two Hellenic gods were then equated with Mithra, who had acquired some solar connotations of his own. Mithra's original association with light and justice make his subsequent evolution in Commagene easy to understand. As a celestial judge and the supernal exponent of ethical behavior, Mithra also championed the sanctity of contracts and covenants. Hermes operated in part as a Greek sponsor of commerce, and the contractual element of mercantile enterprises partly accounts for his inclusion in this multidimensional deity. Mithra's responsibility to protect the "Divine Glory" of the king makes him a defender of sovereignty. He sustains the tenure of righteous leadership.

The third god seated on the summit with Antiochus is Ares-Artagnes-Heracles. Ares was the Greek god of war, and Heracles, a hero and demigod, also enforced will through strength. Artagnes played a comparable role in Persian myth. His name on Nemrut Dag is a variant of Verethragna, who was a warrior spirit of military victory. As he moved west, he acquired the attributes of a hero like Heracles.

Finally, Antiochus included a goddess in the transcendental meeting on the mountain. He called her "my homeland, the Fertile Commagene." Personifying the wealth and productivity of the land, she shared some traits in common with Fortuna, the Roman goddess of happiness, well-being, and good fortune. She probably incorporated the fertility of Venus and the abundant output of Mother Earth. Crowned with pomegranates, grapes, and other fruits and grain, she is the Carmen Miranda of Commagene.

The statues are immense, even though all of their heads have toppled to the ground. Now upright, the heads—some of which are ten feet high—seem to emerge from the rocky terrace. The old gods look like they're up to their necks in mountain. Along with immense statues, including one of himself, Antiochus erected statues of eagles and lions as allusions to divinity and sovereignty, respectively. The words he had carved explain why he, too, is there: "I have placed in the middle of the seats of the benevolent gods, the statue of my own self, associated by good fortune with the ancient majesty of the gods and wishing to show constant preoccupation which I have vowed to the immortal gods, which more than once have visibly helped me and have shown themselves favorable to the undertaking of my reign." According to the gods' wishes and according to their will, Antiochus the king has done the bidding of the divine.

Several panels that illustrate in carved relief the king receiving a handshake from each of the gods accompany these great statues. This camaraderie with divine powers is a common theme in the memorial art of Commagene royalty, and it publicizes the king's own exalted status. The gods greet him as a partner in the pantheon and extend to him their friendship and loyalty. He is confident and comfortable in their company. Divine fellowship was, of course, what Antiochus had in mind when he built his skyscraping tomb. Declaring his divinity in death, he called himself "The great King Antiochus, righteous God," and described his action as worthy of "the superiority of celestial spirits." He also explained his own celestial ambitions: "Thus I justify my intention in erecting close to the celestial thrones and on foundations inaccessible to the ravages of time, this hierothesion where my body, after having aged in the midst of these blessings, will sleep in eternal rest separated from the pious soul flying off toward the celestial regions of Zeus-Oromasdes."

One more terrace relief depicting a star-studded lion informs us that the superior celestial spirits on top of Nemrut Dag possessed planetary counterparts. Three "stars" above the lion's back are named Pyroeis of Heracles, Stilbon of Apollon, and Phaethon of Zeus. We know that these names refer to Mars, Mercury, and Jupiter, the celestial talismans of Ares, Hermes, and Zeus. The unambiguously astronomical

Antiochus I glad-handed the gods on top of Mount Nemrut. These reliefs on his tomb's western terrace show him in the company of three of them. In each case, he is the figure on the left. The panel on the far left joins him with Apollon-Mithras-Helios-Hermes. To the right of that, he greets Zeus-Oromasdes, and still farther to the right, Ares-Artagnes-Hercules offers him five. The lion to the right of the sequence of divine handshakes represents the constellation Leo. The moon is symbolized by the crescent that hangs below the lion's mane. Three stars above its back are named by accompanying inscriptions as the planets Mars, Mercury, and Jupiter. (photograph E. C. Krupp)

character of the relief also tells us the crescent dangling from the lion's neck is the moon, perhaps a celestial stand-in for the fourth deity, the goddess Commagene. Although no lunar dimension for her is known, the crescent may parallel her presence on the mountain, for the moon often carries connotations of women and growth. In any case, celestial powers are gathered in a starry lion like the gods are gathered on the mountaintop, and the lion must be the constellation we know as Leo. Several attempts have been made to date this lunar-planetary configuration in Leo, and the most-favored contender is July 7, 62 B.C. This is believed to correspond to the "coronation" of Antiochus—the date when the Roman general Pompey the Great turned Commagene into a buffer by officially acknowledging its king. Even though the meaning of the celestial event is uncertain, its commemoration on the mountain reaffirms that Antiochus governed with the approval of celestial power.

Antiochus declared two national holidays in perpetuity—his birthday and the anniversary of his coronation. They were to be observed both annually *and* monthly after his death. He provided financially for a cult priest who would celebrate the festivals on the mountaintop with the banquet offering open to all. These ongoing devotions may have been intended to honor the gods, but they also reinforced the tradition of dynastic sovereignty.

After nearly two thousand years of mountaintop exposure, the ruins of the monumental tomb of Antiochus I are not exactly inaccessible to the ravages of time. No one is religiously celebrating his red-letter days, and there is no longer a free lunch on the terraces of Nemrut Dag. The parade of tourists, however, now brings more pilgrims pursuing the cult of celestial kingship than Antiochus could have imagined in even his most grandiose dreams.

PYRAMID SCHEMES

When the Old Kingdom (2686–2181 B.C.) pharaohs of Egypt died, it wasn't just the kingdom of the Nile that was threatened with disruption. The whole universe was at risk, and even in death, the pharaoh had to take charge. According to the *Pyramid Texts,* inscribed upon the interior walls of pyramids of the Fifth and Sixth Dynasties (2498–2181 B.C.), his soul, or *ba,* rose into the sky and became "a living star at the head of his brethren."

In heaven, the pharaoh does what the stars do. He maintains the calendar, administers the seasons, and regulates the hours. When the dead king reaches his celestial destination, he "takes possession of the sky, its pillars and its stars." His pyramid tomb was his point of departure for the sky and the magical device that dispatched him to heaven.

A trip to the dictionary informs us that a pyramid is—first and foremost, in ancient Egypt—"a quadrilateral masonry mass having smooth, steeply sloping sides meeting at an apex, used as a tomb." Scarcely pausing, the dictionary continues, "(in ancient Egypt and pre-Columbian Central America) a quadrilateral masonry mass, stepped and sharply sloping, used as a tomb or a platform for a temple." It is a relatively simple—but somehow evocative—geometric shape.

Massive platforms with steeply sloped sides are actually found in many parts of the world, and many of them incorporate cosmological ideas through symbolic design elements. These cosmological aspects of the pyramid forged a connection with heaven. In mimicking a mountain, the pyramid linked earth and sky. To the elite, it provided symbolic access to the celestial realm.

Certainly the pyramids of Egypt are the best known. All of the pyramids of the Old and Middle Kingdom are located between Abu Rawash (about six miles north of Giza) and the Faiyum (about 70 miles south of Giza), and they are explicitly linked with royal mortuary activities and the cult of the dead king. Altogether, there are more than 70 pyramids in Egypt, and all of them were built in the western desert beyond the west bank of the Nile. Their placement on the west side of the river reflects an obvious directional association with the dead. Celestial objects set in the west, and their disappearances below the horizon are—like death—departures from the realm of the living. In the same way, sunrise and the appearance of a star on the eastern horizon were equated with rebirth.

Monumental architecture is often intended to mirror cosmic order, and the square base of the pyramid and the cardinal orientation of its sides are allusions to the framework that organized and sustained the Egyptian universe. Those "pillars" of the sky mentioned in the *Pyramid Texts* are the cardinal directions.

Egyptian pyramids are conspicuously uniform in orientation to the cardinal directions. Although the accuracy of alignment varies—with the Great Pyramid claiming the highest honors—the intention to conform with true cardinal orientation is evident. The greatest departure of the Great Pyramid from true alignment is 5½ minutes of arc, or about one-sixth the diameter of the full moon. It is possible, of course, to survey a line at this accuracy, even with simple techniques and the unaided eye, but maintaining it in a minimountain of multiton limestone blocks that once towered 481 feet above the desert and measured 756 feet on a side is an engineering, construction, and management achievement.

Most of the pyramids have entry shafts that descend from the north side, and popular descriptions of the pyramids often allege that these corridors were aligned with the north pole of the sky or with the Pole Star. The truth is that they do not, although they are generally oriented toward the zone where the pole and nearby cir-

Three large pyramids dominate the Giza plateau. The one in the distance, on the right, is the largest—the Great Pyramid of Khufu. His son Khafre built the one in the middle, and the pyramid of Menkaure is the smallest of the three. (photograph E. C. Krupp)

cumpolar stars are to be found. These circumpolar stars were known in Egypt as the "imperishable" or "undying" stars because they never set, and a symbolic, if not precise, link with them may have been intended with the entry's placement on the north. The angle of descent was most likely determined by simple geometric principles and not by celestial targets. There is, however, more astronomical alignment in the Great Pyramid, and it may help explain the symbolic function of pyramids in general.

Larger and more famous than the rest of the entire pyramid population, the Great Pyramid was already ancient in the time of the ancient Greeks, who recognized it as one of the wonders of the ancient world. It still claims that status, and of the original Seven Wonders of the World, it is the only one left. It was built a little after 2600 B.C. and has beat every challenger from ancient armies to the Grateful Dead. All kinds of astronomical interpretations have been imposed upon it. Some have involved the arrangement of its interior passages and others have invoked the shadows it casts. It has been an astronomical observatory to some and a prophecy in stone to others. Arcane measurements of its height, perimeter, and inner corridors have supposedly revealed the length of the solar year (which presumably the Egyptians could write on papyrus anytime they wanted to do so) or past events in the parade of history and the shape of things to come. Even the shape of the Great Pyramid has been enlisted into pseudoscientific service as a lens that mysteriously concentrates unknown energy. In a continuing quest to conserve stainless steel, advocates of pyramid power resharpen razor blades through pyramid empowerment.

The interior of the Great Pyramid departs from the simple geometry of its profile on the Giza plateau with an enigmatic arrangement of passages, chambers, and halls, in some cases oddly kiltered and in others, on the level. Although the modern names of the rooms and galleries are somewhat romantic—the Ascending and Descending Corridors, the Queen's Chamber, the Grand Gallery—the titles the ancient Egyptians actually gave to them were probably even more evocative. For example, to us, this entire artificial pharaonic mountain is the "Great Pyramid." To the king who built it, it was the *Akhet Khufu*—or "Horizon of Khufu." The Great Pyramid's connection with Khufu (2589–2566 B.C.), the second pharaoh of the Fourth Dynasty (2613–2498 B.C.), has never been forgotten. Herodotus, the Greek historian of the fifth century B.C., knew whose tomb it was more than two millennia after it was completed, and thanks to Herodotus and to hieroglyphics scratched on its stone blocks, we still know today.

Inside the Great Pyramid, architecture and ritual seem to be focused in the room known as the King's Chamber. Unlike the limestone bulk of the pyramid, this room was built of massive slabs of granite, which is far more durable and harder to cut and shape. The limestone came from local quarries, not far from Cairo. The red granite, on the other hand, was transported from Aswan. Although this was the pyramid builders' closest source, they still had to sail 500 miles upriver to get it. They wouldn't have bothered if the chamber hadn't meant something special to them.

Without the granite sarcophagus that remains in the King's Chamber, that meaning might completely elude us. The room contains no murals or hieroglyphic texts to reveal its purpose. Two odd apertures in the walls, however, provide one more clue. One of these ports is on the north side of the room, and the other is on the south. Modest in size, each could almost be covered by a copy of this book. They are, in fact, one of the most remarkable features of the Great Pyramid, for they are both an interior opening of a duct that first penetrates a few feet horizontally through the pyramid's stone core and then turns sharply upward and adds another odd angle to the Great Pyramid's interior design as it drives through tons of limestone blocks. Finally, each shaft bends horizontally again, and then opens high up on the outside slope. Sometimes explained as ventilators to deliver air-conditioning to the dead, the real meaning of these peculiar shafts is coded in their different and awkward angles of ascent. The one on the north rises at about 31 degrees to the horizontal. The southern shaft is steeper, with an angle of 44 degrees and 5 minutes.

In 1964, the late Professor Alexander Badawy, an Egyptologist who taught at U.C.L.A., challenged conventional wisdom and the "air shaft" interpretation of the odd pair of channels. Collaborating with a student, Virginia Trimble, who is now a well-known astrophysicist, he demonstrated that the shafts were targeted on the stars. Trimble calculated the angles of transiting stars in the era when the Great Pyramid was built. The northern shaft, she realized, was aimed at Thuban, a star in the constellation we know as Draco the Dragon. Between third and fourth magnitude Thuban was not an unusually bright star in the ancient Egyptian night, but in the third millennium B.C. it was the closest thing the Egyptians—or anyone else—had to a North Star. Every 24 hours, it traveled in a small circle around the north celestial pole. Between 2700 and 2600 B.C., the summit of its daily arc coincided with an angle of 31 degrees above the northern horizon and aligned with the northern shaft. To the south, Alnilam, the middle star in Orion's Belt—and the brightest of the three—transited Giza at the same angle as the southern shaft. No other competing stars came close to either alignment, and the significance of both Thuban and the stars of Orion's Belt are confirmed in the *Pyramid Texts.* Both were stellar destinations of the dead pharaoh.

To the Egyptians, Thuban was the leader of the circumpolar Imperishable Stars. They were undying soldiers in a celestial army that defended cosmic harmony. Their eternal marching orders were given by the pharaoh, who flew to the head of their ranks upon his death. He ascended to the sky among the stars to regulate the night, as Thuban and the circumpolar stars did. He joined the Imperishable Stars and governed them, just as he governed the earth.

Orion's Belt, on the other hand, was not part of the immortal cavalcade in the north. It rose and set like the sun, moon, planets, and the rest of the stars. Yet despite the death these celestial objects endured with every passage below the western horizon, they were always reborn in the east. Their cyclic renewal was a different kind of immortality. Manifested among the stars as Orion's Belt, Osiris was the dying god responsible for the world's cyclic renewal and for the resurrection of the dead into the afterlife. In the guise of Orion, Osiris also marched in step with the decans, a set of stars the Egyptians used to keep track of the hours of the night and the passage of the seasons. Sirius, the brightest star in the sky, led the decans into the New Year. It was the stellar incarnation of the goddess Isis, the consort of Osiris as well as the celestial partner of Orion. Orion and all of the other decans returned successively to the predawn sky and, like Sirius, remained to mark nocturnally first one hour, and then another, every ten days during a 120-day tour of timekeeping duty in the night sky.

With the circumpolar, undying stars, the dead pharaoh tirelessly drove the night forward and sent the hours on their way. By joining these eternal stars, he became eternal. By joining Orion, he became Osiris, the resurrected spirit of birth, growth, death, and rebirth. In each of his celestial destinations, the pharaoh upheld the same cosmic order he embraced while alive. His agenda was nothing less than the stability and continuity of the world, and he exerted celestial power to fulfill it.

The *Pyramid Texts* were actually inscribed two centuries after Khufu completed the Great Pyramid, and significant changes occurred in the cult of the king after the time of Khufu and his immediate successors. Despite this evolution of pharaonic power, the orientation of the Great Pyramid's celestial conduits tells us the same themes defined the destiny of the pharaoh in the next life.

Zahi Hawass, the archaeologist in charge of the monuments at Giza and Sakkara for the Egyptian Antiquities Organization, comprehensively reevaluated the Giza funerary complex in 1987. His analysis links the Great Pyramid to a new cult of the deified king. Khufu, he argues, equated himself with the divine sun. His message to the priests of the sun god Re at Heliopolis was unambiguous. As members of the elite and managers of a theological bureaucracy, they controlled land, wealth, and people and shared power with the king. This king, however, put them on notice. He was the living incarnation of the sun—the object, and therefore the beneficiary, of their devotions. They might possess power, but they answered to the king. Inflating his own power at the expense of the Heliopolis priesthood and bureaucracy, Khufu contrived the Great Pyramid not so much as a tomb and a site for the royal funeral but as a monument intended to consolidate and preserve the continuity of royal power. Operating as a temple for a cult of the divine king, returned by death to the highest frontier of power, the pyramid complex retold, in architecture and ritual, the primary myth of kingship—the command of celestial forces and cosmic order.

Hawass believes various components of the Great Pyramid complex reflect this new solar character of the pharaoh. For example, he interprets the boat excavated from one of the pits on the south side of the Great Pyramid as a solar boat. When Re, the sun, made his daily journey through the sky and his nightly journey through the netherworld—the realm of death—he traveled by boat and exchanged boats at key transitions in the 24-hour cycle. According to Hawass, Khufu supplied himself with boats he needed to perform as the sun.

Controversy, of course, surrounds this interpretation of the boats, but it is consistent with other evidence at Giza. The crypts of the first pyramid, Djoser's Step Pyramid at Sakkara, are below ground level, and in subsequent pyramids of the Third and Fourth Dynasties, a subterranean or near-ground-level location for the burial chamber is normal. In the Great Pyramid, however, a deep underground room was abandoned, unfinished in favor of the King's Chamber, which places Khufu above ground, in the dominion of the sun.

A second room, the "Queen's Chamber," was started on the Great Pyramid's central vertical axis a little above ground level, but it, too, was left incomplete. In 1993, however, a robot was sent up the southern "air shaft" of the Queen's Chamber, and although the passage was previously believed to terminate a few feet into the pyramid's core, the robot continued for 213 feet, when it reached a "door." In *The Orion Mystery,* Robert Bauval and Adrian Gilbert discuss this shaft and claim it intentionally points to Sirius. This may be so, but their book also argues that Giza's three big pyramids map Orion's Belt on the desert plateau, a dubious assertion burdened by equivocal evidence. Also, to make the map work, Egypt has to be turned upside down. The logic is labored, even for those uncomfortable with the fact that downriver for the Nile is north. Despite the enthusiastic dust jacket promotion: "Unlocking the Secrets of the Pyramids" and "The Revolutionary Discovery That Rewrites History," consumers in the market for esoteric knowledge should beware. *The Orion Mystery* sold successfully, but it didn't decode the Great Pyramid's astronomy.

Whatever Khufu intended in the Queen's Chamber, he finally decided what he really wanted and elevated what would be his last stop on earth, the King's Chamber, to a point about a third of the way between the base and the summit. Hawass thinks this was Khufu's way of putting himself in the horizon, or *akhet,* of the sun.

The concept of "Khufu's horizon" assumed a new celestial dimension with Mark Lehner's discovery of a possibly intentional solar alignment at Giza. Lehner is an Egyptologist with the University of Chicago's Oriental Institute. Initially, his work at Giza involved a new, comprehensive mapping of every feature on the plateau, and his studies revealed sockets, trenches, and other depressions in the bedrock that would have permitted Khufu's surveyors to level the foundation of the Great Pyramid with water and orient its sides with stakes. While mapping the details of Giza, Lehner also noticed how the summer solstice sun sets. From the Sphinx, the northernmost sunset of the year lodges the sun midway between the silhouetted pyramids of Khufu and Khafre. Khafre was one of Khufu's sons. He built the second large pyramid at Giza, acknowledged his father as the sun, and called himself "son of the sun." The father-and-son pyramids and summer solstice sun together form the akhet symbol, the solar disk framed within a pair of stylized mountains. This symbol is at least as old as the Old Kingdom, and what may be variants of the emblem occur early in fourth millennium B.C. on neolithic ceramics. It is probably rooted in a primordial observational technique that monitored the progress of the year through the sun's displacement along the distant mountainous horizon.

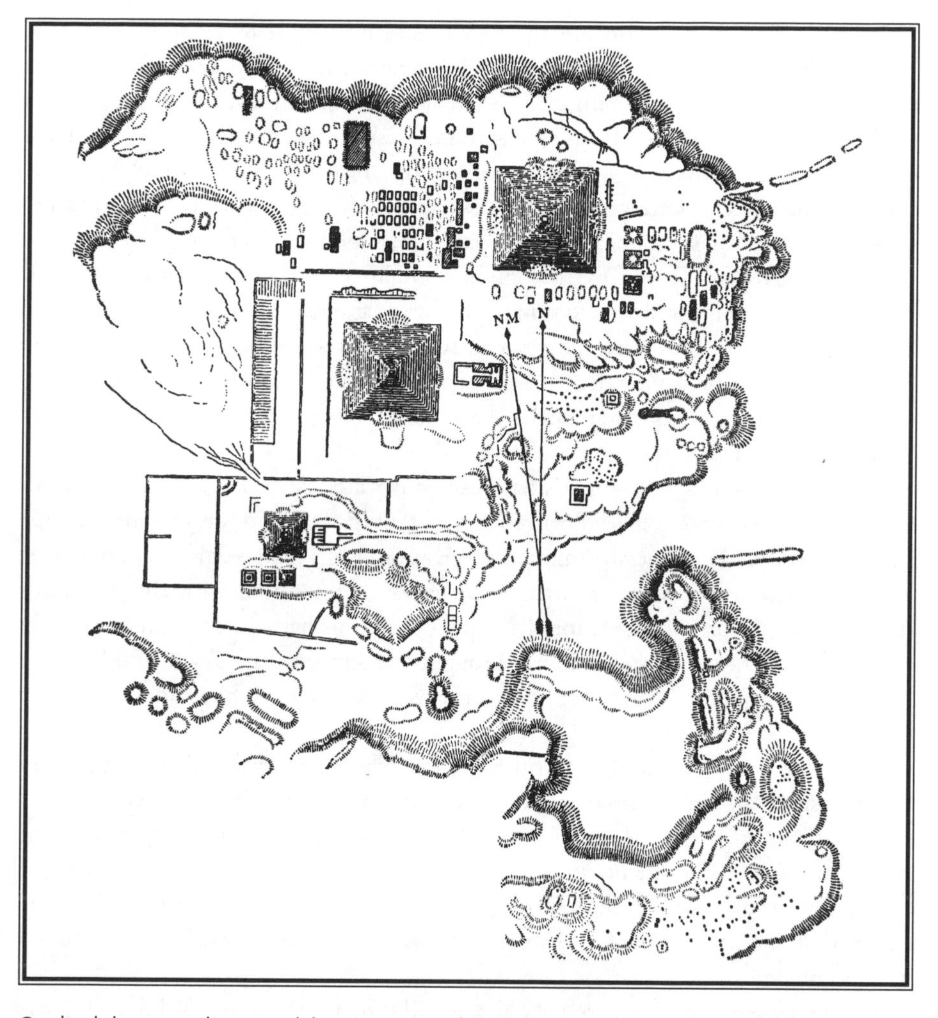

Cardinal directions determined the orientation of the Giza pyramids and the rest of the Giza necropolis. The Great Pyramid is the large one in the upper right. To make the three main pyramids mimic the Belt of Orion, as *The Orion Mystery* claims they were intended to do, you have to turn the map upside down, and that reverses the directionality of the pyramids with respect to north in the sky, to say nothing of the fact that the brightest star is in the middle of the belt and the biggest pyramid is on the wrong end. (from *The Dawn of Astronomy,* J. Norman Lockyer, 1894)

Now, the summer solstice corresponds to the same time of year when Sirius reappeared in the predawn sky to signal the annual Nile flood, the New Year, and the rebirth of life. If this alignment were actually contrived by Khafre, the necropolis of Giza became, in his reign, a stage on which the sun "died" at the horizon among the tombs before the rebirth signaled by the heliacal, or predawn, rising of Sirius in the east. In this cosmic context, the soul of the entombed pharaoh made his pilgrimage to heaven to activate the great cosmic cycles.

It seems likely Khafre also put the Sphinx and its temple to work on behalf of dynastic lineage. For him, a royal solar bloodline must have been a useful myth, and the Sphinx probably helped him turn the story into required reading. The Sphinx is now believed to represent Horus, the son and designated successor of Re. It faces east. At Giza, then, the sun god beams his rays upon the face of his lawful heir as he rises each day. Reciprocating with offerings for his celestial parent, the Sphinx translates the concept of royal lineage into the monumental ritual landscape.

The open-roofed Temple of the Sphinx contains what are probably architectural allusions to the rising and setting sun, to the calendar, and to the systematic measure of time. Details such as these, along with solstitial phenomena associated with the Sphinx, may have been intended to firm up the dynasty's authority through its intimacy with the sun and its command of the seasonal change that accompanies the sun's annual migration.

Solar connotations of pyramids have been advocated since 1912, when historian and Egyptologist James Henry Breasted argued that the pyramid was intended to represent the *benben*, the primordial mound of creation. Both the predawn appearance of Sirius and the morning-star aspect of Venus were associated with the *benu* bird. According to the priests of Heliopolis, the flight of this heronlike bird, momentarily pausing upon the world's only perch, brought the sun into the sky. Heralded by a "Morning Star," the sun ordered the world with celestial rhythm and energized it with life.

I. E. S. Edwards, an authority on Egyptian pyramids, advanced the suggestion that the shape of the pyramid was originally inspired by the dramatic vista of the sun's rays fanning through the clouds to form a towering and glowing pyramid of light that reached into the heavens. There is, however, considerable uncertainty over the derivation of the word for a true pyramid, and there is also skepticism in some circles that the pyramid had anything to do with the benben, which is thought to have been conical.

If the pyramid really represents the world's first mound of solid earth, the pharaoh's tomb is also the place of Creation. In the mortuary environment, however, the allusion to Creation may really refer to the pharaoh's rebirth. Although the solar metaphor puts the pharaoh on another trail to the sky, it does not really exclude the stellar imagery of the *Pyramid Texts* and the Great Pyramid's astral-ducts. As the primary regulator of his world, Khufu could embody the power of the divine sun and still travel to the stars through the transformations of death.

The austere and empty character of pyramid burial chambers has tempted Mark Lehner to speculate, at least informally, that some Old Kingdom pyramids were not even true tombs where the body of the deceased pharaoh was interred, but symbolic tombs, magical houses for the spirits of the dead kings. Whether or not all of the pyramids were genuine tombs, they certainly seem to have had something to do with the departed king's ascent to the sky. One of the *Pyramid Texts* (#267) tells us, "A staircase to heaven is laid for him (the king) so that he may mount up to heaven thereby." The hieroglyph for "ascent" resembles a step pyramid, and that has prompted conjecture that there is etymological support for thinking of the pyramid, at least Djoser's Step Pyramid, "as a stairway to the stars."

An independent scholar, John Charles Deaton, has tried to get a handle on the meaning of the pyramids by taking a second look at the individual names given to them by their makers. He first reminds us that there are celestial associations in the names of two mortuary complexes: "Re loves Khufu" and "the Constellation of Khufu." One implies a close relationship between Khufu and Re, and the other implies a stellar destiny for the pharaoh. Khufu's first successor was his son Djedefre, or "Enduring like Re." For reasons unknown, he located his own pyramid at Abu Rawash. His name for the pyramid—"The Starry Sky of Djedefre"—tells us his afterlife was also played out in the stars.

In the *Pyramid Texts,* the deceased pharaoh grows wings and feathers and becomes a ba, or soul. Portrayed as a bird with a human head and united with the body in life, the ba departs the body at death and hovers close to the spirit double, or *ka,* in which the power of life resides. As a bird-ba, the pharaoh flies to heaven and becomes a star.

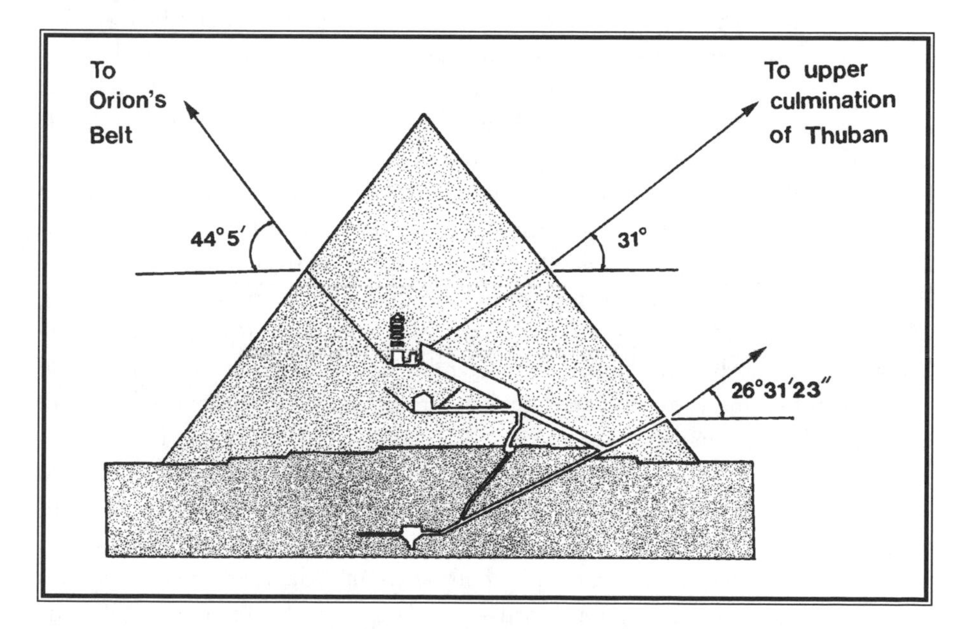

Inside the Great Pyramid, narrow shafts reach from the King's Chamber toward the Thuban, the leader of the Imperishable Stars, on the north, and to the Belt of Orion, the stellar identity of the god Osiris, on the south. The purpose of these shafts—and probably the entire pyramid—was a successful launch of the pharaoh's soul into the sky. Although similar shafts that extend from the Queen's Chamber were thought to end not far into the core of the pyramid, a robot has now traveled more than 200 feet up the southern shaft, which may have been aligned on Sirius. If so, the Queen's Chamber might really have more to do with women than its arbitrary name, for the star Sirius was the celestial incarnation of the goddess Isis, the consort of Orion, which was targeted by the south shaft in the king's burial chamber. (Griffith Observatory drawing)

The ba, Deaton adds, is frequently equated with a star and has, at least at times, luminous properties. Deaton emphasizes there is only one method for celestial ascent mentioned in both the names of the Old Kingdom pyramids and in the *Pyramid Texts:* Ascent by means of the king's ba. Some names for pyramids refer to the pharaoh's ba: "The *ba* of Sahure shines forth," "The *ba* of Neferirikare," "The *ba* of Neferefre," and "The *ba* of Itti." Several other proper pyramid names—such as "Snefru shines forth (the northern, or "red," pyramid at Dahshur)—refer to light, but what kind of light is not entirely clear.

All of these facts persuade Deaton the pyramid shape was intended to capture and reflect sunlight and moonlight on its broad triangular sides. These triangles, Deaton judges represent the luminous, birdlike bas. Their smooth, white surfaces of tura limestone would, in fact, have made the pyramids gleam, and the changing play of light—through the day and in the night—may have symbolized the departure of the ba for the sky. This reasoning is plausible and simple, and if it is true, it says that the pure, geometric shape of the pyramid was exploited not only to dramatize in dancing light and shifting shadow the king's ascent but actually to propel him into heaven.

Even if we are wrong in detail, and there is certainly a reasonable possibility that we are missing much of the true story, the celestial allusions we find in the Egyptian pyramids reinforce what we see in the star-oriented ducts of the Great Pyramid. It and the rest of the Egyptian pyramids reflect a labor mobilization that helped turn Egypt into a nation. Power was centralized through the ideology of kingship. Ever challenged by competitors for power—local governors, the nobility, and the bureaucracy of priests—the pharaoh's influence could grow and decline from dynasty to dynasty and from king to king. History and Khufu's own pyramid hint that he appropriated unprecedented power, enough to enforce his assertion of solar divinity. His Great Pyramid didn't emerge from an architectural vacuum, but it is part of the beginning of that monumental enterprise. Relying on the knowledge and experience acquired by his father Snefru, who built himself two pyramids, and other predecessors, including Djoser, Khufu managed to erect the largest and most accurately constructed pyramid ever built. The task must have impressed the participants at every social level—not just with a remarkable monument but with the acquisition of the power needed to succeed. Whatever else Khufu did—and the record is not clear—he acquired a reputation for high-handed despotism that dogged his memory to the time of Herodotus. We cannot tell if his image as a temple-shutting tyrant is an accurate criticism by a pious—and economically overburdened—population or the fabrication of self-righteous ancient historians philosophically in bed with the displaced priests.

Whatever else they may be, the pyramids are a testimony to the function of ideology in the concentration of power. They were designed and built with the energy of belief, and that belief began with a sense of necessity. Ensuring the cosmic effectiveness of the dead pharaoh through monument and ritual strengthened the institutions of centralized power. In life and death, a powerful monarch stabilized the whole system. Uniting people and territory through a shared belief in values that transcended personal and local priorities, the ruler developed a tool of political and economic power that would help protect his world from domestic turbulence and foreign threat. In Old Kingdom Egypt, the vehicle that gave the pharaoh upward mobility returned to earth with a cargo of national unity.

SOME NEW WORLD ORDER

It is fair to argue that geometrically pristine pyramids occur only in Egypt, in neighboring Sudan, in Rome (the tomb of Cestius, 12–11 B.C.) in Paris (the Louvre), in San Francisco (the Transamerica Building), and in Las Vegas (the Luxor), but the dictionary already loosened Egypt's grip on this kind of monument by mentioning the pyramidal platforms of ancient Mesoamerica. Strictly speaking, these are not true pyramids—which are supposed to have triangular sides that meet in a point—but truncated pyramids, and we also encounter similar structures, built of earth instead of stone, in the eastern half of the United States.

From the top of the largest of these prehistoric mounds you can see the St. Louis Gateway Arch, about eight miles to west. Known as Monks Mound, this multileveled platform is almost a hundred feet high, and it once supported a huge timber building on its flat summit. Now only the earthen mound remains. It belongs to the Mississippian culture of ancient America. Mississippian peoples are named for the Mississippi Valley heartland of a late prehistoric tradition known for its numerous organized settlements. They built large mounds from Wisconsin to Mississippi, from Florida to Oklahoma. Their presence spans the centuries from 750 A.D. to about 1500 A.D. Their wealth and power was based on maize agriculture and long-distance trade, and Cahokia, their greatest city—the main center of their wealth and power, was centered on Monks Mound.

Cahokia is in west-central Illinois, next door to Collinsville and on the floodplain of American Bottom. The presence here of the largest prehistoric urban development north of Mexico and the largest prehistoric monument in the United States is no coincidence. Ideal for intensive agriculture and strategically located on one of the most important river drainages in the country, near the junction of the Missouri with the Mississippi River, Cahokia was well situated to produce wealth and to acquire wealth. As it grew, it became more complex and socially stratified through the redistribution of wealth and through monumental mounds that put the elite on pedestals of earth.

The Mississippian peoples built several kinds of mounds. Some contained burials. Others elevated the homes of the elite. Temples for ritual were built upon others. Access to any of them meant access to the gods, including perhaps mythical ancestors. Those permitted to ascend them transformed status into vertical privilege. They were closer to the sky and to the sources of cosmic power. They could see the territory they controlled, and the power of their elevated perspective was evident to everyone below, who could not miss the huge mounds that defined the social landscape.

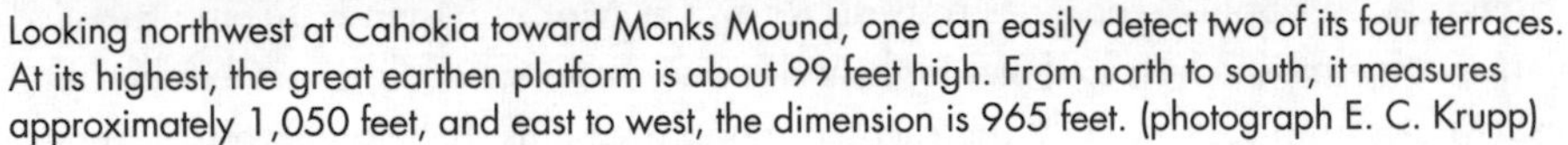
Looking northwest at Cahokia toward Monks Mound, one can easily detect two of its four terraces. At its highest, the great earthen platform is about 99 feet high. From north to south, it measures approximately 1,050 feet, and east to west, the dimension is 965 feet. (photograph E. C. Krupp)

As an emergent chiefdom, Cahokia was probably ruled by a primarily hereditary aristocracy headed by a political leader believed to belong to a celestial bloodline. Mississippian tradition survived to the sixteenth century, and the first European contacts with people like the Natchez and the Creek shed a little light on Mississippian social organization. To a clan-oriented society, lineage is everything, and so it comes as no surprise that rulers of chiefdoms in the southeastern United States had a sibling in the sky. Recognized as the sun's younger brother, the Natchez chief emerged from his lodge every morning and saluted his solar relative. Uplifted by his own mound, the chief enjoyed closer contact with his brother than was permitted the rest of the community. His duties included encouraging the real sun to follow its expected course. As the sun's counterpart on earth and fraternal partner in power, the chief was known as "Great Sun." We don't know if Cahokia's Great Sun actually resided on Monks Mound or fulfilled ritual responsibilities in a temple that topped it, but its summit could have put Cahokia's high chief closer to the sun than any other brother the sun might have.

Mississippian ceremonial paraphernalia is saturated with rayed disks and quartered circles, symbols believed to mirror the divine priority of the sun and the directionally ordered world. Agricultural ritual tempered by the sun's association with seasonal change was also part of Mississippian religious life. Appreciative of the sun's role in renewing the earth's productivity and cognizant of the order and stability its behavior in the sky conferred on the world, the Mississippians must have done more than worship the sun. They also probably paid close attention to it.

Systematic astronomical observations probably were performed at Cahokia to coordinate the complexities of urban life, to unify the community through public ceremony, and to maximize the yield of increasingly intensive cultivation. Archaeologist Warren Wittry realized he might have found Cahokia's solar observatory after he discovered, in 1961, several large rings and arcs of postholes in a field about 3,000 feet west of Monks Mound. The size of the holes—about a foot and half across—and the adjacent trenches that were used to prop the poles into place imply the use of tall posts, perhaps 30 feet high. One of the postcircles, now called Woodhenge 3 (or Circle 2), provided sunrise alignments for both solstices and the equinoxes between a post slightly offset from the center of the ring and posts on its perimeter. Another pair of posts defined the circle's north-south axis. Over the years, research on the "American Woodhenge" or "Sun Circle" continued to add details and modify our knowledge of the structure. At a special symposium, held at the Cahokia Mounds Museum in 1994, Wittry and several other experts reexamined the astronomical, cosmological, and symbolic potential of Cahokia's woodhenges. Wittry confirmed that Woodhenge 3, with a diameter of 410 feet, originally hosted 48 uniformly-spaced posts, and also explained that the insertion trenches were radially oriented toward the circle's center. The trenches not only clarified which holes were actually part of the ring, they underscored the significance of the circular figure to the builders.

With posts alone, Woodhenge 3 can't discern the date of solstice to better than a few days or even a week. As an astronomical instrument, it could indicate the turning points of the solstice sun and approximate seasonally important dates, but it would be calendrically challenged in any effort to establish something like the exact number of days in a year. That doesn't mean the skywatchers of old Cahokia didn't know the answer. It just means the woodhenge by itself couldn't tell them.

In most respects, Woodhenge 3 looks more like a ceremonial enclosure that incorporates astronomical and cosmographical alignments. It has been viewed as part of the monumental symbolic vocabulary that supplemented the mechanisms of political power. Competing interests of separate lineages and factions in the Mississippian chiefdoms likely put power in perpetual contention, but visible demonstrations of the authority of the elite and the hierarchy of power—great mounds, esoteric ritual, and astronomically tuned woodhenges—ordain political and social stratification and endorse its leadership. Matching the social pyramid with the architecture of the cosmos through the analogy of monumental and astronomical architecture, Cahokia's elite must have allied itself with celestial gods, primordial Creation, cosmic order, and the power of the sky.

Melvin L. Fowler, the archaeologist who organized the 1994 Cahokia Woodhenge conference, found new support for the Sun Circle's place in Cahokia's symbolism of power. He reported his discovery of another, similar "woodhenge" to the south of Monks Mound and closely associated with Mound 72, a ridgetop mound in which

Fowler had uncovered unprecedented and extraordinarily rich elite burials in 1973. He had already demonstrated that other, larger ridgetop mounds seem to establish the primary cardinal axes of Cahokia and mark the city's limits. Mound 72 was also speared by Cahokia's central north-south axis, and the mound itself was oriented to summer solstice sunset and winter solstice sunrise. Fowler's new circle, Woodhenge 72, incorporates the large post sockets he had exposed in his excavation of Mound 72 and therefore also occupies the city's north-south axis. With a diameter that matches Woodhenge 3, a probable posthole population of 48, and one confirmed astronomical post alignment (summer solstice sunrise/winter solstice sunset), Woodhenge 72 compels us to accept the importance of these attributes in any interpretation of the timber circles of Cahokia. With Woodhenge 3 on the city's east-west axis and Woodhenge 72 on the north-south axis, we should be asking if the placement of these monuments also had something to do with commemorating the cardinal plan of Cahokia. Both of these axes converge at Monks Mound, on the southwest corner of its first terrace. Earlier excavations there by Fowler revealed a sequence of intense construction as well as isolated postholes that appeared, like the ridgetop mounds and the woodhenges, to have something to do with preserving the fundamental reference lines of the city. Most of this story is missing, lost like the mounds that have eroded away or been intentionally destroyed and replaced with giant retail stores and parking lots. In that respect, Cahokia is still functioning as a center of redistribution. Although the real story of its symbols of authority will always be incomplete, Monks Mound alone tells us its rulers secured status and power through closer contact with the sky.

Maize cultivation, pyramidal platforms, and perhaps even the Mississippian version of the sun cult were Mexican imports. Long before the Mississippians started piling up dirt to lift their chiefs closer to heaven, people in Mesoamerica were building pyramids for access to the gods, and some of their ideas along with their trade goods traveled north. The details, complexities, and mechanisms of this diffusion are not really sorted out, but there is enough evidence to persuade us it happened.

You can, however, take cross-cultural comparisons too far. The superficial resemblance of American pyramids to the monuments of Egypt, for example, has inspired dedicated but undisciplined and uninformed advocates of cultural diffusion to find the origin of the New World monuments in those of the Old. In the late nineteenth century Ignatius Donnelly, the world's most energetic promoter of Atlantis, cited the presence of pyramids in Egypt and pyramids in Mexico as evidence of the influence in both hemispheres of the celebrated lost continent. Time alone, however, undermines Donnelly's pyramid scheme. The earliest known pyramids in Mesoamerica (at Nakbé, Guatemala, 500 B.C.) were built more than a thousand years after the last big pyramid in Egypt and more than two thousand years after the inspired era of Giza. In addition, Mesoamerican pyramids are topped by temples, not points. Like the Mississippian platform mounds, they are really truncated pyramids. Like all pyramids, however, they had something to do with upward mobility. They lifted their owners closer to the sky and were equipped with stairways to do it.

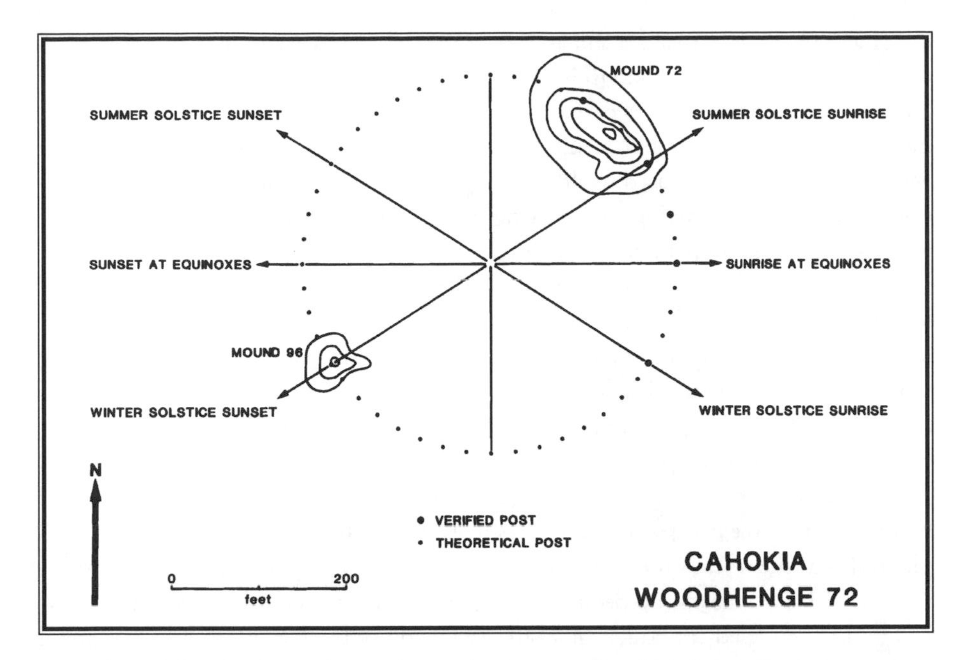

An analysis of the placement of Mound 96 with respect to Mound 72, famous for its extravagant, elite burials, led Melvin L. Fowler to propose that both features were related by an undiscovered circle of timber posts. Subsequent excavation between 1991 and 1993 revealed the presence of additional postholes matching the perimeter of the hypothetical ring, and this new element of Cahokia is now known as Woodhenge 72. It seems to incorporate solstitial and equinox alignments more or less like those identified at Woodhenge 3 to the west of Monks Mound. Both of these apparently astronomical post circles are intersected by a prime cardinal axis of the entire city. (Griffith Observatory drawing, Joseph Bieniasz, after Melvin L. Fowler)

Most Mesoamerican pyramids were artificial cosmic mountains, and they were also said to symbolize the realms of heaven and the levels of the underworld in the stack of platforms they pushed into the sky. They were, in fact, pictures of the cosmos organized around a world axis. Sometimes this axis was defined by the zenith and the nadir, and in other circumstances it was the polar axis, centered on the sky's rotation. These pyramids could also represent the place of Creation, permit communication with the gods, and mark the center of the world. The Aztecs, for example, symbolized Coatepec (Serpent Hill), the sacred mountain, in Templo Mayor, the great double pyramid at the heart of their capital, Tenochtitlán. Coatepec was the place where the order of the Aztec world was first installed through the violent birth of Huitzilopochtli, a warrior avatar of the sun. At Templo Mayor, the Aztecs reenacted this myth of tribal sovereignty, and the ritual they performed symbolized the cosmic origin of Aztec power.

Modeling pyramids on cosmological concepts, the Maya raised stages on which their kings could perform in the company of the gods. Linda Schele and David Freidel, both specialists on ancient Maya civilization, show the process was at work at least by the Late Preclassic era and as early as the first century B.C. In their description of the evolution of Maya kingship, *A Forest of Kings,* we see Cerros, a modest settlement on the southwest shore of Chetumal Bay, in what is now Belize, inaugurate an innovative program of public architecture to accommodate royal ritual as the village was transformed by increased wealth from agriculture, fishing, and trade into a chiefdom. After clearing, leveling, and burying most of their houses, they refurnished the place with plazas and pyramids that turned a village into a ceremonial center.

The first temple at Cerros, known to archaeologists as Structure 5C, was a two-level pyramid. Its central stairway faced south, and it had its back to the sea. More or less oriented cardinally, it straddled the site's new north-south axis. When the king went up the stairs, he was headed north and so climbed a polar axis into the sky and the otherworld of the gods. The roof of the temple on top of the pyramid was supported by four immense poles. Two were lodged in sockets that framed the front door. The other two echoed the front pair against the inside of the back wall. Schele and Freidel regard these posts as the four trees that support the sky—in this case the temple roof—at the world's four directions.

Four elaborate masks molded in plaster put celestial gods on the south face of the pyramid. They filled the panels of both terraces on both sides of the stairs. Careful analysis of the details in these four masks revealed who they portrayed. On the first level, stylized jaguars snarled at any crowd assembled in the plaza, and a symbol for the sun on each jaguar's cheek verifies the jaguar is the sun. Whether the sun traveled through the blue sky of the daytime or through the underworld at night, it could be depicted with the features of a jaguar. In this case, the jaguar on the east side is the rising sun, and the western jaguar naturally is the setting sun. In both cases, the people of Cerros may have intended the rising and setting Jaguar Suns on their first pyramid to represent the sun in underworld, for the masks on the second level architecturally hover above the faces of the Jaguar Sun like Venus, as the Morning Star in the east and as the Evening Star in the west, hangs above the rising and setting sun. Details of these two masks led Schele and Freidel to conclude they really do represent these two aspects of Venus.

In architecture and ornament, then, the Cerros pyramid seems to tell us the king entered the otherworld spirit realm of night when he disappeared inside the temple. In the dark recesses of that sanctuary, he perforated himself, perhaps with a stingray spine, as a sacrificial offering intended to feed the gods with his own blood. Bloodletting may have also been a part of the quest for visions. Through hallucinatory trance, the king could chat with his ancestors and with the gods and so acquire knowledge and inspiration that would benefit his community through his acquisition and exercise of power. The temple was a portal to the territory of the divine, a place of magical power where the curtain between worlds parted. When the king reappeared in the doorway or stood in the middle of the stairway surrounded by celestial masks, people could see he was in the territory of gods. Those two sky gods that flanked him—Venus and the sun—conveyed another message to the people below. They were the celestial identities of the Hero Twins, the mythical brothers whose adventures in the underworld cultivated the first dawn out of the fields of time and established the tradition of kingship that made cosmological sense out of the new social order. The myth, the pyramid, and the ritual performances of the king all verified the same thing. Maya society was congruent with cosmic order. That's the kind of news that gives kingship a good name.

Maya twin pyramid complexes, like Tikal's Complex Q, are also thought to merge royal power and cosmological symbolism in their architecture. This pair of pyramids was constructed in 771 A.D. by Ah Cacau, Tikal's ruler at the time. Each pyramid has four stairways, one for each key direction. There is a building with nine doorways on the south and an enclosed area for an inscribed stela and altar on the north. These stelae usually carry a portrait of the king and inscriptions pertaining to dynastic events. The nine doorways are said to represent the underworld, with one door for each Lord of the Night. The ruler, equated with the sun, occupies a focal point in the complex, and the enclosure for the royal stela is believed to represent the sky, from which the king derives authority and power. The sun's passage from east to west is signified by the east-west arrangement of the twin pyramids.

Neither the Cerros pyramid nor the pyramid pair at Tikal came with a user's guide to the monument's software. We have to infer the meaning from the character of the architecture. The kings of Palenque, on the other hand, paneled the walls of their temples with elliptical commentary on dynastic events.

Palenque is in southern Mexico, in the state of Chiapas, and it probably dominated the western Maya lands during the Late Classic era and the seventh century A.D. Lodged against the last low ridges on the sierra's northern flank, Palenque persuades the eye it can see all the way to the Gulf of Mexico. To the south of the plazas, palaces, and pyramids, the land rolls out like a carpet as it turns into the flat Plain of Tabasco. Through it threads the Usumacinta River, one of the great highways of the western Maya.

As the home of the first real tomb discovered inside a Mesoamerican pyramid, Palenque transformed our ideas about Maya kingship and royal burial. The tomb was found in 1952 by Mexican archaeologist Alberto Ruz Lhuillier inside the pyramid known as the Temple of the Inscriptions. The ruler of Palenque was interred in a stone crypt at ground level. His name was Pacal, or "Shield," and he was one of Palenque's most prominent kings. He was responsible for ambitious programs of public construction, including most of the monumental Palace and his own pyramid tomb. The lid of his sarcophagus is periodically promoted by pseudoscientific advocates of ancient astronauts like Erich von Däniken as a portrait of one of antiquity's space aliens blasting back to the planet he came from. During the last 20 years or so, however, we have lived through the golden age of decipherment of Maya hieroglyphics. Now we can read most of the inscriptions, and that is why we know who was buried in Pacal's tomb. We also know that the richly symbolic carving on the lid of his coffin shows him falling, with the dying sun, into the underworld. Identifying himself with the setting sun, Pacal injected himself into a celestial metaphor of royal succession. Although his trip to the underworld was as inevitable as the sunset, his son, Chan-Bahlum (Snake-Jaguar), would revive kingship, if not the dead king, by rising to the occasion like the rising sun. Chan-Bahlum's own monuments at Palenque, in the group of the Temple of the Cross, commemorate his accession and other significant events of his reign. One passage of hieroglyphics documents his heir-designation ceremony on June 17, 641 A.D., and its accompanying pictorial relief illustrates Pacal's transfer of power to his son. Five days later, on the summer solstice, Chan-Bahlum "became the sun."

Dedicatory inscriptions on the Temple of the Inscriptions reveal the meaning Chan-Bahlum read into his accession monument. Reminding posterity of the edifice erected by First Father at the beginning of the world, Chan-Bahlum gave his pyramid the same name. First Father's first project was actually the celestial dome supported by the cosmic axis world tree. That means First Father really established the fundamental order of the cosmos, and Chan-Bahlum was just letting people know that what was good for the gods was just fine for the king. According to the texts, Chan-Bahlum performed the same consecration ritual for his architectural replica of cosmic order as was celebrated by First Father when he "entered into the sky" and fabricated the real thing.

Chan-Bahlum—the son, heir, and successor of Pacal, Palenque's famed entombed king—built the Temple of the Cross to the east of his father's burial pyramid and through it memorialized the inauguration of an ordered cosmos by First Father at the beginning of time. (photograph E. C. Krupp)

Pacal had established his divinity through a calendrical affiliation with the goddess known as First Mother. She gave birth to the men and gods at the time of Creation, including the three gods associated with royal sovereignty known as the Palenque Triad. Later, when Chan-Bahlum elected to dedicate his pyramid on July 23, 690 A.D., he was ensuring that the triple gods of kingship would be on hand for the ceremony, along with First Mother. That day coincided with a gathering of three planets—Jupiter, Saturn, and Mars—in the company of the moon in the constellation we see as Scorpius the Scorpion. Noticing this suggestive celestial assembly, anthropologist Floyd Lounsbury brought it to the attention of Dieter Dütting and astronomer Anthony F. Aveni. Aveni had already helped shape the modern study of ancient astronomy with more than two decades of field research, and together, he and Dütting realized Chan-Bahlum had timed the grand opening of his temple with a family reunion. As the moon, First Mother was reenacting the birth of her children—the three planets. In an astronomical configuration occurring seven years after Pacal's death, his acknowledged heir renewed the primordial covenant between king and gods. Symbolically propelled back to the time of the world's beginning by the completion of his pyramid, Chan-Bahlum reinvigorated the divine ancestry of the king and demonstrated his commitment to the continuity of cosmic order through royal duty. Two days after the Creation-mimicking dedication ritual, Chan-Bahlum ritually pierced himself—probably with an obsidian blade—and offered his own blood to his gods. The rite occurred just three days short of the 75th anniversary of his father's installation as king and established a parallel between Chan-Bahlum and the original mythic sacrifice that the dawn of time brought for the gods and the cosmically ordained line of Maya kings.

KING OF THE MOUNTAIN

Khmer rulers of ancient Cambodia also built pyramidal monuments that imitated the central world-axis mountain. Officially titled "King of the Mountain," they governed on earth in accordance with heaven's harmonies, and they traced their lineage to the merging of two "celestial" dynasties. According to legend, the royal bloodline of a wealthy kingdom in southern Cambodia originated with the nuptials of an aristocrat and a local girl named Soma, a Hindu word for offering. The same word is sometimes used for the elixir of immortality and also for the moon. In this case, Soma was the daughter of a Naga, one of the serpent lords of the Indochinese soil, and her husband was a Brahmin from India. Their marriage established the "lunar dynasty" of Funan, as early Chinese historians called the country. Angkor scholar George Coedès asserts this Chinese name derives from the Khmer word *phnom*, which means "mountain." If so, it suggests a longstanding association of national sovereignty with the power of the mountain. The story also implies influence from India, and it is likely that Hindu tradition was transplanted to Indochina from the west.

While the moon's clan was turning the territory around the Mekong delta into a wealthy city-state, the sun's realtives were busy in the north. There, the legends claim, the ascetic Kambu Svayambhuya decided to abstain from abstinence when he met Mera, a celestial nymph. Their "solar dynasty" ruled Chenla, which may have been a tribute state of Funan, until the descendants of the sun and moon got together and unified the land and its people with a national identity. Their name—Khmer—is probably a combination of Kambu and Mera.

With a language that belongs to the Austro-Asiatic linguistic family, the Khmers are not closely related ethnically to either the Sanskrit-speaking Hindus of India or to the Chinese. Belief and tradition from both cultures, however, migrated into Indochina. The Khmer modified these ideas into their own cosmology and institutions of kingship, and the synthesis is evident in their monumental architecture.

Certainly the most famous Khmer monument is Angkor Wat. It was built in the first half of the thirteenth century by King Suryavarman II, and it is one of many relics of the old Khmer capitals near the modern town of Siem Reap, a little northeast of the north end of Cambodia's great lake, Tonle Sap. The lake has enriched Cambodia with fish and other aquatic resources, and the floodplain of the Mekong River and its tributaries ensured Cambodia's agricultural success. Fertility of the land and productivity of the waters were supplemented with active international trade. The economy subsidized military defense and territorial expansion as well as a vigorous campaign of public architecture. In sanctifying kingship, buildings like Angkor Wat helped solidify centralized power in an environment of constant contention for control.

Angkor Wat is really a nested set of enclosures—galleries, walls, and a moat—centered on an imitation of Mount Meru, the polar axis world peak. The heavenly character of all of Angkor Wat is communicated by the *apsaras*, or celestial nymphs, that cover its walls. The entire complex replicates the structure and behavior of the Hindu cosmos, and in that way it harmonizes the terrestrial realm inhabited by people with the celestial realm where the gods dwell. The goal is access to the gods and the transfer of some of the eternal and pure reality of heaven to the flawed and spiritually fogged world in which we live.

Like every ritualized mountain, Angkor Wat reflects a simple notion: To get to the gods, you have to get to the sky. It also exploits the first principle of cosmo-magical power. Stability and prosperity on earth result from congruence with heaven. The match is made by incorporating the design of the universe and the rhythms of its change into places of power on earth. In a traditional society, where the sacred is a shared vehicle of cultural identity, the community's power is maximized when the belief system is meshed with the lattice of the state. For that reason, King Suryavarman II made sure that Angkor Wat—a sacred instrument of the state—would mimic the cosmos.

Angkor Wat is cardinally oriented and huge. The outer dimensions, including the moat, are about 4,260 feet north-south and 4,920 feet east-west, almost a mile in

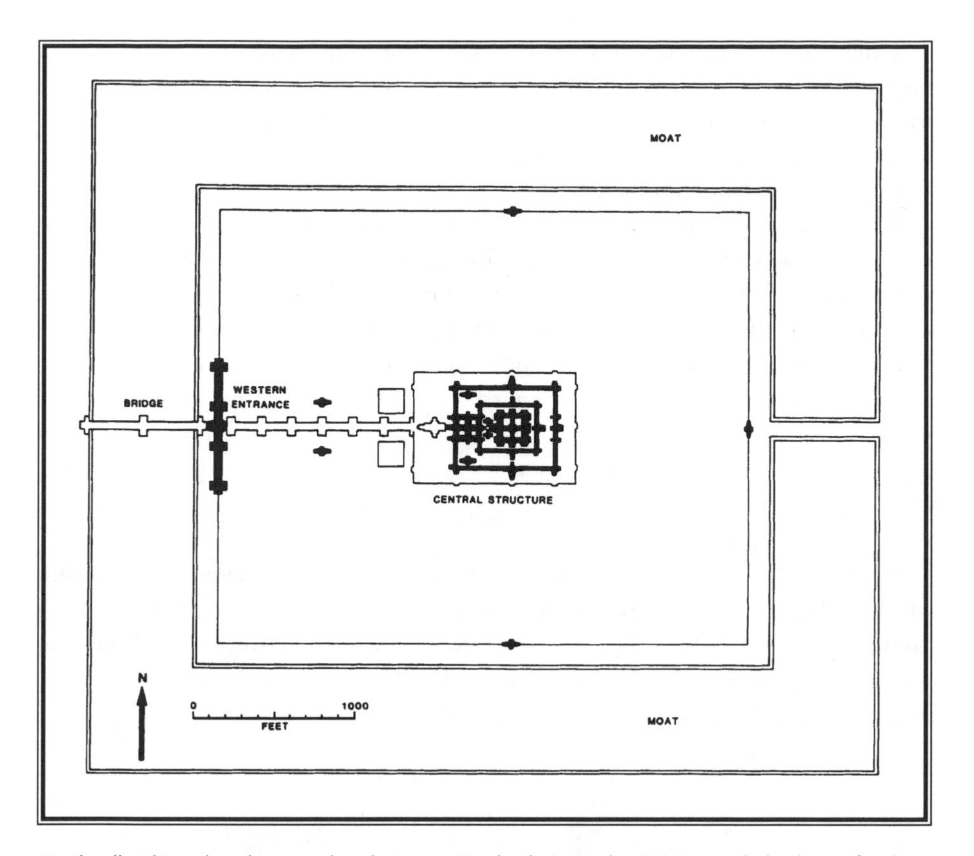

Cardinally aligned and oriented to the west, Cambodia's Angkor Wat provided a home for the deified spirit of its builder, Suryavarman II, the Khmer king. Its central structure was designed to mirror the structure and character of Mount Meru, the world-axis mountain that centered the cosmos, and the dead king, transformed into the god Vishnu, who resided there. The first enclosure (bold outline) of the central structure is the Second Gallery. Its finely carved mythological and historical reliefs completely surround the inner monument. (Griffith Observatory drawing, Joseph Bieniasz)

both directions. The moat and outermost wall correspond to the edge of the world and to the ocean that encircles the world's central continent. In that sense, the outer limits of Angkor Wat also symbolize the frontier of the Khmer empire, the periphery of its influence.

The sanctuary in the middle of Angkor Wat touches heaven 213 feet above the ground. Its summit is the earthly incarnation of the Brahma's celestial city, the apartments of the gods. This heavenly high-rise is actually surrounded by four lower towers at the corners. They represent the subordinate mountains that flank Meru and restate the cardinal reference frame. Tapered to their tops like rising stacks of diminishing doughnuts, elaborately glazed, all five towers resemble pagodas. Their alliance with the sky is obvious, and one attempt to convey their vertical reach likens them to "so many rockets aimed at heaven."

Unlike most of the Angkor monuments, however, Angkor Wat's principal gate is on the west, and the elevated promenade that conducted the privileged to the innermost temple-mountain sanctuary starts on the far west bank of the moat. It penetrates eastward through gates, over terraces, and up the mountainous slopes of the inner sanctuary. All of this emphasis on the west led some to interpret Angkor Wat as Suryavarman II's funerary monument. To an extent that may be so, for most of the other cosmic mountain temple complexes like it probably did shelter the final remains of the king and members of the royal family. George Coedès argued effectively, however, that neither Angkor Wat nor the other great Khmer sanctuaries were strictly sepulchres. They were homes for the gods—the spirits of the divine kings.

The spirit of the Khmer king occupied a central pyramid in a capital city at the center of an empire, and his power, like the power of cosmic order, radiated to all eight points of the compass. In that way, he behaved like the source of cosmic power at the high throne of the universe, and Angkor Wat mirrored the structure of the cosmos and the capital.

Khmer civilization was introduced to the concept of divine kingship with the help of Java, where the tradition was already entrenched by the seventh century A.D., when Jayavarman II, who belonged to a noble Khmer family in exile, returned to Cambodia. Jayavarman II founded the first capital at Angkor. He is credited with unifying Cambodia, creating the Khmer empire, and laying the groundwork for the ideology of the

The view east from about halfway along the elevated promenade between Angkor Wat's western entrance and the west side of the Second Gallery keeps the center tower of the sanctuary on-axis. Two of the four lower corner towers around it are also evident, but the second pair is hidden behind them. (photograph E. C. Krupp)

kings who followed him. When those kings died, they were transmuted into a high god of the Hindu pantheon, usually Siva. By the time Suryavarman II occupied the Khmer throne, however, Vishnu, the deity who manages the periodic destructions and regenerations of the cosmos, was the choice of kings. Transfigured through death into Vishnu, Suryavarman II was believed to live in Angkor Wat, and Vishnu's power of creation was centered there.

An interdisciplinary study of the measurements of Angkor Wat led Robert Stencel, Fred Gifford, and Eleanor Morón to conclude that fundamental dimensions of Angkor Wat were intended to symbolize numerically the vast intervals of time which Hindu cosmology assigned to successive world ages. For example, the distance in traditional Khmer units from the first step of the outer western bridge to the center of the monument closely approximated one millionth of the duration of the first—and golden—age, the Krita Yuga. It lasted 1,728,000,000 years, and other distances along the main axis were judged to correspond to the three shorter ages that followed. Other elements of Angkor were associated with various astronomical and calendrical cycles, and alignments with significant risings of the sun and moon were argued for various pairs of the monument's key architectural components. Although it is difficult to prove these congruences, they are consistent with the objective of harmonizing earth with heaven in a royal monument inhabited by the gods.

The famed wraparound wall reliefs of Angkor Wat are as compelling as its architectural presence, and they, too, echo cosmological themes. Located on the outer walls of the Second Gallery, they carry the imagery of war and creation to the cardinal directions. On the east side, Suryavarman II's artists illustrated the myth of the Churning of the Ocean of Milk. It is a story of Creation in the first world age. Attempting to curdle the lost elixir of immortality out of the milky waters of primordial time, the gods and the demons collaborated by making the mother of all butter churns out of the great mythical mountain known as Mandara. Coiling the monstrous serpent Vasuki around the massive peak, the two teams engaged in a cooperative tug-of-war with the snaky rope to agitate the mountain and emulsify the sea, but as they swiveled it back-and-forth, it started to slip and sink into the creamy ooze on the ocean floor. Vishnu reacted quickly to preserve their investment. Transforming himself into Kurma the cosmic tortoise, he submerged himself under the mountain and supported it upon his shell. With a world turtle to steady the churn-stick mountain, the gods and demons kept on whipping up the ocean for over a thousand years. In time, marvelous and precious things emerged from the clotting waters—the sun, the moon, the goddess of fortune, the goddess of wine, the first apsara, the tree of celestial paradise, the magic cow, the white horse, the milk-white elephant, the bow, the conch of victory, the pearl of gems, and more, including the beverage of eternal life.

Suryavarman II allocated 150 feet of stone wall for this scene and so allied himself with Vishnu's creative energy and the wealth it bequeathed to the world. Because it is on the east side, the relief greets the rising sun. This analogy of creation and

Although this relief is almost six feet high, it is only a small piece of Angkor Wat's half-gallery depiction of the Churning of the Ocean of Milk. This is the very center of the entire scene, and it illustrates Mount Mandara stabilized on the back of the great tortoise with Vishnu himself assisting in the foreground. Celestial nymphs churned from the sea are in the upper register. Part of the serpent Vasuki can be seen extending to the left and in the grasp of a demon. (photograph E. C. Krupp and Robin Rector Krupp)

beginnings finds closure in the battle relief on the gallery's western wall. There, where they can catch the light of the setting sun, the images of war and destruction symbolize the other side of the cycle of life and death. Although the world is inevitably destroyed at the end of a world age, complete annihilation is averted by Vishnu's participation in the regeneration of time. He awakens from his dreaming at the end of each cosmic cycle. On earth, the king is the same kind of anchoring influence who responds to the threat to world order.

These analogies help define what Robert von Heine-Geldern, a German scholar, identified as Angkor's "magic relation between microcosm and macrocosm, between the human world and the universe, between terrestrial manifestations on the one hand and the points of the compass and constellations on the other." Everything, he added, had its own "magic position" in the structure of the cosmos and its own "magic moment" in the rhythms of the sky. Cosmic forces controlled the world, and "nothing could prosper unless it was in harmony with these universal forces."

Harmony between heaven and earth continued to preoccupy the Khmer kings. They each created their own central-world-mountain monuments to proclaim their own divine credentials, and although the basic design conformed to the same principles that enfranchised earlier kings, they added personal touches congruent with their own times. At the end of the twelfth century, Jayavarman VII enlarged the Khmer empire and rebuilt the capital as the city of Angkor Thom—square, cardinal, and centered on his shrine, the Bayon. The Bayon was constructed a century after the completion of Angkor Wat. It has 54 towers, and they all champion cardinality with a face of Jayavarman VII transformed into Avalokiteshvara—the Buddha of the future—in each of the four directions. After ambling through the Bayon's third and highest level, you are convinced nothing escapes the view of all of these faces charged with monitoring the world's four directions.

Adding Buddhist icing to the Hindu cosmological cake, Jayavarman VII appropriated the myth of the churning of the milky ocean from Hindu epic and embedded it in the design of his fortified royal capital. Each of the bridges that connect the city's gates with the outside world has a guardrail of gods on one side of the causeway and a palisade of demons on the other. In single file, each team of these supernaturals is gripping a serpent banister, its head at the front of the parade of gods and its tail in the hands of the demons. They are united to retrieve the sacred treasures lost in an

Teams of demons and gods are lined across the bridge to the South Gate of Angkor Thom, the cosmic, royal city built by Jayavarman VII. The flaring figure in the foreground is the multiheaded serpent whose coils were put to work to manufacture the elixir of immortality from the Ocean of Milk, but here the symbolism all refers to the king. He is the source of the world's prosperity, and the churn-stick of the universe is his world-mountain monument. (photograph E. C. Krupp)

earlier age through the magical churning of the sea. What they really generate is an unflagging renewal of royal power in the layout of Angkor Thom.

Paul Wheatley, in *The Pivot of the Four Quarters,* his book about the cosmo-magical properties of sacred world-axis cities, imagined the snake railing was conceptually coiled around the central tower of the Bayon. George Coedès, on the other hand, put the churning pole in another location. There are five bridges and tower gates into the city—one for each direction and a fifth, also on the east, for the route known as the Avenue of Victory. Coedès imagined each tower gate was the world-mountain churn stick. As a terrestrial manifestation of Mount Mandara, each tower is crowned, like the towers of the Bayon, with cardinally directed faces of the future Buddha.

Because the center tower of the Bayon was equated with Mount Meru, the central mountain of the world, it may be more sensible to think of each gate as the second cosmic peak, offset, like Mount Mandara, from the world's true center. It is also reasonable, however, to suppose the two mountains were merged in the center of the city and in all five gates. In any case, we have a royal monument that represents the creative and stabilizing power of the divine king. The message is clear. The king himself is the cosmic pivot that articulates the land with productivity and strength.

Infiltrating architecture with myth, King Jayavarman VII told his empire that he was the source of its victory and prosperity. Through his shrine, at the heart of the Khmer capital, he was the foundation—the tortoise Vishnu—that steadied the world and churned out its treasures. Opposing factions were united like the gods and demons under the king's effective leadership. Through dynastic succession, he ensured the wealth and well-being of the empire. He didn't just climb the cosmic mountain for power from the gods. He was the cosmic mountain, and his monuments celebrated his divinity.

Of course, not everybody built pyramids, and those who did put them to different use. In the language of symbolic architecture, however, their fundamental meaning is obvious. They are artificial cosmic mountains. The resources required to construct them are extraordinary. It takes an upwardly mobile chiefdom or a wealthy sacred kingship to get the idea off the ground. Centralized power is a requirement, but so is ideology. Ideology is the duct tape that holds the monument together, and the monument reinforces the centralized power that got it built in the first place.

Where power is sanctioned by the sacred, pyramidal monuments are public displays of divine cosmic mandate. They legitimize lineage and authenticate authority. Frequently these monuments served as homes for the dead . . . but not just any dead. These were the exalted and celestially transfigured dead—the highest elite, the divine kings of societies that could afford to be ruled by gods. Although our catalog is not complete, the evidence we have implies that pyramidal temple-tombs were intended to do the obvious: Lift the honored dead to the sky. In detail, these pyramids probably meant many different things, but by invoking the sky, they provided a little boost up to the gods in the sky, a kind of footstool to heaven—and even a divine king can use a step in the right direction.

WHY THEY NEVER LAND ON THE WHITE HOUSE LAWN

People wouldn't build astronomically tuned pyramids and cosmic cities, and they wouldn't elevate rulers to the top of a social pyramid, if centralized power weren't useful. Its usefulness depends, of course, on the environment in which people live and on the challenges they face. Competition, aggression, and economic stress favor the concentration of power, for the ability to direct resources is an organizational advantage. For all of our complaints about faceless bureaucracies, remote government, and abusive authority, we have continued to return to these tools again and again. We have invented ways to temper influence and limit power, but despite our nostalgia for Paradise, most of us don't willingly reduce the scale of our cultural impact and abandon useful technology and knowledge to return to the Eden of the small hunting/foraging band. And as egalitarian as some of those bands may appear, they also seem to be perfectly capable of developing complicated systems of belief and ritual interaction. The scale is different, but the imagination still has reach.

Centralization of power is driven by the economic and political advantages it confers. It can do some things that are otherwise impossible, and when it is managed effectively, it is a potent weapon in a community's arsenal for survival. That does not mean, however, that it's the free lunch of cultural evolution. It comes with factory-standard costs. Centralized power inevitably introduces social stratification and uneven distribution of wealth. If no one cared about status and opportunity, this wouldn't be an issue, but people generally can judge where their personal interests lie, and they usually don't lie low on the social ladder. So those personal interests are not always congruent with the goals of the group. Rank and resources can, then, also be powerful agents of social subversion. The family loyalty that naturally bonds people despite their differences, through recognition of shared kinship, operates on behalf of social cohesion at the level of the bloodline, but when the society gets big and complicated, it can only remain stable if its mixed population somehow believes in its own distinct identity. Language contributes to the cause, and people are also persuaded by the belief system that they belong to a defined and worthy group. Power in the broadly extended group relies on ideology.

Ideology, we have seen, legitimizes leadership and status by establishing a clear connection between human affairs and the cosmos as a whole. If society reflects the character of the universe, it can be regarded as an inherent element of nature. It is, after all, endorsed by the gods. Canny interaction with the gods invests the ruler with their patronage, and the successful ruler works out an ideological mechanism to pump divine will through the chain of command.

We know, then, why the expressions of power are so saturated with celestial themes. Ideas are the negotiable currency in the marketplace of power, and the sky is where the gods do most of their high-end business. With the highest order of the universe revealed overhead, celestial metaphors are inevitable components of the belief system.

Maintenance of the ideology requires investment in visible emblems of its validity. Recognition of the sanctity of high places is the beginning of the process, but chiefs, kings, and emperors—and their staffs—are compulsively attracted toward monuments that incorporate and advertise cosmology, divinity, and royal sovereignty. This, too, contains its own destabilizing forces, however, for the centralized economy has a finite surplus. It can only afford to invest so much. If the leadership outspends the assets, and can't sustain the wager of its debt, the system fragments, and the game folds. That is, in part, what happened in Old Kingdom Egypt—too many pyramids for the gross national product. Similar pressures may have contributed to the collapse of the Classic Maya and the decline of Khmer civilization. Capital investment in ideology, by itself, probably doesn't trigger the plunge, but combined with other factors—like the expense of warfare and environmental disruption, it may break the bank.

Ideology is conservative. That's its job, but in fulfilling its function, it also restricts invention and experimentation. That doesn't mean ideologically unified societies are not inventive and culturally rich. They are, but the important factor is their ability to adapt to change. Because change is inevitable, adaptation—like centralized power—is a useful skill.

Popular culture reiterates ancient myth. The sky is charged with transcendental power, and some of it now and then is transferred to earth. (Griffith Observatory drawing)

Institutions that proliferate ideas, tools, technology, and problem-solving options seem to add an adaptive edge. One experiment in that direction is our own era's secularization of power. The separation of church and state has decoupled the sacred from the social contract. The celestial seal of divine approval is not conferred on presidents, prime ministers, parliaments, courts, or Congress, no matter what oaths we swear with a hand on the Bible. The effectiveness of celestial ideology in merchandising political institutions has evaporated wherever cosmovision no longer issues the writ of governance.

So far the experiment shows promise, but the baseline of evidence is still short. Abandoning the sacred covenant between the social order and the divine also compromises social cohesion, and social cohesion was the point behind banking on ideology in the first place.

Even if the government derives its power from the people, and not from the gods, the universe is no modest off-Broadway production. It is a cosmic performance piece that dazzles and daunts our minds with its energy and scale. People still understand that most of what happens in the world is out of their control and that great forces activate the physical universe. It may be gravity and particle physics calling the shots instead of ancestral spirits and rain gods. Redshifted galaxies and the curvature of space may define heaven instead of hierarchies of angels. But whatever governs the nature of the cosmos is still more powerful than the sentient organisms of planet earth.

People are reluctant to relinquish the sacred entirely, for they still have to come to personal terms with their place in the cosmos. A physical interpretation of nature does a lot for predictive understanding, but it doesn't necessarily make people feel at home. Belief in transcendental power once fortified the social order. Science delivers its own brand of power, but some people still feel there is something missing. They wish we could still count on heaven the way our ancestors did. Reacting to the loss, traditionalists, fundamentalists, and, frankly, anyone looking for the linchpin of a universe without a center or an edge, embrace mechanisms that reinstate archaic ideology and restore familiar values. These responses may reach back to an earlier system of belief in which the universe and its sources of power made more sense, or they may adapt the imagery of modern scientific discovery to our old habit of personalizing power. In the first case, divine power—now not so literally lodged in the sky—is back in the driver's seat. In the second case, supernatural power is actually the unimaginably advanced technology of aliens from space.

In the late twentieth century, factions of fundamentalist movements now supplement acceptance of the literal meaning of sacred texts with attempts to enlist the coercive power of the state on behalf of belief. In the United States, that is what the warfare over prayer in public schools, over Darwinian evolution in science classes, and over the introduction of Creation pseudoscience into the curriculum is all about. Establishment of so-called Islamic republics is another facet of the same beguiling gem of true belief.

Traditionalists, on the other hand, operate, to the extent that they can, as if the world has not changed. Marginalized by the structure of power, they skirt the fields of outside influence and fulfill ritual duties they recognize as still essential even though the greater world in which they live neither recognizes the value of what they undertake nor is grateful for their effort. Even now, on behalf of an oblivious planet, the Elderly Elders of the village of Hotevilla, on Third Mesa in northern Arizona's Hopi reservation, battle with other Hopi factions over their responsibility to fulfill a sacred covenant with Maasaw, the Guardian Spirit of the Earth. According to Hopi prophecy, the planet is on the eve of catastrophic crisis, and the Traditionals of Hotevilla hope to avert it, or minimize its destructive impact, by perpetuating time-honored Hopi lifeways and protecting the integrity of the ancient charter with the gods. For the Traditionals, this means forbidding the introduction of water pipes and electrical lines in the village. Their aversion to the convenience of modern utilities is not just a Luddite rejection of modern technology. There is an object of power, a "marker," hidden underground in Hotevilla. Ritually deposited some time after the village was founded, it is linked to the ancient Hopi compact. Excavation, the Elderly Elders fear, could disturb it, with tragic consequences for the planet. These traditionalists deprive themselves to steward the threatened belief system. Acting autonomously, they rely on their own limited resources and unbending commitment.

Both fundamentalists and traditionalists make room for the miraculous in a scientific age. This is not easy when systems of belief and knowledge collide, but even the rest of the population seems intent on identifying an ideology for miracles that can be accommodated by the secularization of power and a scientific engagement with nature. We want the world to retain a capacity for miracle because the miraculous is an expression of transcendental power.

Now that there is no transcendental center of the world, the sacred is no longer embedded in the architecture of public power. Because the arena of nation-states no longer operates on sacred power, miracles and the power that makes them possible must be encountered in personal life, not in public forums. That is why flying saucers never land on the White House lawn. Space aliens and UFOs are a modern myth of transcendental power, but that power no longer has any charter with the government. The saucer folk may say, "Take me to your leader," but they don't really mean it. Instead, the aliens seem to be fans of ordinary people. Stories of contact and abduction are essentially visionary or religious experiences—and not necessarily always pleasant. With the focus of extraterrestrial attention, the people who have these encounters become extraordinary, but only individuals are transformed. The state is untouched and actually is targeted by believers as the agency of official denial and duplicitous knowledge. For believers, then, there may be a secret empire of institutional power linked to restricted knowledge of the activity of alien spacecraft, but this is no cosmic mandate to run the country.

The new geography of the physical universe allows us to find a home for miraculous power in a system of knowledge that bypasses the gods. Mysterious apparitions and encounters are now instead accepted by many as evidence of spacecraft from other worlds and alien contact. Marvels are fit for polite company if they are really superscience from the stars. We don't even have to accept the reality of visiting extraterrestrials to find the concept useful in allowing us to understand the perceptions of those who do believe that space aliens float and glow among us.

Whether or not we embrace the tales of interaction with visitors from other worlds, most of us still have an urge to see something uncanny in the nature of the cosmos. Because we no longer all share the same world view, however, we can't rely on cosmovision to contain it. That doesn't mean there is no longer any traffic in symbols. Cosmovision has been replaced by television. The images and commentary that appear in the flames of the electronic hearth are stories that help create and reinforce the mythic structure of the world. As passive—and often uncritical—consumers of televised data, we are influenced not just by verifiable fact, but by repeated exposure. Abduction by UFOs and alien autopsies are as commonplace as commercials and as compelling as pulp fiction. Whether we believe in them or not, they are part of our vocabulary of cosmic wonder.

We demand, however, a personal and human dimension in our miracles, and that may help explain sightings of Elvis and the enshrining of Graceland. Merging the beatification of Elvis with mystery in space, *The Sun,* a sensationalist tabloid, headlined the September 20, 1988, edition with "Statue of Elvis Found on Mars." The Elvis shrine in my own house is dedicated to the proposition that a religion is gradually and quietly condensing around the "King." Skeptics understandably reject the idea as absurd, but Elvis is becoming a source of personal power and inspiration. A neighbor, at first skeptical of this notion, was singing in Brazil as backup for an Elvis impersonator. On that occasion, in a conversation with a woman from the audience, she was shown a picture of Elvis in the woman's wallet, and the Brazilian woman said, "I feel much safer with this with me." For those who prefer a more public expression of the Elvis influence, I direct your attention to the route that connects the Israeli port of Ashdod with Jerusalem. A gas station and cafe on Highway 1, at the summit of the last hill before descending to the city that is a holy cosmic center for three of the world's great living religions, greets visitors with nonstop Elvis recordings and Elvis memorabilia covering every wall and column. The sign outside proclaims "I saw Elvis at the Elvis Inn," and the future may judge it is no coincidence this place of Elvis power is near Jerusalem. Anyone still doubtful about the ubiquity of Elvis and the evolution of his myth should take a look at Rowland Scherman's book of photographs *Elvis Is Everywhere.*

The transfiguration of Elvis and inklings of technology imported from the stars may be understandable mythic consequences of the transformation of the belief system and the secularization of power, but for now, at least, they are a little too wacky for a mainstream ideology of transcendental power. It is reasonable to ask, however, what, if anything, might reignite a cross-cultural ideology in a world that has separated the sacred from the state and whose fortunes depend on scientific analysis and testable knowledge. What reservoirs of power can mobilize a society of Balkanized belief, skeptical of authority but absolutely dependent on centralized power? There are, perhaps, two concepts that have emerged as agents of ideological unification—the environmental sanctity of the earth and the cosmic mystery of outer space. Our fear of fatally fouling our own nest and the elevating imagery of exotic and distant frontiers have great emotional power, and emotional response stokes the furnace of belief. These energizers of ideology are maybe not so different after all from what preoccupied our ancestors. The planet's well-being may now be a global issue, but prosperity and survival are still our concerns. Demonstrations of cosmic power, on the other hand, are now evident in a comet's crash into Jupiter and in Space Telescope views of vast smoky pillars of interstellar gas that cradle newborn stars. These landscapes and unprecedented events become the property of the entire earth through television and the World Wide Web. Centers of public astronomy, like Griffith Observatory in Los Angeles, put more people than ever eyeball to the universe with in-person views through a telescope and with the cosmic verisimilitude of the planetarium. The cosmos of our ancestors may seem more cohesive, but ours is unmatched for spectacle, mystery, and awe.

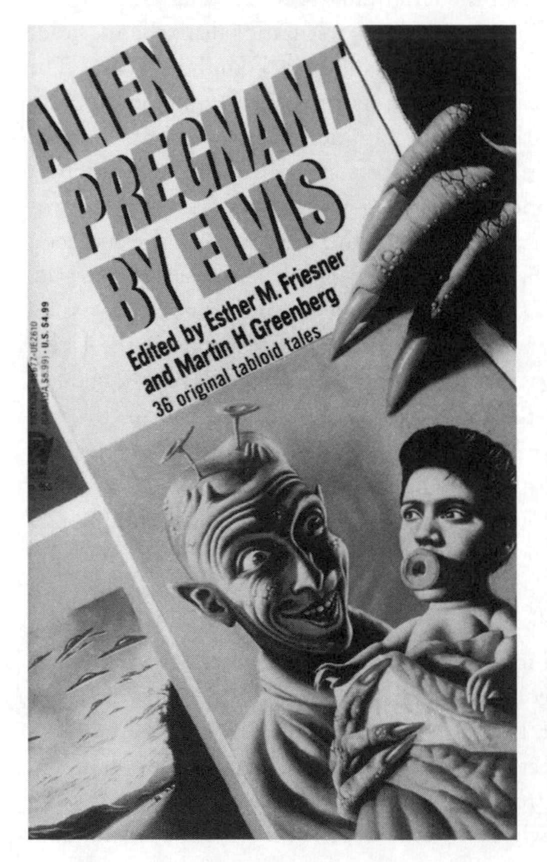

Supermarket tabloids sensationalize the news and even sometimes make it up. Although their content is consumed at the expense of critical thought and real facts, the tabloids, like the rest of popular culture, reflect the mythology of our age. Here, however, a collection of fake tabloid stories sensationalizes the tabloids themselves in the interest of entertainment. It is significant that these cheap thrills are packaged with an Elvis headline. Clearly that's what gets attention, and in this case the power of Elvis goes cosmic. Inside, art imitates art, for the contents include the tale "Martian Memorial to Elvis Sighted," a fictional echo of an earlier tabloid report of Elvis on Mars. (photograph Griffith Observatory, cover by DAW Books, Inc.)

We may, then, be seeing a glimpse of a new ideology of power. It naturally emerges from a new vision of nature that is too compelling for most of us to reject, although there will always be some who persist in the anachronistic quest for the center of the world. We, however, remain on alert, thanks to the Irish poet William Butler Yeats, for that new rough beast at last slouching toward Bethlehem. But as its ordained hour comes round, we should keep in mind that it could look something like Elvis.

BIBLIOGRAPHY

The broad conceptual character of the following references, or their cross-cultural content, made them useful sources throughout this book. For that reason, they are collected together here and are not referenced again with the other sources listed according to the individual sections they supported.

ANCIENT AND PREHISTORIC ASTRONOMY AND COSMOLOGY

Aveni, Anthony F. *Ancient Astronomers,* Exploring the Ancient World Series. Washington, D.C.: Smithsonian Books, 1993.

———. *Conversing with the Planets.* New York: Times Books, 1992.

———. *Empires of Time.* New York: Basic Books, Inc., Publishers, 1989.

———. *Skywatchers of Ancient Mexico.* Austin: University of Texas Press, 1980.

———, ed. *Archaeoastronomy in the New World.* Cambridge: Cambridge University Press, 1982.

———, ed. *Archaeoastronomy in Pre-Columbian America.* Austin: University of Texas Press, 1975.

———, ed. *The Lines of Nazca.* Philadelphia: The American Philosophical Society, 1990.

———, ed. *Native American Astronomy.* Austin: University of Texas Press, 1977.

———, ed. *New Directions in American Archaeoastronomy.* Oxford: B.A.R., 1988.

———, ed. *The Sky in Mayan Literature.* New York: Oxford University Press, 1992.

———, ed. *World Archaeoastronomy.* Cambridge: Cambridge University Press, 1989.

Aveni, Anthony F., and Gordon Brotherston. *Calendars in Mesoamerica and Peru and Native American Computations of Time.* Oxford: B.A.R., 1983.

Aveni, Anthony F., and Gary Urton, eds. *Ethnoastronomy and Archaeoastronomy in the American Tropics.* New York: New York Academy of Sciences, 1982.

Baity, Elizabeth Chesley. "Archaeoastronomy and Ethnoastronomy So Far." *Current Anthropology* 14 (1973): 389–449.

Bauer, Brian S., and David S. Dearborn. *Astronomy and Empire in the Ancient Andes: The Cultural Origins of Inca Sky Watching.* Austin: University of Texas Press, 1995.

Benson, Arlene, and Tom Hoskinson, eds. *Earth and Sky—Papers from the Northridge Conference on Archaeoastronomy.* Thousand Oaks, California: Slo'w Press, 1985.

Blacker, Carmen, and Michael Loewe, eds. *Ancient Cosmologies.* London: George Allen & Unwin Ltd., 1975.

Blake, John F. *Astronomical Myths.* London: Macmillan and Co., 1877.

Brecher, Kenneth, and Michael Feirtag, eds. *Astronomy of the Ancients.* Cambridge, Massachusetts: MIT Press, 1979.

Broda, Johanna, Stanislaw Iwaniszewski, and Lucrecia Maupomé, eds. *Arqueoastronomía y Etnoastronomía en Mesoamérica.* Mexico City: Universidad Nacional Autónoma de México, 1991.

Burl, Aubrey. *Prehistoric Astronomy and Ritual.* Aylesbury, Bucks., U.K.: Shire Publications Ltd., 1983.

———. *Rites of the Gods.* London: J. M. Dent & Sons Ltd., 1981.

Carlson, John B. "America's Ancient Skywatchers," *National Geographic Magazine* (March 1990): 76–107.

Carlson, John B., and W. James Judge. *Astronomy and Ceremony in the Prehistoric Southwest.* Albuquerque: Maxwell Museum of Anthropology, 1987.

Cook, Roger. *The Tree of Life—Image for the Cosmos.* New York: Avon Books, 1974.

Cornell, James. *The First Stargazers.* New York: Charles Scribner's Sons, 1981.

de Santillana, Giorgio, and Hertha von Dechend. *Hamlet's Mill.* Boston: Gambit, Incorporated, 1969.

Dicks, D. R. *Early Greek Astronomy to Aristotle.* London: Thames and Hudson, 1970.

Eliade, Mircea. *Cosmos and History: The Myth of the Eternal Return.* 1954. Reprint. New York: Harper & Row, 1959.

———. *The Sacred & the Profane: The Nature of Religion.* 1959. Reprint. New York: Harcourt Brace Jovanovich, no date.

Frankfort, H. et al. *The Intellectual Adventure of Ancient Man.* Chicago: University of Chicago Press, 1946.

Griffin-Pierce, Trudy. *Earth Is My Mother, Sky Is My Father.* Albuquerque: University of New Mexico Press, 1992.

Hadingham, Evan. *Early Man and the Cosmos.* New York: Walker and Company, 1984.

Hawkins, Gerald S. *Beyond Stonehenge.* New York: Harper & Row, Publishers, Inc., 1973.

———. *Mindsteps to the Cosmos.* New York: Harper & Row, Publishers, Inc., 1983.

Heath, Thomas L. *Greek Astronomy.* London: J. M. Dent & Sons Ltd., 1932.

Hodson, F. R., ed. "The Place of Astronomy in the Ancient World." *Philosophical Transactions of the Royal Society of London* 276, no. 1257:1–276.

Iwaniszewski, Stanislaw, ed. *Readings in Archeoastronomy.* Warsaw: State Archaeological Museum and Department of Historical Anthropology, Institute of Archaeology, Warsaw University, 1992.

Iwaniszewski, Stanislaw, Arnold Lebeuf, Andrzej Wiercinski, and Mariusz Ziolkowksi, eds. *Time and Astronomy at the Meeting of Two Worlds.* Warsaw: Warsaw University Center for Latin American Studies, 1994.

James, E. O. *The Tree of Life.* Leiden: E. J. Brill, 1966.

Jaschek, Carlos, ed. *European Meeting on Archeoastronomy & Ethnoastronomy.* Strasbourg, France: l'Observatoire Astronomique de Strasbourg, 1992.

Krupp, E. C. "Ancient Watchers of the Sky," *1980 Science Year. World Book Science Annual.* Chicago: World Book-Childcraft International, 1979, pp. 98–113.

———. *Beyond the Blue Horizon—Myths and Legends of the Sun, Moon, Stars, and Planets.* New York: HarperCollins, 1990 (now only available in paperback from Oxford University Press, 1991).

———. *Echoes of the Ancient Skies.* New York: Harper & Row, Publishers, Inc., 1983 (now only available in paperback from Oxford University Press, 1994).

———, ed. *Archaeoastronomy and the Roots of Science.* Boulder, Colorado/Washington, D.C.: Westview Press/American Association for the Advancement of Science, 1984.

———, ed. *In Search of Ancient Astronomies.* Garden City, New York: Doubleday & Company, Inc., 1978.

———. "Light and Shadow," *Griffith Observer* 54, no. 6 (June 1983): 12–20.

Lewis, Sir George Cornewall. *An Historical Survey of the Astronomy of the Ancients.* London: Parker, Son, and Bourn, West Strand, 1862.

Lloyd, G. E. R. *Early Greek Science: Thales to Aristotle.* New York: W. W. Norton & Company, Inc., 1970.

———. *Greek Science After Aristotle.* New York: W. W. Norton & Company, Inc., 1973.

Lyle, Emily. *Archaic Cosmos—Polarity, Space and Time.* Edinburgh: Polygon, 1990.

———, ed. *Sacred Architecture* (*Cosmos,* volume 8, the Yearbook of the Traditional Cosmology Society). Edinburgh: Edinburgh University Press, 1992.

Malville, J. McKim, and Claudia Putnam. *Prehistoric Astronomy in the Southwest.* Boulder, Colorado: Johnson Publishing Company, 1989.

Marshack, Alexander. *The Roots of Civilization.* New York: McGraw-Hill Book Company, 1972.

McCoy, Ron. *Archaeoastronomy—Skywatching in the Native American Southwest.* Flagstaff: The Museum of Northern Arizona Press, 1992.

Mercer, Samuel A. B. *Earliest Intellectual Man's Idea of the Cosmos.* London: Luzac & Co., 1957.

Michell, John. *At the Center of the World.* New York: Thames and Hudson, Inc., 1994.

———. *Secrets of the Stones—the Story of Astro-archaeology.* Harmondsworth, Middlesex, England: Penguin Books Ltd., 1977.

Neugebauer, O. *The Exact Sciences in Antiquity.* 2d edition. 1957. Reprint. New York: Harper & Row, Publishers, Inc., 1962.

———. *A History of Ancient Mathematical Astronomy, Part 2.* New York: Springer-Verlag, 1975.

O'Neil, W. M. *Early Astronomy from Babylon to Copernicus.* Sydney, Australia: Sydney University Press, 1986.

———. *Time and the Calendars.* Sydney, Australia: Sydney University Press, 1975.

Pennick, Nigel. *The Cosmic Axis.* Cambridge, England: Runestaff Publications, 1985.

Ruggles, C. L. N., ed. *Archaeoastronomy in the 1990s.* Loughborough, U.K.: Group D Publications Ltd., 1993.

———, ed. *Astronomy and Cultures.* Niwot, Colorado: University Press of Colorado, 1993.

———. *Records in Stone—Papers in Memory of Alexander Thom.* Cambridge: Cambridge University Press, 1988.

Sarton, George. *A History of Science, Volume 1* (Ancient Science Through the Golden Age of Greece) and *Volume 2* (Hellenistic Science and Culture in the Last Three Centuries, B.C.). 1952 and 1959. Reprint. New York: W. W. Norton & Company, Inc., 1970.

Schiffman, Robert A., ed. *Visions of the Sky—Archaeological and Ethnological Studies of California Indian Astronomy* (Coyote Press Archives of California Prehistory no. 16). Salinas, California: Coyote Press, 1988.

Swarup, G., A. K. Bag, and K. S. Shukla. *History of Oriental Astronomy.* Cambridge: Cambridge University Press, 1987.

Thurston, Hugh. *Early Astronomy.* New York: Springer-Verlag, Inc., 1994.

van der Waerden, B. L. *Science Awakening II: The Birth of Astronomy.* Leyden: Noordhoff International Publishing, 1974.

Warren, William Fairfield. *The Earliest Cosmologies.* New York: Eaton & Mains, 1909.

Williamson, Ray A., ed. *Archaeoastronomy in the Americas.* Los Altos, California, and College Park, Maryland: Ballena Press and Center for Archaeoastronomy, 1981.

———. *Living the Sky—the Cosmos of the American Indian.* Boston: Houghton Mifflin Company, 1984.

Williamson, Ray A., and Claire R. Farrer, eds. *Earth & Sky—Visions of the Cosmos in Native American Folklore.* Albuquerque: University of New Mexico Press, 1992.

Worthen, Thomas D. *The Myth of Replacement—Stars, Gods, and Order in the Universe.* Tucson: The University of Arizona Press, 1991.

Zinner, Ernst. *Die Geschichte der Sternkunde.* Berlin: Verlag von Julius Springer, 1931.

Zinner, Ernst. *The Stars Above Us.* New York: Charles Scribner's Sons, 1957.

Ziółkowski, Mariusz S., and Robert M. Sadowski, eds. *Time and Calendars in the Inca Empire.* Oxford: B.A.R., 1989.

RELATED JOURNALS

Archaeoastronomy, Bulletin of the Center for Archaeoastronomy. Center for Archaeoastronomy, Post Office Box X, College Park, Maryland, 20740.

Archaeoastronomy, Supplement to the Journal for the History of Astronomy. Science History Publications Ltd., Halfpenny Furze, Mill Lane, Chalfont St. Giles, Buckinghamshire, England, HP8 4NR, U.K.

Cosmos. Traditional Cosmology Society, School of Scottish Studies, University of Edinburgh, 27 George Square, Edinburgh EH8 9LD, Scotland, U.K.

OTHER GENERAL REFERENCES

Berger, Peter L. *The Sacred Canopy.* Garden City, New York: Doubleday & Company, Inc., 1967.

Bonnefoy, Yves, compiler. *Mythologies, Volumes One and Two* (trans. Wendy Doniger). Chicago: The University of Chicago Press, 1991.

Carlyon, Richard. *A Guide to the Gods.* New York: William Morrow and Company, Inc., 1982.

Campbell, Joseph. *Historical Atlas of World Mythology, Volume I: The Way of the Animal Powers.* New York: Alfred van der Marck Editions, 1983.

Campbell, Joseph. *Historical Atlas of World Mythology, Volume II: The Way of the Seeded Earth, Part 1: The Sacrifice.* New York: Harper & Row, Publishers, Inc., 1988.

———. *Historical Atlas of World Mythology, Volume II: The Way of the Seeded Earth, Part 2: Mythologies of the Primitive Planters: The Northern Americas.* New York: Harper & Row, Publishers, Inc., 1989.

———. *Historical Atlas of World Mythology, Volume II: The Way of the Seeded Earth, Part 3: Mythologies of the Primitive Planters: The Middle and Southern Americas.* New York: Harper & Row, Publishers, Inc., 1989.

———. *The Inner Reachers of Outer Space.* New York: Alfred van der Marck Editions, 1986.

———. *The Mythic Image.* Princeton, N.J.: Princeton University Press, 1975.

Chetwynd, Tom. *Dictionary of Sacred Myth.* London: The Aquarian Press, 1986.

Cotterell, Arthur, ed. *The Encyclopedia of Ancient Civilizations.* New York: Mayflower Books, Inc., 1980.

———, ed. *The Penguin Encyclopedia of Classical Civilizations.* New York: Viking Penguin Group, 1993.

Devereux, Paul. *Secrets of Ancient and Sacred Places.* London: Blandford Press, 1992.

———. *Symbolic Landscapes.* Glastonbury, England: Gothic Images Publications, 1992.

Earle, Timothy K. "Chiefdoms in Archaeological and Ethnohistorical Perspective," *Annual Reviews of Anthropology* 16. Annual Reviews, Inc., 1987, pp. 279–308.

Editor et al. *Vanished Civilisations.* Sydney: Reader's Digest, 1983.

Eliade, Mircea. *The History of Religious Ideas. Vol. 1: From the Stone Age to the Eleusinian Mysteries.* Chicago: University of Chicago Press, 1978.

———. *The History of Religious Ideas. Vol. 2: From Gautama Buddha to the Triumph of Christianity.* Chicago: University of Chicago Press, 1982.

———. *The History of Religious Ideas. Vol. 3: From Muhammad to the Age of Reforms.* Chicago: University of Chicago Press, 1985.

———. *Myth and Reality.* 1963. Reprint. New York: Harper & Row, 1975.

———. *Patterns in Comparative Religion.* 1958. Reprint. New York: New American Library, 1974.

———. *Rites and Symbols of Initiation: The Mysteries of Birth and Rebirth.* 1958. Reprint. New York: Harper & Row, 1975.

Ferm, Vergilius, ed. *Ancient Religions.* New York: The Philosophical Library, Inc., 1950.

Frankfort, H. *Kingship and the Gods.* Chicago: University of Chicago Press, 1948.

Frazer, Sir James George. *The Worship of Nature, Volume 1* (Volume 2 never published). London: Macmillan and Co., Limited, 1926.

Harpur, James. *The Atlas of Sacred Places.* New York: Henry Holt and Company, 1994.

Harpur, James, and Jennifer Westwood. *The Atlas of Legendary Places.* New York: Weidenfeld & Nicholson, 1989.

Hooke, S. H. *The Labyrinth—Further Studies in the Relation between Myth and Ritual in the Ancient World.* London: Society for Promoting Christian Knowledge, 1935.

———. *Myth and Ritual.* London: Oxford University Press, 1933.

———. *Myth, Ritual, and Kingship.* Oxford: Oxford University Press, 1958.

Johnson, Allen, W., and Timothy Earle. *The Evolution of Human Societies—From Foraging Group to Agrarian State.* Stanford, California: Stanford University Press, 1987.

Kearney, Michael. *World View.* Novato, California: Chandler & Sharp Publishers, Inc., 1984.

Kramer, Samuel Noah, ed. *Mythologies of the Ancient World.* Garden City, New York: Doubleday & Company, Inc./Anchor Books, 1961.

L'Orange, H. P. *Studies in the Iconography of Cosmic Kingship.* New Rochelle, New York: Caratzas Brothers, Publishers, 1982.

Lovin, Robin W., and Frank E. Reynolds, eds. *Cosmogony and the Ethical Order.* Chicago: The University of Chicago Press, 1985.

Lowry, Shirley Park. *Familiar Mysteries.* New York: Oxford University Press, 1982.

Mercatante, Anthony S. *The Facts On File Encyclopedia of World Mythology and Legend.* New York: Facts On File, 1988.

Molyneaux, Brian Leigh. *The Sacred Earth.* Boston: Little, Brown and Company, 1995.

Ries, Julien. *The Origins of Religions.* Grand Rapids, Michigan: William B. Eerdmans Publishing Company, 1994.

Savill, Sheila. *Pears Encyclopaedia of Myths and Legends, Volumes 1–4* ("The Ancient Near and Middle East and Classical Greece and Rome," "Western and Northern Europe and Central and Southern Africa," "The Orient," and "Oceania and Australia and the Americas"). London: Pelham Books, 1976.

Shapiro, Max S., and Rhoda H. Hendricks. *Mythologies of the World.* Garden City: Doubleday & Company, Inc., 1979.

Townsend, Richard F., ed. *The Ancient Americas—Art from Sacred Landscapes.* Chicago: The Art Institute of Chicago, 1992.

Traube, Elizabeth G. *Cosmology and Social Life.* Chicago: The University of Chicago Press, 1986.

Westwood, Jennifer, ed. *The Atlas of Mysterious Places.* New York: Weidenfeld & Nicholson, 1987.

ADDITIONAL REFERENCES BY CHAPTER

INTRODUCTION—FINGERPRINTS OF COSMIC POWER

Chamberlain, Von Del. *A Guide to Archaeoastronomy at Hovenweep, Chaco Canyon, & Canyon de Chelly.* Privately distributed, no date.

Frazier, Kendrick. *People of Chaco.* New York: W. W. Norton & Company, 1986.

Gabriel, Kathryn. *Roads to Center Place.* Boulder, Colorado: Johnson Books, 1991.

Gay, Carlo. *Chalcacingo.* Graz, Austria: Akademische Druck- u. Verlagsanstalt, 1971.

Grove, David C. *Chalcatzingo—Excavations on the Olmec Frontier.* New York: Thames and Hudson, Inc., 1984.

Krupp, E. C. "The Cosmic Temples of Old Beijing." *World Archaeoastronomy* (A. F. Aveni, ed.). Cambridge, England: Cambridge University Press, 1989, pp. 65–75.

———. "Cosmos on Parade—A Year, More-or-Less, of Sky Jinks and Earth Tremors, Part One, Eclipses and Earthquakes," *Griffith Observer* 58, no. 12 (December 1994): 2–17.

———. "The Mandate of Heaven," *Griffith Observer* 46, no. 6 (June 1982): 8–17.

———. "Shadows Cast for the Son of Heaven," *Griffith Observer* 46, no. 8 (August 1982): 8–18.

Lister, Robert H., and Florence C. Lister. *Chaco Canyon.* Albuquerque: University of New Mexico Press, 1981.

Meyer, Jeffrey F. *The Dragons of Tiananmen—Beijing as a Sacred City.* Columbia, South Carolina: University of South Carolina Press, 1991.

Needham, Joseph. *Science and Civilisation in China, Vol. 3,* "Mathematics and the Sciences of the Heavens and the Earth." Cambridge: Cambridge University Press, 1959.

Polo, Marco. *The Travels of Marco Polo* (edited by Milton Rugoff). New York: The New American Library of World Literature, 1961.

Thurston, Hugh. "The Length of the Year," *Griffith Observer* 57, no. 6 (June 1993): 2–11.

CHAPTER ONE—THE CENTER OF THE WORLD

Adams, Richard E. W. "Río Azul, Lost City of the Maya," *National Geographic Magazine* 169, no. 4 (April 1986): 420–451.

Aveni, Anthony F., and Giuliano Romano. "Orientation and Etruscan Ritual," *Antiquity* 68, no. 260 (September 1994): 545–563.

Batchelor, Stephen. *The Tibet Guide.* London: Wisdom Publications, 1987.

Baum, L. Frank. *The Wonderful Wizard of Oz.* 1900. Reprint. New York: Books of Wonder/William Morrow and Company Inc., 1987.

Brown, Joseph Epes, ed. *The Sacred Pipe—Black Elk's Account of the Seven Rites of the Oglala Sioux.* Norman, Oklahoma: University of Oklahoma Press, 1953.

Burkhardt, V. R. *Chinese Creeds and Customs.* Hong Kong: South China Morning Post Ltd., 1982.

Destenay, Anne L. *Nagel's Encyclopedia Guide China.* Geneva: Nagel Publishers, 1979.

Editor et al. *Etruscans: Italy's Lovers of Life,* Lost Civilizations Series. Alexandria, Virginia: Time-Life Books, 1995.

Freidel, David, Linda Schele, and Joy Parker. *Maya Cosmos—Three Thousand Years on the Shaman's Path.* New York: William Morrow and Company, Inc., 1993.

Graham, Ian. "Looters Rob Graves and History," *National Geographic Magazine* 169, no. 4 (April 1986): 452–461.

Hamblin, Dora Jane. *The Etruscans,* The Emergence of Man Series. New York: Time-Life Books, 1975.

Herberger, Charles. "Matching Microcosm with Macrocosm: The Urge to Order," *Griffith Observer* 48, no. 4 (April 1984): 2–15.

Hoppál, Mihály, ed. *Shamanism in Eurasia, Parts 1 and 2.* Göttingen, Germany: Edition Herodot, 1984.

Kuan Yu-chien. *Magnificent China—A Guide to Its Cultural Treasures.* San Francisco: China Books & Periodicals, Inc., 1987.

Manguel, Alberto, and Gianni Guadalupi. *The Dictionary of Imaginary Places.* New York: Macmillan Publishing Co. Inc., 1980.

Meyer, Jeffrey F. "Chinese Buddhist Temples as Cosmograms," *Sacred Architecture* (ed. Emily Lyle. *Cosmos,* volume 8, the Yearbook of the Traditional Cosmology Society). Edinburgh: Edinburgh University Press, 1992.

Morrell, Elizabeth. *A Visitor's Guide to China.* London: Michael Joseph, 1983.

Neihardt, John G. *Black Elk Speaks.* Lincoln, Nebraska: University of Nebraska Press, 1961.

Noriyuki, Duane. "Stories, Traditions and the Baby White Buffalo," *Los Angeles Times.* September 22, 1994.

Parsons, Elsie Clews. *Pueblo Indian Religion, Volume I, Parts 1 & 2, and Volume II, Parts 1 & 2.* 1939. Reprint. Chicago: The University of Chicago Press, 1974.

Pettazzoni, Raffaele. *The All-knowing God.* London: Methuen & Co., Ltd., 1956.

Post, J. B. *An Atlas of Fantasy.* Baltimore: The Mirage Press, Ltd., 1973.

Shirokogoroff, S. M. *Psychomental Complex of the Tungus.* 1935. Reprint. New York: AMS Press, Inc., 1982.

Stephen, Alexander M. *Hopi Journal of Alexander M. Stephen* (ed. Elsie Clews Parsons). New York: Columbia University Press, 1936.

Tedlock, Barbara. "The Road of Light: Theory and Practice of Mayan Skywatching," *The Sky in Maya Literature* (ed. Anthony F. Aveni). New York: Oxford University Press, 1992, pp. 18–42.

Tylor, Edward B. *Primitive Culture* (two volumes). New York: G. P. Putnam's Sons, 1920.

———. *Researches in the Early History of Mankind and the Development of Civilization.* London: John Murray, 1878.

Watson, William. *Art of Dynastic China.* New York: Harry N. Abrams, Inc., 1981.

Werner, E. T. C. *A Dictionary of Chinese Mythology.* 1932. Reprint. New York: The Julian Press, Inc., 1961.

Werner, E. T. C. *Myths and Legends of China.* London: George G. Harrap & Co., Ltd., 1922.

Williams, C. A. S. *Outlines of Chinese Symbolism and Art Motives.* 1941. Reprint. Rutland, Vermont: Charles E. Tuttle Company, Inc., 1974.

CHAPTER TWO—PLUGGING IN TO COSMIC POWER

Bean, Lowell John. "Power and Its Applications in Native California," *California Indian Shamanism* (ed. Lowell John Bean). Menlo Park, California: Ballena Press, 1992.

Benítez, Fernando. *In the Magic Land of Peyote.* Austin: University of Texas Press, 1975.

Berrin, Kathleen, ed. *Art of the Huichol Indians.* New York: Harry N. Abrams, Inc., Publishers, 1978.

Bord, Janet. *Mazes and Labyrinths of the World.* London: Latimer New Dimensions Limited, 1976.

Brodzky, Anne Trueblood, Rose Danesewich, and Nick Johnson, eds. *Stones, Bones and Skin—Ritual and Shamanic Art.* Toronto: The Society for Art Publications, 1977.

Clarke, D. V., T. G. Cowie, and Andrew Foxon. *Symbols of Power at the Time of Stonehenge.* Edinburgh: Her Majesty's Stationery Office, 1985.

Diószegi, V. *Tracing Shamans in Siberia.* Oosterhout, The Netherlands: Anthropological Publications, 1968.

Diószegi, V., and Hoppál, M., eds. *Shamanism in Siberia.* Budapest: Akadémiai Kiadó, 1978.

Drury, Nevill. *The Elements of Shamanism.* Longmead, Shaftsbury, Dorset: Element Books, 1989.

Eliade, Mircea. *Shamanism: Archaic Techniques of Ecstasy.* Princeton, New Jersey: Princeton University Press, 1964.

Fienup-Riordan, Ann. *The Real People and the Children of Thunder.* Norman, Oklahoma: University of Oklahoma Press, 1991.

Furst, Peter T. *Hallucinogens and Culture.* San Francisco: Chandler & Sharp Publishers, 1976.

———, ed. *Flesh of the Gods—the Ritual Use of Hallucinogens.* New York: Praeger Publishers, Inc., 1972.

Geertz, Clifford. *Negara.* Princeton, New Jersey: Princeton University Press, 1980.

Halifax, Joan. *Shaman—the Wounded Healer.* New York: The Crossroad Publishing Company, 1982.

Herberger, Charles F. "The Labyrinth as an Emblem of the Womb, the Tomb, and Lunisolar Cyclical Time," *Griffith Observer* 55, no. 3 (March 1991): 2–19.

Hoppál, Mihály, and Keith D. Howard, ed. *Shamans and Cultures.* Budapest: Akadémiai Kiadó, and Los Angeles/Fullerton: International Society for Trans-Oceanic Research, 1989.

Hoppál, Mihály, and Juha Pentikäinen, ed. *Northern Religions and Shamanism.* Budapest: Akadémiai Kiadó, and Heksinki: Finnish Literature Society, 1992.

Hoppál, Mihály, and Otto von Sadovszky, ed. *Shamanism Past and Present, Parts 1 and 2.* Budapest: Ethnographic Institute, Hungarian Academy of Sciences, and Los Angeles/Fullerton: International Society for Trans-Oceanic Research, 1993.

Kraft, John. *The Goddess in the Labyrinth.* Åbo, Finland: Åbo Akademi, 1985.

Lonsdale, Steven. *Animals and the Origins of Dance.* New York: Thames and Hudson Inc., 1981.

Myerhoff, Barbara G. *Peyote Hunt.* Ithaca, New York: Cornell University Press, 1974.

Nelson, Edward William. *The Eskimo About Bering Strait.* 1899. Reprint. Washington, D.C., Smithsonian Institution Press, 1983.

Oswalt, Wendell H. *Bashful No Longer—an Alaskan Eskimo Ethnohistory, 1778–1988.* Norman, Oklahoma: University of Oklahoma Press, 1990.

Partridge, Eric. *Origins—A Short Etymological Dictionary of Modern English.* 1966. Reprint. New York: Greenwich House, 1983.

Pennick, Nigel. *Labyrinths: Their Geomancy and Symbolism.* Cambridge, England: Runestaff Publications, 1984.

———. *Mazes and Labyrinths.* London: Robert Hale Limited, 1990.

Rogers, Malcolm J., et al. *Ancient Hunters of the Far West.* San Diego: The Union-Tribune Publishing Company, 1966.

Purce, Jill. *The Mystic Spiral—Journey of the Soul.* New York: Avon Books, 1974.

Siikala, Anna-Leena, and Mihály Hoppál. *Studies on Shamanism.* Budapest: Akadémiai Kiadó, and Heksinki: Finnish Literature Society, 1992.

VanStone, James W. "Mainland Southwest Alaska Eskimo," *Handbook of North American Indians, Volume 5* (ed. David Damas). Washington, D.C.: Smithsonian Institution, 1984, pp. 224–242.

von Werlhof, Jay. "Construction of Earthen Art," *Rock Art Papers, Volume 5* (ed. Ken Hedges). San Diego: San Diego Museum of Man, 1987, pp. 37–42.

———. *Spirits of the Earth, Volume 1, The North Desert.* El Centro, California: Imperial Valley College Museum Society, 1987.

Walsh, Roger N. *The Spirit of Shamanism.* Los Angeles: Jeremy P. Tarcher, Inc., 1990.

Wosien, Maria-Gabriele. *Sacred Dance—Encounter with the Gods.* New York: Avon Books, 1974.

CHAPTER THREE—CENTERS OF CREATION

Alexander, Hartley Burr. *The Mythology of All Races, Volume X—North American.* 1916. Reprint. New York: Cooper Square Publishers, Inc., 1964.

Bancroft-Hunt, Norman, and Werner Forman. *People of the Totem.* New York: G. P. Putnam's Sons, 1979.

Bernbaum, Edwin. *Sacred Mountains of the World.* San Francisco: Sierra Club Books, 1990.

Blackman, Margaret B. "Haida: Traditional Culture," *Handbook of North American Indians, Volume* 7 (ed. Wayne Settles). Washington, D.C.: Smithsonian Institution, 1990, pp. 240–260.

Blofeld, John. *The Tantric Mysticism of Tibet.* New York: E. P. Dutton & Co., Inc., 1970.

Bruggmann, Maximilien, and Peter Gerber. *Indians of the Northwest Coast.* New York: Facts on File, Inc., 1989.

Buckley, Michael, and Robert Strauss. *Tibet—A Travel Survival Kit.* Berkeley, California: Lonely Planet Publications, 1986.

Chan, Victor. *Tibet Handbook.* Chico, California: Moon Publications, Inc., 1994.

Chodag, Tiley. *Tibet—The Land and the People.* Beijing: New World Press, 1988.

Clark, Ella Elizabeth. *Indian Legends of Canada.* Toronto: McClelland and Stewart Limited, 1960.

Daniélou, Alain. *The Gods of India.* New York: Inner Traditions International Ltd., 1985.

Douglas, Nik. *Tibetan Tantric Charms & Amulets.* New York: Dover Publications, Inc., 1978.

Dunham, Carroll, and Ian Baker. *Tibet—Reflections from the Wheel of Life.* New York: Abbeville Press, Inc., 1993.

Editor et al. *Ancient Tibet.* Berkeley, California: Dharma Publishing, 1986.

———. *Keepers of the Totem,* The American Indians Series. Alexandria, Virginia: Time-Life Books, 1993.

———. *Tibet.* New York: McGraw-Hill Book Company, 1981.

Einarsen, John, ed. *The Sacred Mountains of Asia.* Boston: Shambhala Publications, Inc., 1993.

Fausboll, V. *Indian Mythology—According to the Mahabharata.* London: Luzac & Co., 1903.

Feldman, Susan, ed. *African Myths and Tales.* New York: Dell Publishing Company, Inc., 1963.

Frawley, David. *Gods, Sages and Kings.* Salt Lake City: Passage Press, 1991.

Hyde-Chambers, Fredrick, and Audrey Hyde-Chambers. *Tibetan Folk Tales.* Boston: Shambhala Publications, Inc., 1981.

Ions, Veronica. *Indian Mythology.* London: The Hamlyn Publishing Group Limited, 1967.

Knappert, Jan. *The Aquarian Guide to African Mythology.* Wellingborough, England: Thorsons Publishing Group, 1990.

Leach, Maria. *The Beginnings—Creation Myths Around the World.* New York: Funk & Wagnalls Company, 1956.

Leakey, Richard E. *The Making of Mankind.* New York: E. P. Dutton, 1981.

Lewis-Williams, J. David. *Discovering Southern African Rock Art.* Cape Town, South Africa: David Philip Publishers (Pty) Ltd., 1990.

———. "A Dream of Eland: an Unexplored Component of San Shamanism and Rock Art," *World Archaeology* 19, no. 2 (October 1987): 165–177.

———. *The Rock Art of Southern Africa.* Cambridge, England: Cambridge University Press, 1983.

Lewis-Williams, David, and Thomas Dowson. *Images of Power—Understanding Bushman Rock Art.* Johannesburg, South Africa: Southern Book Publishers (Pty) Ltd., 1989.

Lewis-Williams, J. D., and T. A. Dowson. *Rock Paintings of the Natal Drakensberg.* Pietermartizburg, South Africa: University of Natal Press, 1992.

Li An-che. "Tibetan Religion," *Ancient Religions* (ed. Vergilius Ferm). New York: The Philosophical Library, Inc., 1950, pp. 251–269.

McCrea, Barbara, and Tony Pinchuck. *Zimbabwe and Botswana—the Rough Guide.* London: Rough Guides Ltd., 1993.

Radin, Paul, and James Johnson Sweeney. *African Folktales and Sculpture.* New York: Pantheon Books, 1952.

Rawson, Philip. *Sacred Tibet.* New York: Thames and Hudson, Inc., 1991.

Smelcer, John E. *A Cycle of Myths—Native Legends from Southeast Alaska.* Anchorage: Salmon Run, 1993.

Sproul, Barbara C. *Primal Myths—Creating the World.* New York: Harper & Row, Publishers, Inc., 1979.

Stein, R. A. *Tibetan Civilization.* Stanford, California: Stanford University Press, 1972.

Tucci, Giuseppe. *The Religions of Tibet.* London: Routledge & Kegan Paul Ltd., 1980.

van der Post, Laurens. *The Heart of the Hunter.* New York: William Morrow and Company, Inc., 1961.

———. *The Lost World of the Kalahari* 1958. New edition. New York: William Morrow and Company, Inc., 1988.

Waddell, L. Austine. *Tibetan Buddhism.* 1895. Reprint. New York: Dover Publications, Inc., 1972.

Wannenburgh, Alf. *The Bushmen.* New York: Mayflower Books, 1979.

CHAPTER FOUR—MOTHER EARTH

Angel, Myron. *La Piedra Pintada—The Painted Rock, a Legend.* 1910. Reprint. San Luis Obispo, California: Padre Productions, 1979.

Bean, Lowell John, and Thomas F. King, eds. *'Antap—California Indian Political and Economic Organization.* Ramona, California: Ballena Press, 1974.

Brady, James E., George Hasemann, and John H. Fogarty. "Harvest of Skulls & Bones," *Archaeology* 48, no. 3 (May/June 1995): 36–40.

Breuil, Abbé H. *Four Hundred Centuries of Cave Art.* New York: Hacker Art Books, 1979.

Burenhult, Goran, ed. *The First Humans.* San Francisco: HarperSanFrancisco, 1993.

Clottes, Jean. "Rhinos and Lions and Bears (Oh, My!)," *Natural History* 104, no. 5 (May 1993): 64–70.

Clottes, Jean, and Jean Courtin. "Neptune's Ice Age Gallery," *Natural History* 102, no. 4 (April 1995): 30–34.

Daubisse, Paulette, et al. *La Grotte de Font-de-Gaume.* Périgueux, France: Pierre Fanlac Éditeur, 1984.

Díaz, Bernal. *The Conquest of New Spain.* Harmondsworth, England: Penguin Books Ltd., 1963.

Du Bois, Constance Goddard, "The Religion of the Luiseño Indians of Southern California," *University of California Publications in American Archaeology and Ethnology* 8, no. 3 (June 1908): 69–186.

Gallenkamp, Charles. *Maya.* Third edition. New York: Viking Penguin, Inc., 1985.

Giedion, S. *The Eternal Present: The Beginnings of Art.* New York: Bollingen Foundation, 1962.

Hadingham, Evan. *Secrets of the Ice Age.* New York: Walker and Company, 1979.

Hamayon, Roberte N. "Shamanism in Siberia: From Partnership in Supernature to Counter-power in Society," *Shamanism, History, and the State* (eds. Nicholas Thomas and Caroline Humphrey). Ann Arbor, Michigan: The University of Michigan Press, 1994, pp. 76–89.

Hudson, Travis, and Ernest Underhay. *Crystals in the Sky: An Intellectual Odyssey Involving Chumash Astronomy, Cosmology and Rock Art.* Socorro, New Mexico: Ballena Press, 1978.

Hudson, Travis, Thomas Blackburn, Rosario Curletti, and Janice Timbrook, eds. *The Eye of the Flute—Chumash Traditional History and Ritual.* Santa Barbara, California: Santa Barbara Museum of Natural History, 1977.

James, E. O. *Prehistoric Religion.* New York: Frederick A. Praeger, 1957.

Kühn, Herbert. *The Rock Pictures of Europe.* Fair Lawn, New Jersey: Essential Books, Inc., 1956.

Laird, Carobeth. *The Chemehuevis.* Banning, California: Malki Museum Press, 1976.

———. *Mirror and Pattern—George Laird's World of Chemehuevi Mythology.* Banning, California: Malki Museum Press, 1984.

LeRoi-Gourhan, André. *Treasures of Prehistoric Art.* New York: Harry N. Abrams, Inc., Publishers, no date.

Marshack, Alexander. "Exploring the Mind of Ice Age Man," *National Geographic* 147, no. 1 (January 1975): 64–89.

———. "Images of the Ice Age," *Archaeology* 48, no. 4 (July/August 1995): 28–39.

McGowan, Charlotte. *Ceremonial Fertility Sites in Southern California.* San Diego: San Diego Museum of Man, 1982.

Rafter, John. "The Bernasconi Hills Discoveries," *Rock Art Papers, Volume* 7 (ed. Ken Hedges). San Diego: San Diego Museum of Man, 1990, pp. 33–40.

———. "More Sunlight/Petroglyph Interaction at Counsel Rocks," *Rock Art Papers, Volume 8* (ed. Ken Hedges). San Diego: San Diego Museum of Man, 1991, pp. 65–74.

———. "Shelter Rock of the Providence Mountains," *Rock Art Papers, Volume 5* (ed. Ken Hedges). San Diego: San Diego Museum of Man, 1989, pp. 27–50.

Ruspoli, Mario. *The Cave of Lascaux—The Final Photographs.* New York: Harry N. Abrams, Inc., Publishers, 1987.

Sieveking, Ann. *The Cave Artists.* New York: Thames and Hudson, Inc., 1979.

Solecki, Ralph S. *Shanidar, the First Flower People.* New York: Alfred A. Knopf, Inc., 1971.

Stringer, Christopher, and Clive Gamble. *In Search of Neanderthals.* New York: Thames and Hudson, Inc., 1993.

Trinkaus, Erik, and Pat Shipman. *The Neandertals.* New York: Alfred A. Knopf, 1993.

Whitley, David S. "Shamanism, Natural Modeling and the Rock Art of Far Western North American Hunter-Gatherers," *Shamanism and Rock Art in North America* (ed. Solveig A. Turpin). San Antonio, Texas: Rock Art Foundation, Inc., 1994, pp. 1–43.

Zigmond, Maurice L. *Kawaiisu Mythology.* Socorro, New Mexico: Ballena Press, 1980.

———. "The Supernatural World of the Kawaiisu," *Flowers of the Wind* (ed. Thomas C. Blackburn). Socorro, New Mexico: Ballena Press, 1977: 59–95.

CHAPTER FIVE—AGENTS OF RENEWAL

Akurgal, Ekrem. *The Art of the Hittites.* New York: Harry N. Abrams, Inc., no date.

Alexander, Robert L. *The Sculpture and Sculptors of Yazilikaya.* Newark, New Jersey: University of Delaware Press, 1986.

Austen, Hallie Iglehart. *The Heart of the Goddess—Art, Myth and Meditations of the World's Sacred Feminine.* Berkeley, California: Wingbow Press, 1990.

Baring, Anne, and Jules Cashford. *The Myth of the Goddess.* London: Viking Penguin Group, 1991.

Bittel, Kurt. *Hattusha—the Capital of the Hittites.* New York: Oxford University Press, 1970.

Brennan, Martin. *The Stars and the Stones—Ancient Art and Astronomy in Ireland.* London: Thames and Hudson Ltd., 1983.

Burenhult, Goran, ed. *People of the Stone Age.* San Francisco: HarperSanFrancisco, 1993.

Cunliffe, Barry, ed. *The Oxford Illustrated Prehistory of Europe.* Oxford, England: Oxford University Press, 1994.

Editor et al. *Anatolia: Cauldron of Cultures,* Lost Civilizations Series. Alexandria, Virginia: Time-Life Books, 1995.

———. *Early Europe: Mysteries in Stone,* Lost Civilizations Series. Alexandria, Virginia: Time-Life Books, 1995.

Eogan, George. *Knowth and the Passage-tombs of Ireland.* London: Thames and Hudson Ltd., 1986.

Ergener, Resit. *Anatolia Land of Mother Goddess.* Ankara: Hitit Publications Ltd., 1988.

Evans, J. D. *The Prehistoric Antiquities of the Maltese Islands.* London: The Athlone Press, The University of London, 1971.

Formosa, Gerald J. *The Megalithic Monuments of Malta.* Vancouver: Skorba Publishers, 1975.

Frymer-Kensky, Tikva. *In the Wake of the Goddesses.* New York: The Free Press, 1992.

Gadon, Elinor W. *The Once & Future Goddess.* San Francisco: Harper & Row, 1989.

Getty, Adele. *Goddess—Mother of Living Nature.* New York: Thames and Hudson Ltd., Inc., 1990.

Gimbutas, Marija. *The Civilization of the Goddess.* New York: HarperCollins, 1991.

———. *The Gods and Goddesses of Old Europe 7000–3500 B.C.* London: Thames and Hudson Ltd., 1974.

———. *The Language of the Goddess.* New York: Harper & Row, Publishers, Inc., 1989.

Güterbrock, Hans Gustav. "Hittite Mythology," *Mythologies of the Ancient World* (ed. Samuel Noah Kramer). Garden City, New York: Doubleday & Company, Inc./Anchor Books, 1961, pp. 139–179.

———. "Hittite Religion," *Ancient Religions* (ed. Vergilius Ferm). New York: The Philosophical Library, Inc., 1950, pp. 81–109.

Hicks, Jim. *The Empire Builders,* The Emergence of Man Series. New York: Time-Life Books, 1974.

Jacobsen, Thorkild. *The Treasures of Darkness—a History of Mesopotamian Religion.* New Haven, Connecticut: Yale University Press, 1976.

James, E. O. *From Cave to Cathedral.* London: Thames and Hudson Ltd., no date.

———. *The Cult of the Mother Goddess.* New York: Frederick A. Praeger, 1959.

———. *Seasonal Feasts and Festivals.* London: Thames and Hudson Ltd., 1961.

Johnson, Buffie. *Lady of the Beasts.* San Francisco: Harper & Row, Publishers, 1988.

Lehmann, Johannes. *The Hittites—People of a Thousand Gods.* New York: The Viking Press, 1977.

Malone, Caroline, Anthony Bonanno, Tancred Couder, Simon Stoddart, and David Trump. "The Death Cults of Prehistoric Malta," *Scientific American* 269, no. 6 (December 1993): 110–117.

Mann, A. T. *Sacred Architecture.* Shaftesbury, England: Element Books Limited, 1993.

McCrickland, Janet. *Eclipse of the Sun.* Glastonbury, England: Gothic Image Publications, 1990.

Mellaart, James. *Çatal Hüyük—A Neolithic Town in Anatolia.* New York: McGraw-Hill Book Company, 1967.

Micallef, Paul I. *Mnajdra Prehistoric Temple—A Calendar in Stone.* Malta: Paul I. Micallef, 1989.

Neumann, Erich. *The Great Mother.* Princeton, New Jersey: Princeton University Press, 1963.

O'Kelly, Michael J. *Early Ireland—An Introduction to Irish Prehistory.* Cambridge, England: Cambridge University Press, 1989.

———. *Newgrange—Archaeology, Art and Legend.* London: Thames and Hudson Ltd., 1982.

Ó Ríordáin, Sean P., and Glyn Daniel. *New Grange and the Bend of the Boyne.* London: Thames and Hudson Ltd., 1964.

Pallis, Svend Aage. *The Babylonian Akitu Festival* (Historisk-Filologiske Meddelelser Volume 12). Copenhagen: Andr. Fred. Host & Son, 1926–27.

Ray, T. P. "The Winter Solstice Phenomenon at Newgrange, Ireland: Accident or Design?" *Nature* 337 (January 26, 1989): 343–345.

Renfrew, Colin. *Before Civilization.* New York: Alfred A. Knopf, Inc., 1973.

Stone, Merlin. *When God Was a Woman.* New York: The Dial Press, 1976.

Temizer, Raci. *Museum of Anatolian Civilizations.* Ankara: Bahçelievler, 1981.

Temizsoy, I. *The Anatolian Civilisations Museum.* Ankara: no publisher, no date.

Trump, D. H. *Malta: An Archaeological Guide.* London: Faber and Faber Limited, 1972.

Wolkstein, Diane, and Samuel Noah Kramer. *Inanna, Queen of Heaven and Earth—Her Stories and Hymns from Sumer.* New York: Harper & Row, Publishers, Inc., 1983.

CHAPTER SIX—SHAMANS, CHIEFS, AND SACRED KINGS

Ananikian, Mardiros, and Alice Werner. *The Mythology of All Races, Volume VII—Armenian and African.* 1925. Reprint. New York: Cooper Square Publishers, Inc., 1964.

Atmore, Anthony, and Gillian Stacey. *Black Kingdoms Black Peoples.* New York: G. P. Putnam's Sons, 1979.

Beckwith, Martha Warren. *The Kumulipo—A Hawaiian Creation Chant.* Chicago: The University of Chicago Press, 1951.

Blackburn, Thomas C. *December's Child—a Book of Chumash Oral Narratives.* Berkeley and Los Angeles: University of California Press, 1975.

Blitzer, Charles. *Age of Kings,* Great Ages of Man Series. New York: Time-Life Books, 1967.

Cable, Mary, ed. *The African Kings,* Treasures of the World Series. Chicago: Stonehenge Press, Inc., 1983.

Carlson, Robert G. "Hiearchy and the Haya Divine Kingship: A Structural and Symbolic Reformulation of Frazer's Thesis," *American Ethnologist* 20, no. 2 (1993): 312–335.

Davidson, Basil. *African Kingdoms,* Great Ages of Man Series. New York: Time-Life Books, 1966.

Dudley, Michael Kioni. *Man, Gods, and Nature.* Honolulu: Na Kane O Ka Malo Press, 1990.

Editor et al. *Africa's Glorious Legacy,* Lost Civilizations Series. Alexandria, Virginia: Time-Life Books, 1994.

———. *The First Americans,* The American Indians Series. Alexandria, Virginia: Time-Life Books, 1992.

———. *Mound Builders & Cliff Dwellers,* Lost Civilizations Series. Alexandria, Virginia: Time-Life Books, 1992.

Galloway, Patricia, ed. *The Southeastern Ceremonial Complex: Artifacts and Analysis.* Lincoln, Nebraska: University of Nebraska Press, 1989.

Garlake, Peter. *The Kingdoms of Africa,* The Making of the Past Series. Oxford, England: Elsevier-Phaidon, 1978.

Griaule, M. *Conversations with Ogotemmêli—an Introduction to Dogon Religious Ideas.* Oxford: Oxford University Press, 1965.

Griaule, M., and G. Dieterlen. *The Pale Fox.* Chino Valley, Arizona: Continuum Foundation, 1986.

Guidoni, Enrico. *Primitive Architecture.* New York: Harry N. Abrams, Inc., Publishers, 1978.

Hudson, Charles. *The Southeastern Indians.* Knoxville, Tennessee: The University of Tennessee Press, 1976.

Hudson, Travis, ed. *Breath of the Sun—Life in Early California as Told by a Chumash Indian, Fernando Librado, to John P. Harrington.* Banning, California: Malki Museum Press, 1980.

Hudson, Travis, and Ernest Underhay. *Crystals in the Sky: An Intellectual Odyssey Involving Chumash Astronomy, Cosmology and Rock Art.* Socorro, New Mexico: Ballena Press, 1978.

Hudson, Travis, Thomas Blackburn, Rosario Curletti, and Janice Timbrook, eds. *The Eye of the Flute—Chumash Traditional History and Ritual.* Santa Barbara, California: Santa Barbara Museum of Natural History, 1977.

Huet, Michel. *The Dance, Art and Ritual of Africa.* New York: Pantheon Books, 1978.

Hull, Richard. W. *African Cities and Towns Before the European Conquest.* New York: W. W. Norton & Company, Inc., 1976.

Ingoglia, Gina. *Disney's The Lion King.* New York: Disney Press, 1994.

Johnson, Rubellite Kawena. *Kumulipo—Hawaiian Hymn of Creation, Volume I.* Honolulu: Topgallant Publishing Co., Ltd., 1981.

Johnson, Rubellite Kawena, and John Kaipo Mahelona. *Na Inoa Hoku—a Catalogue of Hawaiian and Pacific Star Names.* Honolulu: Topgallant Publishing Company, Ltd., 1975 (updated unpublished manuscript 1989).

Johnson, Rubellite Kawena. *Mo'Olelo Hawai'i* (World of the Hawaiians). Honolulu: unpublished manuscript, 1993.

Kalakaua, King David. *The Legends and Myths of Hawaii.* 1888. Reprint. Rutland, Vermont: Charles E. Tuttle Company, 1972.

Kirch, Patrick Vinton. *Feathered Gods and Fishhooks.* Honolulu: University of Hawaii Press, 1985.

Knappert, Jan. *The Aquarian Guide to African Mythology.* Wellingborough, England: Thorsons Publishing Group, 1990.

Krupp, E. C. "Hiawatha in California," *The Astronomy Quarterly,* 7, no. 4 (1990).

Makemson, Maud Worcester. *The Morning Star Rises—an Account of Polynesian Astronomy.* New Haven: Yale University Press, 1941.

Parrinder, Geoffrey. *African Mythology.* London: The Hamlyn Publishing Group Limited, 1967.

Phillipson, David W. *African Archaeology.* Second edition. Cambridge, England: Cambridge University Press, 1993.

Shaw, Thurstan. *Nigeria—Its Archaeology and Early History.* London: Thames and Hudson Ltd., 1978.

South, Malcolm, ed. *Topsell's Histories of Beasts.* Chicago: Nelson-Hall, 1981.

Swanton, John R. *Indian Tribes of the Lower Mississippi Valley and Adjacent Coast of the Gulf of Mexico. Bureau of American Ethnology Bulletin 43.* Washington, D.C.: Government Printing Office, 1911.

Valeri, Valerio. *Kingship and Sacrifice—Ritual and Society in Ancient Hawaii.* Chicago: The University of Chicago Press, 1985.

Vastokas, Joan M., and Romas K. Vastokas. *Sacred Art of the Algonkians.* Peterborough, Ontario: Mansard Press, 1973.

The Walt Disney Company. *The Lion King* (videotape). Burbank, California: The Walt Disney Company, 1994.

Willet, Frank. *Ife in the History of West African Sculpture.* New York: McGraw-Hill Book Company, 1967.

CHAPTER SEVEN—CELESTIAL EMPIRES

Aston, W. G. *Shinto: The Way of the Gods.* London: Longmans, Green and Co., 1905.

———, trans. *Nihongi—Chronicles of Japan from the Earliest Times to A.D. 697.* 1896. Reprint. Rutland, Vermont: Charles E. Tuttle Company, 1972.

Berger, Patricia, and Terese Tse Bartholomew. *Mongolia—The Legacy of Chinggis Khan.* New York: Thames and Hudson, Inc., 1995.

Black, Jeremy, and Anthony Green. *Gods, Demons, and Symbols of Ancient Mesopotamia—an Illustrated Dictionary.* Austin, Texas: University of Texas Press, 1992.

Chamberlain, Basil Hall, trans. *The Kojiki—Records of Ancient Matters.* 1920. Reprint. Rutland, Vermont: Charles E. Tuttle Company, 1982.

Cleaves, Francis Woodman, trans. and ed. *The Secret History of the Mongols.* Cambridge, Massachusetts: Harvard University Press, 1982.

Collcutt, Martin, Marius Jansen, and Isao Kumakura. *Cultural Atlas of Japan.* New York: Facts On File, Inc., 1988.

Davis, F. Hadland. *Myths and Legends—Japan.* Boston: David D. Nickerson & Company, Publishers, no date.

Editor et al. *Baedeker's Japan.* New York: Prentice Hall, 1993.

Ferguson, John C., and Masaharu Anesaki. *The Mythology of All Races, Volume VIII—Chinese and Japanese.* 1928. Reprint. New York: Cooper Square Publishers, Inc., 1964.

Grousset, René. *The Empire of the Steppes—a History of Central Asia.* New Brunswick, New Jersey: Rutgers University Press, 1970.

Hackin, J. *Asiatic Mythology.* New York: Crescent Books, no date.

Heissig, Walter. *The Religions of Mongolia.* Berkeley and Los Angeles: University of California Press, 1970.

Humphrey, Caroline. "Shamanic Practices and the State in Northern Asia: Views from the Center and the Periphery," *Shamanism, History, and the State* (eds. Nicholas Thomas and Caroline Humphrey). Ann Arbor, Michigan: The University of Michigan Press, pp. 191–228.

Ji Zhi and Monhnasen. *Folk Customs of the Ordos Mongolian People.* Neimenggu Autonomous Region (Inner Mongolia), China: no publisher, 1991.

Kessler, Adam T. *Empires Beyond the Great Wall—the Heritage of Genghis Khan.* Los Angeles: Natural History Museum of Los Angeles County, 1993.

Kidder, Edward. *Ancient Japan,* The Making of the Past Series. Oxford, England: Elsevier-Phaidon, 1977.

Kidder, J. Edward. *Japan before Buddhism.* New York: Praeger Publishers, Inc., 1966.

Kuan Yu-chien. *Magnificent China—A Guide to Its Cultural Treasures.* San Francisco: China Books & Periodicals, Inc., 1987.

Legg, Stuart. *The Barbarians of Asia.* New York: Dorset Press, 1970.

Lister, R. P. *Genghis Khan.* New York: Dorset Press, 1969.

Mackenzie, Donald A. *Myths of China & Japan.* London: The Gresham Publishing Company Ltd., no date.

Marshall, Robert. *Storm from the East.* Berkeley, California: The University of California Press, 1993.

Morgan, David. *The Mongols.* Oxford, England: Basil Blackwell Ltd., 1986.

Morrell, Elizabeth. *A Visitor's Guide to China.* London: Michael Joseph, 1983.

O'Neill, John. *The Night of the Gods, vol. 1.* London: Harrison & Sons and Bernard Quaritch, 1893.

Phillips, E. D. *The Mongols.* London: Thames and Hudson Ltd., 1969.

Picken, Stuart D. B. *Shinto—Japan's Spiritual Roots.* Tokyo: Kodansha International Ltd., 1980.

Piggott, Juliet. *Japanese Mythology.* London: The Hamlyn Publishing Group Limited, 1969.

Ratchnevsky, Pau. *Genghis Khan—His Life and Legacy.* Oxford, England: Basil Blackwell Ltd., 1991.

Rossabi, Morris. *Khubilai Khan—His Life and Times.* Berkeley, California: The University of California Press, 1988.

Schoenberger, Karl. "Akahito in Final Ritual of Passage," *Los Angeles Times,* November 23, 1990.

Severin, Tim. *In Search of Genghis Khan.* New York: Atheneum, 1991.

Storey, Robert. *Mongolia—a Travel Survival Kit.* Berkeley, California: Lonely Planet Publications, 1993.

Storry, Richard. *The Way of the Samurai.* New York: Mayflower Books, 1978.

Tamburello, Adolfo. *Monuments of Civilization: Japan.* New York: Grosset & Dunlap, Inc., 1973.

Vitebsky, Piers. *The Shaman.* Boston: Little, Brown and Company, 1995.
Watanabe, Teresa. "Ancient Rites Unite Prince, Ex-Diplomat," *Los Angeles Times,* June 9, 1993.
———. "Royal Wedding in Japan," *Los Angeles Times,* June 7, 1993.

CHAPTER EIGHT—ENLIGHTENED SELF-INTEREST AND ULTERIOR MOTIVES

Abell, George O. "Astrology," *Science and the Paranormal* (ed. George O. Abell and Barry Singer). New York: Charles Scribner's Sons, 1981, pp. 70–94.
Allen, James P. *Genesis in Egypt—the Philosophy of Ancient Egyptian Creation Accounts.* New Haven, Connecticut: Yale Egyptological Seminar, Yale University, 1988.
Baigent, Michael. *From the Omens of Babylon: Astrology and Ancient Mesopotamia.* London: Penguin Books Ltd./Arkana, 1994.
Baines, John, and Jaromír Málek. *Atlas of Ancient Egypt.* New York: Facts on File Publications, 1980.
Barton, Tamsyn. "Astrology and the State in Imperial Rome," *Shamanism, History, and the State* (eds. Nicholas Thomas and Caroline Humphrey). Ann Arbor, Michigan: The University of Michigan Press, pp. 146–163.
Bunson, Matthew. *Encyclopedia of the Roman Empire.* New York: Facts On File, Inc., 1994.
Calvin, William H. *How the Shaman Stole the Moon.* New York: Bantam Books, 1991.
Cazeau, Charles J. "Prophecy: The Search for Certainty," *The Skeptical Inquirer* VII, no. 1 (fall 1982): 220–229.
Eberhard, Wolfram. "The Political Function of Astronomy and Astronomers in Han China," *Chinese Thought and Institutions* (ed. John K. Fairbank). Chicago: The University of Chicago Press, 1957, pp. 33–70.
Editors of Time-Life Books. *Cosmic Connections,* Mysteries of the Unknown Series. Alexandria, Virginia: Time-Life Books, 1988.
———. *Visions and Prophecies,* Mysteries of the Unknown Series. Alexandria, Virginia: Time-Life Books, 1988.
Eddy, John A. "Astronomical Alignment of the Big Horn Medicine Wheel," *Science* 184 (June 7, 1974): 1,035–1,043.
Eddy, John A. "Medicine Wheels and Plains Indian Astronomy," *Native American Astronomy* (ed. Anthony F. Aveni). Austin, Texas: University of Texas Press, 1977, pp. 147–169.
———. "Medicine Wheels and Plains Indian Astronomy," *Astronomy of the Ancients* (ed. Kenneth Brecher and Michael Feirtag). Cambridge, Massachusetts: MIT Press, 1979, pp. 1–24.
Ermans, Adolf. *The Literature of the Ancient Egyptians.* 1927. Reprint. New York: Arno Press, 1977.
Flaherty, Gloria. *Shamanism and the Eighteenth Century.* Princeton, New Jersey: Princeton University Press, 1992.

Goodrich, Norma Lorre. *Priestesses.* New York: Franklin Watts, 1989.

Grinnell, George Bird. *By Cheyenne Campfires.* 1926. Reprint. Lincoln, Nebraska: University of Nebraska Press, 1971.

Grinnell, George Bird. *The Cheyenne Indians, Volume I, History and Society.* 1923. Reprint. Lincoln, Nebraska: University of Nebraska Press, 1972.

Grinnell, George Bird. *The Cheyenne Indians, Volume II, War, Ceremonies, and Religion.* 1923. Reprint. Lincoln, Nebraska: University of Nebraska Press, 1972.

Harner, Michael J., ed. *Hallucinogens and Shamanism.* New York: Oxford University Press, 1973.

Hart, George. *A Dictionary of Egyptian Gods and Goddesses.* London: Routledge & Kegan Paul, 1986.

———. *Egyptian Myths.* Austin, Texas: University of Texas Press, 1990.

Hobson, Christine. *The World of the Pharaohs.* New York: Thames and Hudson, Inc., 1987.

Hook, Brian, ed. *The Cambridge Encyclopedia of China.* Cambridge, England: Cambridge University Press, 1982.

Ions, Veronica. *Egyptian Mythology.* London: The Hamlyn Publishing Group Limited, 1968.

Jastrow, Morris. *Aspects of Religious Belief and Practice in Babylonia and Assyria.* 1911. Reprint. New York: Benjamin Blom, Inc., 1971.

Jastrow, Morris. *The Religion of Babylonia and Assyria.* Boston: Ginn & Company, 1898.

Lindsay, Jack. *Origins of Astrology.* New York: Barnes & Noble, Inc., 1971.

Loewe, Michael, and Carmen Blacker, eds. *Divination and Oracles.* London: George Allen & Unwin Ltd., 1971.

Lyons, Albert S. *Predicting the Future.* New York: Harry N. Abrams, Inc., Publishers, 1990.

Manilius. *Astronomica,* Loeb Classical Library. Cambridge, Massachusetts: Harvard University Press, 1977.

Mercatante, Anthony S. *Who's Who in Egyptian Mythology.* New York: Clarkson N. Potter, Inc., 1978.

Murnane, William J. *The Guide to Ancient Egypt.* New York: Facts on File Publications, 1983.

Needham, Joseph. *Science and Civilisation in China, Vol. 3,* "Mathematics and the Sciences of the Heavens and the Earth." Cambridge: Cambridge University Press, 1959.

Needham, Joseph, Gwei-djen Lu, John H. Combridge, and John S. Major. *The Hall of Heavenly Records—Korean Astronomical Instruments and Clocks 1380–1780.* Cambridge, England: Cambridge University Press, 1986.

Needham, Joseph, Ling Wang, Derek J. de Solla Price. *Heavenly Clockwork—The Great Astronomical Clocks of Medieval China.* Second edition. Cambridge, England: Cambridge University Press, 1986.

Saggs, H. W. F. *Babylonians.* Norman, Oklahoma: University of Oklahoma Press, 1995.

Sayce, A. H. *Astronomy and Astrology of the Babylonians.* 1874. Reprint. San Diego: Wizards Bookshelf, 1981.

Schafer, Edward H. *Pacing the Void.* Berkeley and Los Angeles: University of California Press, 1977.

Schlesier, Karl H. *The Wolves of Heaven—Cheyenne Shamanism, Ceremonies, and Prehistoric Origins.* Norman, Oklahoma: University of Oklahoma Press, 1987.

Temple, Robert K. G. *Conversations with Eternity.* London: Rider & Company, 1986.

Thompson, R. Campbell. *The Reports of the Magicians and Astrologers of Nineveh and Babylon in the British Museum, Volumes 1 and 2.* 1900. Reprint. New York: AMS Press Inc., 1977.

Vandenberg, Philipp. *The Mystery of the Oracles.* New York: Macmillan Publishing Co., Inc., 1979.

van der Waerden, B. L. *Science Awakening II: the Birth of Astronomy.* Leyden: Noordhoff International Publishing, 1974.

Vogt, David E. "Medicine Wheel Astronomy," *Astronomies and Cultures* (ed. Clive L. N. Ruggles and Nicholas J. Saunders). Niwot, Colorado: University Press of Colorado, 1993, pp. 163–201.

Walsh, Roger N. *The Spirit of Shamanism.* Los Angeles: Jeremy P. Tarcher, Inc., 1990.

Waltham, Clae, ed. *Shu Ching—Book of History.* Chicago: Henry Regnery Company, 1971.

Watterston, Barbara. *The Gods of Ancient Egypt.* New York: Facts on File Publications, 1984.

Wheatley, Paul. *The Pivot of the Four Quarters.* Edinburgh: Edinburgh University Press, 1971.

CHAPTER NINE—IT PAYS TO ADVERTISE

Arlington, L. C., and W. Lewisohn. *In Search of Old Peking.* 1935. Reprint. New York: Paragon Book Reprint Corp., 1967.

Basilov, Vladimir N., ed. *Nomads of the Eurasian Steppe.* Los Angeles: Natural History Museum of Los Angeles County, 1989.

Benedict, Ruth. *Zuni Mythology, Vol. II.* New York: Columbia University Press, 1935.

Berrin, Kathleen, and Esther Pasztory, eds. *Teotihuacán—Art from the City of the Gods.* New York: Thames and Hudson, Inc., 1993.

Bredon, Juliet. *Peking.* 1931. Reprint. Hong Kong: Oxford University Press, 1982.

Bredon, Juliet, and Igor Mitrophanow. *The Moon Year.* Shanghai: Kelly & Walsh, 1927.

Brodzky, Anne Trueblood, Rose Danesewich, and Nick Johnson, eds. *Stones, Bones and Skin—Ritual and Shamanic Art.* Toronto: The Society for Art Publications, 1977.

Bunson, Matthew. *Encyclopedia of the Roman Empire.* New York: Facts On File, Inc., 1994.

Carlson, John B. "Rise and Fall of the City of the Gods," *Archaeology* 46, no. 6 (November/December 1993): 58–69.

Carlson, John B. "Venus-Regulated Warfare and Ritual Sacrifice in Mesoamerica," *Astronomies and Cultures* (ed. Clive L. N. Ruggles and Nicholas J. Saunders). Niwot, Colorado: University Press of Colorado, 1993, pp. 202–252.

———. *Venus-regulated Warfare and Ritual Sacrifice in Mesoamerica: Teotihuacán and the Cacaxtla "Star Wars" Connection.* College Park, Maryland: Center for Archaeoastronomy, 1991.

Coggins, Clemency Chase. "A New Sun at Chichén Itzá," *World Archaeoastronomy* (ed. A. F. Aveni). Cambridge, England: Cambridge University Press, 1989, pp. 260–275.

Coggins, Clemency Chase, and Orrin C. Shane III, eds. *Cenote of Sacrifice—Maya Treasures from the Sacred Well at Chichén Itzá.* Austin, Texas: University of Texas Press, 1984.

Cohadas, Marvin. "Astronomical Associations of the Chichén Itzá Castillo and Its Relationship to Other Maya Astronomical Observatories," unpublished paper, presented at "Archaeoastronomy in the Americas" conference, Santa Fe, New Mexico, 1979.

Conway, Thor. *Painted Dreams—Native American Rock Art.* Mineocqua, Wisconsin: NorthWord Press, 1993.

Daniels, Les. *DC Comics.* Boston: Little, Brown and Company, 1995.

———. *Marvel.* New York: Harry N. Abrams, Inc., Publishers, 1991.

Diehl, Richard A., and Janet Catherine Berlo, eds. *Mesoamerica after the Decline of Teotihuacán A.D. 700–900.* Washington, D.C.: Dumbarton Oaks Research Library and Collection, 1989.

Diószegi, V., ed. *Popular Beliefs and Folklore Tradition in Siberia.* Bloomington, Indiana: Indiana University Press, 1988.

Diószegi, V., and M. Hoppál, eds. *Shamanism in Siberia.* Budapest: Akadémiai Kiadó, 1978.

Editor et al. *Rome: Knopf Guides.* New York: Alfred A. Knopf, Inc., 1994.

Fitzhugh, William W., and Aron Crowell, eds. *Crossroads of Continents: Cultures of Siberia and Alaska.* Washington, D.C.: Smithsonian Institution Press, 1988.

Garlake, Peter. *The Hunter's Vision—The Prehistoric Rock Art of Zimbabwe.* Seattle: University of Washington Press, 1995.

Habachi, Labib. *The Obelisks of Egypt—Skyscrapers of the Past.* New York: Charles Scribner's Sons, 1977.

Hoppál, Mihály, ed. *Shamanism in Eurasia, Parts 1 and 2.* Göttingen, Germany: Edition Herodot, 1984.

Hoppál, Mihály, and Keith D. Howard, eds. *Shamans and Cultures.* Budapest: Akadémiai Kiadó, and Los Angeles/Fullerton: International Society for Trans-Oceanic Research, 1989.

Krupp, E. C. "The Cosmic Temples of Old Beijing." *World Archaeoastronomy* (ed. A. F. Aveni). Cambridge, England: Cambridge University Press, 1989, pp. 65–75.

———. "The Mandate of Heaven," *Griffith Observer* 46, no. 6 (June 1982): 8–17.

Lounsbury, Floyd. "Astronomical Knowledge and Its Uses at Bonampak, Mexico," *Archaeoastronomy and the New World* (ed. A. F. Aveni). Cambridge, England: Cambridge University Press, 1982, pp. 143–168.

Lanciani, Rodolfo. *The Ruins and Excavations of Ancient Rome.* 1897. Reprint. New York: Bell Publishing Company, 1979.

Macadam, Alta, ed. *Blue Guide Rome and Environs.* London: Ernest Benn Limited, 1979.

Meyer, Jeffrey F. *The Dragons of Tiananmen—Beijing as a Sacred City.* Columbia, South Carolina: University of South Carolina Press, 1991.

Milbrath, Susan. "Astronomical Images and Orientations in the Architecture of Chichén Itzá," *New Directions in American Archaeoastronomy* (ed. A. F. Aveni). Oxford, England: B.A.R., 1988, pp. 57–80.

Miller, Mary Ellen. *The Murals of Bonampak.* Princeton, New Jersey: Princeton University Press, 1986.

Millon, René. "Teotihuacán Studies from 1950 to 1990 and Beyond," *Art, Ideology, and the City of Teotihuacán* (ed. Janet Catherine Berlo). Washington, D.C.: Dumbarton Oaks Research Library and Collection, 1992, pp. 339–429.

Rovin, Jeff. *The Encyclopedia of Superheroes.* New York: Facts On File Publications, 1985.

Ruscito, Stefano. "Un Orologio per Tutte le Stagioni," *Archeologia Viva* I, no. 7 (October 1987), pp. 18–25.

Schaafsma, Polly. "Imagery and Magic: Petroglyphs at Comanche Gap, Galisteo Basin, New Mexico," *Archaeology, Art, & Anthropology: Papers in Honor of J. J. Brody* (ed. Meliha S. Duran and David T. Kirkpatrick). Albuquerque: The Archaeological Society of New Mexico, 1993, pp. 157–194.

Schele, Linda, and David Freidel. *A Forest of Kings: The Untold Story of the Ancient Maya.* New York: William Morrow and Company, Inc., 1990.

Schele, Linda, David Freidel, and Joy Parker. *Maya Cosmos—Three Thousand Years on the Shaman's Path.* New York: William Morrow and Company, Inc., 1993.

Schele, Linda, and Mary Ellen Miller. *The Blood of Kings: Dynasty and Ritual in Maya Art.* Fort Worth, Texas: Kimbell Art Museum, 1986.

Shirokogoroff, S. M. *Psychomental Complex of the Tungus.* 1935. Reprint. New York: AMS Press, Inc., 1982.

Sprajc, Ivan. "The Venus-Rain-Maize Complex in the Mesoamerican World View: Part I," *Journal for the History of Astronomy* 24, parts 1/2 (February/May, 1993): 17–70.

Sprajc, Ivan. "The Venus-Rain-Maize Complex in the Mesoamerican World View: Part II," *Archaeoastronomy* (*JHA Supplement* 18), 1993: S27–S54.

Thompson, J. Eric S. *Maya History and Religion.* Norman, Oklahoma: University of Oklahoma Press, 1970.

Vitebsky, Piers. *The Shaman.* Boston: Little, Brown and Company, 1995.

Wechsler, Howard J. *Offerings of Jade and Silk—Ritual and Symbol in the Legitimation of the T'ang Dynasty.* New Haven, Connecticut: Yale University Press, 1985.

Wheatley, Paul. *The Pivot of the Four Quarters.* Edinburgh: Edinburgh University Press, 1971.

Wild, Fiona, ed. *Eyewitness Travel Guides: Rome.* New York: Dorling Kindersley, 1993.

Zanker, Paul. *The Power of Images in the Age of Augustus.* Ann Arbor: The University of Michigan Press, 1988.

CHAPTER TEN—UPWARD MOBILITY

Akurgal, Ekrem. *Ancient Civilizations and Ruins of Turkey.* Istanbul: Haset Kitabevi, 1973.

Badawy, Alexander. "The Stellar Destiny of Pharaoh and the So-Called Air-Shafts of Cheops' Pyramid," *Mitteilungen des Instituts für Orientforschung,* Band X (1964): 189–206.

Bauval, Robert, and Adrian Gilbert. *The Orion Mystery.* London: William Heinemann Ltd., 1994.

Clancy, Flora Simmons. *Pyramids,* Smithsonian Exploring the Ancient World Series. Washington, D.C.: Smithsonian Books, 1994.

Coèdes, George. *Angkor—An Introduction.* Oxford/Singapore: Oxford University Press, 1963.

Colledge, Malcolm A. R. *The Parthians.* New York: Frederick A. Praeger, 1967.

Curtis, Vesta Sarkhosh. *Persian Myths.* Austin, Texas: University of Texas Press, 1993.

Dagens, Bruno. *Angkor—Heart of an Asian Empire.* New York: Harry N. Abrams, Inc., Publishers, 1995.

Danielou, Alain. *The Gods of India.* New York: Inner Traditions International, Ltd., 1985.

Deaton, John Charles. "The Old Kingdom Evidence for the Function of the Pyramids," *Varia Agyptiaca* 4 (1988): 193–200.

———. "The Post Old Kingdom Evidence for the Function of the Pyramids." Unpublished manuscript.

Dütting, Dieter, and Anthony F. Aveni. "The 2 Cib 14 Mol Event in the Palenque Inscriptions." *Zeitschrift für Ethnologie* 107 (1982).

Editor et al. *Anatolia: Cauldron of Cultures,* Lost Civilizations Series. Alexandria, Virginia: Time-Life Books, 1995.

———. *Southeast Asia: A Past Regained,* Lost Civilizations Series. Alexandria, Virginia: Time-Life Books, 1995.

Edwards, I. E. S. *The Pyramids of Egypt.* New York: The Viking Press, Inc., 1972.

Eicher, Peter. *The Elvis Sightings.* New York: Avon Books, 1993.

Eliot, Joshua, ed., et al. *1995 Vietnam, Laos, & Cambodia Handbook.* Lincolnwood, Chicago: Passport Books, 1995.

Fakhry, Ahmed. *The Pyramids.* Chicago: The University of Chicago Press, 1961.

Faulkner, R. O. *The Ancient Egyptian Pyramid Texts.* Oxford: Oxford University Press, 1969.

———. "The King and the Star-Religion in the Pyramid Texts," *Journal of Near Eastern Studies* XXV (1966): 153–161.

Fowler, Melvin. *The Cahokia Atlas.* Springfield, Illinois: Illinois Historic Preservation Agency, 1989.

———. "A Pre-Columbian Urban Center on the Mississippi," *Scientific American* 233, no. 2 (August 1975): 92–101.

———. *The Woodhenge 72 Project: A Summary Report of the 1991, 1992, and 1993 Field Work.* In press.

Freeman, Michael, and Roger Warner. *Angkor—The Hidden Glories.* Boston: Houghton Mifflin Company, 1990.

Freidel, David, and Linda Schele. "Symbol and Power: A History of the Lowland Maya Cosmogram," *Maya Iconography* (eds. Elizabeth P. Benson and Gillett G. Griffin). Princeton, New Jersey: Princeton University Press, 1988, pp. 44–93.

Friesner, Esther M., and Martin H. Greenberg, eds. *Alien Pregnant by Elvis.* New York: DAW Books, Inc., 1994.

Gökovali, Sadan. *Commagene and Nemrut.* Izmir, Turkey: Ofis Ticaret Matbaacilik Ltd., 1988.

Hacklin, J., ed., et al. *Asiatic Mythology.* New York: Crescent Books, no date.

Hamblin, Dora Jane. "A Unique Approach to Unraveling the Secrets of the Great Pyramids," *Smithsonian* 16, no. 1 (April 1986): 78–93.

Hawass, Zahi A. "The Funerary Establishments of Khufu, Khafra and Menkaura." Doctoral thesis. Philadelphia, Pennsylvania: The University of Pennsylvania, 1987, (University Microfilms, Ann Arbor, Michigan).

Heilman, James M., and Roger R. Hoefer. "Possible Astronomical Alignments in a Fort Ancient Settlement at the Incinerator Site in Dayton, Ohio. *Archaeoastronomy in the Americas* (ed. Ray A. Williamson). Los Altos, California: Ballena Press/Center for Archaeoastronomy, 1981, pp. 157–171.

Higham, Charles. *The Archaeology of Mainland Southeast Asia.* Cambridge, England: Cambridge University Press, 1989.

Ions, Veronica. *Indian Mythology.* 1967. Reprint. New York: Peter Bedrick Books, 1983.

Krupp, E. C. "Cahokia: Corn, Commerce, and the Cosmos," *Griffith Observer* 41, no. 5 (May 1977): 10–20.

———. "Egyptian Astronomy: The Roots of Modern Timekeeping," *New Scientist* 85, (January 3, 1980): 24–27.

———. "Great Pyramid Astronomy," *Griffith Observer* 42, no. 3 (March 1978): 1–18.

———. "How Much Sun Can a Woodhenge Catch?" Paper presented at "The Ancient Skies and Sky Watchers of Cahokia: Woodhenges, Eclipses, and Cahokia Cosmology" symposium, May 9, 1994. In press.

———. "Light in the Temples," *Records in Stone* (ed. Clive L. N. Ruggles). Cambridge, England: Cambridge University Press, 1988, pp. 473–499.

Lewis, James R., ed. *The Gods Have Landed—New Religions from Other Worlds.* Albany, New York: State University of New York Press, 1995.

Lloyd, Seton. *Ancient Turkey.* London: British Museum Publications Ltd., 1989.

Lockyer, J. Norman. *The Dawn of Astronomy.* London: Cassell and Company, 1894.

Mabbet, Ian, and David Chandler. *The Khmers.* Oxford, England: Basil Blackwell Ltd., 1995.

Mails, Thomas E., and Dan Evehema. *Hotevilla—Hopi Shrine of the Covenant, Microcosm of the World.* New York: Marlowe & Company, 1995.

Marcus, Greil. *Dead Elvis.* New York: Doubleday, 1991.

Mazzeo, Donatella, and Chiara Silvi Antonini. *Monuments of Civilization: Ancient Cambodia.* New York: Grosset & Dunlap, Inc., 1978.

McCluskey, Stephen C. "Astronomies and Rituals at the Dawn of the Middle Ages," *Astronomies and Cultures* (ed. Clive L. N. Ruggles and Nicholas J. Saunders). Niwot, Colorado: University Press of Colorado, 1993, pp. 100–123.

Mehling, Marianne, ed. *Turkey—A Phaidon Cultural Guide.* Oxford, England: Phaidon Press Limited, 1989.

Mink, Claudia Gellman. *Cahokia—City of the Sun.* Collinsville, Illinois: Cahokia Mounds Museum Society, 1992.

Morón, Eleanor. "Configuration of Time and Space at Angkor Wat," *Studies in Indo-Asian Art and Culture* 5 (1977), pp. 217–267.

Parmentier, Henri. *Henri Parmentier's Guide to Angkor.* Saigon: Albert Portail, no date.

Petrie, W. M. Flinders. *The Pyramids and Temples of Giza.* With an update by Zahi Hawass. 1885. Reprint. London: Mysteries & Mysteries of Man Ltd., 1990.

Robinson, Daniel, and Tony Wheeler. *Cambodia—A Travel Survival Kit.* Berkeley, California: Lonely Planet Publications, 1992.

Rooney, Dawn F. *Angkor.* Lincolnwood, Chicago: Passport Books, 1994.

Schele, Linda, and David Freidel. *A Forest of Kings: The Untold Story of the Ancient Maya.* New York: William Morrow and Company, Inc., 1990.

Scherman, Rowland. *Elvis Is Everywhere.* New York: Clarkson Potter/Publishers, 1991.

Schrag, Carl. "Graceland's Not as Far from Jerusalem as You Think," *The Northern California Jewish Bulletin,* August 24, 1990.

Stencel, Robert, Fred Gifford, and Eleanor Morón. "Astronomy and Cosmology at Angkor Wat," *Science* 193 (July 23, 1976): 281–287.

Stierlin, Henri. *The Cultural History of Angkor.* London: Aurum Press Ltd., 1984.

Taylor, Lily Ross. *The Divinity of the Roman Emperor.* 1931. Reprint. New York: Garland Publishing, Inc., 1979.

Trimble, Virginia, "Astronomical Investigation Concerning the So-Called Air-Shafts of Cheops' Pyramid," *Mitteilungen des Instituts für Orientforschung,* Band X (1964): 183–187.

Wallace, Amy. "Scholarly Study of Elvis Has Academia All Shook Up," *Los Angeles Times,* September 2, 1995.

Wheatley, Paul. *The Pivot of the Four Quarters.* Edinburgh: Edinburgh University Press, 1971.

Wittry, Warren L. *Summary Report on 1978 Investigations of Circle No. 2 of the Woodhenge, Cahokia Mounds State Historic Site.* Chicago: Department of Anthropology, University of Illinois at Chicago Circle, March 1980.

Yaman, Berker. *Andiyaman-Nemrut Dagi-Commagene.* Istanbul: Minyatür Publications, no date.

Yaraborough, Trin. "Elvis Lives! In Many Shapes and Colors," *Los Angeles Times,* August 16, 1993.

Yörükoglu, Ömer, and Kemal Soyer. *Nemrud Dag Kommaagene.* Ankara: no publisher, 1988.

Illustration Credits

Page 27: In *Rio Azul Reports,* ed. R. E. W. Adams, no. 3, *The 1985 Season.* Property of Rio Azul Archaeology Project. Reprinted with the permission of Rio Azul Project.

Page 28: Reprinted with the permission of R. E. W. Adams.

Page 29: Reprinted with permission of Akademische Druk-u Verlagsanstalt, Universitats-Buchdruckerei u. Universitats-Verlag, Graz, Austria.

Page 62: Illustration by Jay Von Werholf; copyright © Imperial Valley Museum, P.O. Box 2495, El Centro, CA 92243.

Page 72: Reprinted with the permission of Cambridge University Press.

Page 77: Copyright © Nancy G. Cattell. Reprinted with the permission of Nancy G. Cattell.

Page 90: Copyright © Dover Publications, Inc. Reprinted with the permission of Dover Publications, Inc.

Page 114: Copyright © Gary Wirth. Reprinted with the permission of Gary Wirth.

Pages 123 and 124: Copyright © A. Marshack. Reprinted with the permission of A. Marshak.

Page 141: Copyright © 1996 Yorkshire Television Ltd.

Pages 165, 167, and 193: Copyright © Peter C. Keller. Reprinted with the permission of Peter C. Keller.

Page 170: Reprinted with the permission of Yale University Press.

Page 181: © Frank Willett.

Page 190: Copyright © Raymond Turvey. Reprinted with the permission of Raymond Turvey.

Page 232: Copyright © Cambridge University Press. Reprinted with the permission of Cambridge University Press.

Page 316: Cover art by Jim Warren. Reprinted with the permission of Alan Lynch, DAW Books.

Index

Adams, R. E. W., and Río Azul, 26, 27
African cultures
  Ife, 179–82
  San, 69–79
agriculture
  and calendric control, 4, 231
  Dogon granaries, 163–64
  in Mississippian culture, 296–97
  symbolism of, in Shinto Japan, 204–5
  Tibetan, origin of, 85–86
  and women, importance of, 142–43
  See also fertility
akitu ceremony, 144–45
Alexander the Great, astrologers of, 240
Allen, James P., and Egyptian creation myths, 226–27
All-Knowing God, The (Pettazzoni), 34
Altamira cave art, 116–17
altars
  directional, 28–32
  Dogon, 163
Altay shamans, costumes of, 249
altazimuth, 8–11
altered states of consciousness, 56–60
  of Chumash shamans, 155–56
  and dance, 63–65, 69–70, 73
  of Maya, 300
  San trance dancing, 69–70, 73
Amaterasu, 200, 206
Amazing Spider-man, 246
American Woodhenge, 297
Amma's Pillar, 163–64
Amun, 225–26
Anasazi, 1
Anatolia, land of mothers, 142–43
Angkor Thom, 309–10
Angkor Wat, 304–9
  cosmic associations, 307–9
  dimensions, significance of, 304–7
animism, 36
  creation myth of Tibet, 80–81
'antap, in Chumash society, 155–58
Antiochus I
  national holidays of, 284–85
  tomb of, 280–85
Apollon-Mithras-Helios-Hermes, 282
Ares-Artagnes-Heracles, 283
armilla, ecliptic, 8–11
armilla, equatorial, 8–11
Art of Deification, The, 25
Ashurbanipal, library of, 227–28
Assyria, astronomy of, 228–30
Aston, William George, and Shintoism, 207
astrology
  to advise the ruler, 227–36
  in ancient Rome, 240–43
  invention of, 230

astrology *(Continued)*
for Khubilai Khan, 194–95
skepticism of, 236–40
Astronomica (Manilius), 230
astronomical knowledge, and power, 13, 154
astronomy
advising the ruler, 227–36
in ancient China, 7–11
astronomers, 231–36
and the Church, 10–11
and funerary practices, 136–41
of the Mesopotamians, 228–30
restricting information, 238–40
symbolic value, 7–8
at Wijiji ruin, 1
Atlantis, 298
Augustan Altar of Peace, 276–77
Augustus Caesar
deification of, 279
power displays of, 274–80
Austen, Hallie Iglehart, and the Maltese caves, 129
Avalokitesvara (Chenrezi), 86
Aveni, Anthony F., and Chan-Bahlum, 303

Babylon, skepticism of astrology, 236–37
Badawy, Alexander, and the Great Pyramid, 288
Bai Ülgän, 53
Baigent, Michael
astrology, skepticism of, 237
From the Omens of Babylon, 229–30
Bali, hierarchy of power of, 47
Batman, 245–46
Bayon, 309–10
Bean, Lowell John, and character of power, 50–51
Bear Butte, as center of the world, 219
bears, as symbols of rebirth, 111
Beijing
cosmological layout of, 193
Old Beijing Observatory, 7–11
Benin, center of power, 181–82
Bighorn medicine wheel, 218–19
birth imagery, 79, 107–10
at Çatal Höyük, 142–43
See also cyclic renewal; Mother Earth
Black Elk, and Sioux cosmology, 38–39
Bladder Festival of Yupik Eskimos, 56
Bleek Wilhelm and Dorothea, and /Xam Bushmen, 72
bloodletting, of Cerros kings, 300, 303
Bodhisattvas, 86
Bonampak mural, 259–62
Bon religion. *See* Tibetans
Brahmaputra River (Tsangpo), 95
Brochtorff Circle, 132–33
Brown, Joseph Epes, and the White Buffalo Cow Woman, 38–39
Buddhism
Bodhisattvas, 86
creation myth of, 89
Lamaist Buddhism, 20–22
in Tibet, introduction of, 81, 88–89
Buga, 36–37
burials
bear symbolism, 111
at Brochtorff Circle, 132–33
in caves, 110–15
at center of the world, 26–28
in Mississippian culture, 297–98
in Mongol culture, 188
at Shanidar Cave, 110
See also funeral imagery; pyramids; tombs
Burkhan Khaldun, sacred mountain, 186–87
Bushmen. *See* San

Cacaxtla, murals of, 262–66
Cahokia, mounds of, 295–98
caiman, significance of, 5–6
cairns
and Cheyenne spirit wheels, 218
as shrines, 187
Calakmul, sacred center at, 48–49
calendars
at Chichén Itzá, 269–70
control of, 231

Roman, 274–78
Tong tian, 7, 192
California
Indians of, 154–55
See also Chumash
Calvin, William H., and *How the Shaman Stole the Moon,* 209–10
Cambodia, monumental architecture, 303–10
Canopus, significance, to San, 68
Captain Marvel, 246
cardinal directions. *See* directions, cardinal
Carlson, John B.
and Venus iconography, 261
and Venus warfare symbolism, 265–66
Carmen Miranda of Commagene, 283
Çatal Höyük, 142–43
cave art
at Altamira, 116–17
cyclic renewal themes of, 120–25
of the Maya, 112
See also rock art
Cave of Balamkanché, 111–13
caves
evolution of symbolism, 127–28
Mother Earth, access to, 100
ritual significance of, 110–16
shamanic uses, 119–20
celestial globe, 8–11
celestial power. *See* power
celestial symbolism and power, 154
Celtic peoples, passage tombs, 133–41
center of the world, 15–16, 20
burial of kings at, 26–28
caves as, 110
of the Cheyenne, 213–20
for Hittites, 146–52
of San, 76
for Tibetans, 81
See also world axis; world-axis mountain
central lake
feminine qualities of, 93
in Navajo tradition, 99
and the underworld, 93
and the world-axis mountain, 93
Cerros, temples of, 300–301
Chaco Canyon, 1
Chalcatzingo, 2–3
caves, facial symbolism of, 113–15
topography, significance of, 5–7
water supply, 5
Chan-Bahlum, pyramid of, 301–3
change, cosmic power of, 17
Chauvet, Jean-Marie, 117
Chauvet Cave, 117
Chemehuevi, creation myth of, 108
Chenrezi (Avalokitesvara), 86
Cheyenne, 212–20
astronomical knowledge of, 217–20
directions, cardinal, symbolism of, 215–19
Massaum ceremony, 213–17
medicine wheels, 217–19
Pleiades, significance of, 214–15
wolf, significance of, 215–17
world view of, 213
Yellow-haired Woman, 215–17
Chichén Itzá
calendars in public ritual, 269–70
monumental architecture, 267–70
war and sacrifice at, 266–67
chiefdoms
of Cahokia, 295–98
Chumash variation of, 155–61
contact with the gods, 140
Dogon variation of, 162–67
Hawaiian, 168–73
power associations in, 153–54, 168–73
sacred kings and shamans, similarities of, 178–79
China, ancient
astronomy and the Jesuit missionaries, 9–11
astronomy specialists of, 231–36
celestial phenomena documentation, 232
color symbolism of, 272–74
and directions, cardinal, 20–25
emperor, ritual duties of, 271–74
rituals of power, 271–74
Chi Ni Tzu, 231

Christianity, in ancient China, 10
Chumash, 154–61
  center of the world, 103–4, 158–61
  chiefs, 155
  Painted Rock site, 102–3
  power distribution, 157
  sacred places of, 167–68
  shamans, 156–61
  sunstick, 158–61
  tribelet organization, 154–57
  winter solstice ceremony, 158–61
Cicero, skepticism of astrology, 237
circle
  in ancient China, 10
  symbolism of, 2
clouds, Olmec symbolism, 4. *See also* rain
Coedès, George, and the Bayon, 310
Cohadas, Marvin, and Kukulcán's pyramid, 268
color-direction links
  Four Celestial Kings, 25
  in Hollywood, California, 40–42
  in Hopi culture, 31–32
color symbolism
  in ancient China, 272–74
  Cheyenne, 216–17
  in Mongol ideology, 184–85
Columbus, Christopher, and knowledge, 209–210
Comanche Gap petroglyphs, 252–55
Commentary (Macrobius), 230
community, and cosmic order, 50
Conway, Thor, and sacred art, 252
cosmic center
  and sacred mountains in Tibet, 85
  See also center of the world; world axis; world-axis mountains
cosmic order
  control over, by Chinese emperor, 7–8
  and Creation, 145
  enforced by Four Celestial Kings, 24
cosmic symbolism, 17. *See also* center of the world; directions, cardinal; world axis; world-axis mountains
cosmos, structure of, 15–20
cosmovision, 17
  and kingship, 49
  modern-day, 315
  See also world view
Cosquer, Henri, and Grotte Cosquer, 117–18
Cosquer Cave, 117
Counsel Rocks, 108
Crab supernova, account of, 232
crane dance, 63
Creation
  and the central place, 16–17
  and cosmic power, 145
  and shamanic power, 69–72, 80
creation myths
  of Dogon, 163
  of Egyptians, 220–27
  of Haida, 77–79
  of Hawaiian chiefdoms, 171–72
  from Heliopolis, 221–23
  from Hermopolis, 224–25
  of imperial Japan, 198–99
  of Kawaiisu Indians, 101
  of Khmer, 307
  from Memphis, 221–22
  personification of the supernatural, 80–81
  from Thebes, 225–26
  of Tibetans, 80–81, 85–86
  of Yamato, 198–99
  of Yoruba culture, 180
cyclic renewal, 34, 50–51
  at Çatal Höyük, 142–43
  cave art symbolism, 120–125
  Dogon ceremonies for, 165
  in Egypt, 288–89
  Hawaiian symbolism, 171
  images of, 97–99
  in Maltese temples, 128–31
  masculine role in, 143–52
  at Newgrange, 140–41
  ritual sex of mother goddess and father god, 144–52
  in Shinto Japan, 200–203
  See also Mother Earth

Daghda, 134
dance, ritual
 the crane dance, 63
 San trance dance, 69–70, 73
Daniel, Glyn, and Newgrange, 138
datura (jimsonweed), Chumash use of, 155–56
dead, the
 in Maltese temples, 131–32
 spirit realm, 35–36
 See also soul, the
death and renewal theme
 in Irish passage graves, 133–41
 in Maltese caves and tombs, 128–33
Deaton, John Charles, and the pyramids, 293–94
Devi, 93
Dhritarashra, 21
Diamond Kings of Heaven, 24–25
Diana, and fertility, 143–44
directional geometry, 17–18
directions, cardinal
 at Angkor Wat and Angkor Thom, 307–10
 Dogon society, symbolism of, 165–66
 and Four Celestial Kings, 20–25
 Hopi culture, symbolism of, 28–32
 Maya civilization, symbolism of, 26–28
 north pole, 19
 and Pueblo Bonito, 2
 Tibetans, symbolism of, 82
 value of, 17–18
directions altar, 28–32
Disneyland, and traditional cosmology, 40
divination
 of Dogon, 165–67
 of Etruscans, 18
divine kingship, 178
 in Yoruba culture, 179–82
 See also kings
Dogon, 162–67
 astronomy of, 164–67
 and the cardinal directions, 165–66
 creation myth, 163
 granaries, 163–64
 rain, symbolism of, 162
 rock painting, 166–67
 sacred places of, 167–68
 social order, 162–63
 world view, 162–67
Donnelly, Ignatius, and the pyramids, 298
Dowson, Thomas, and San rock art, 71
drums
 and shamanic trance, 64–65
 of Siberian shamans, 250–52

Earle, Timothy
 and Hawaiian chiefdoms, 168
 and the symbolic use of the heavens, 140
Eddy, John A., and Cheyenne medicine wheels, 217–19
Edwards, C.E.S., and pyramids, 292
Egypt
 creation myths of, 220–27
 cyclic renewal, 288–89
 death of pharaohs, 285–95
 priesthood rivalries, 220
 pyramids, significance of, 285–95
 Thuban, significance of, 288
 water, symbolism of, 99–100
Eliade, Mircea, 37
 shamanism, description of, 52–53
Elvis, influence of, 315
Epinomis (Plato), 230
Erketü Tengri, Mongol god, 187–88
Esarhaddon, skepticism of astrology, 237–38
Etruscan civilizations, and the cardinal directions, 18
Evenks, 36
Evolution of Human Societies, The (Johnson and Earle), 176

Fajada Butte, 1
female aspect
 and the central lake, 93
 of Maltese caves, 128–32
 power of renewal, 142–43

fertility
  feminine symbols of, 127
  masculine role in, 143–52
  symbols of, 104–10
  See also Mother Earth; Mother Goddess
fertility stones, 104
feudalism, 168
First Father of the Maya, 26–28
First People
  of the Hopi, 30
  of the Huichol, 57–58
Font-de-Gaume cave paintings, 121–22
Forest of Kings, A (Freidel and Schele), 300
four, the number, significance of, 39–40, 165
Four Celestial Kings, and the cardinal directions, 20–25
Fowler, Melvin L., and Woodhenge 72, 297–98
Frazer, Sir James George, and divine kingship, 178–79
Freidel, David
  Maya Cosmos, 26
  Maya pyramids, 299–300
Frobenius, Leo, and Ife, 179
From the Omens of Babylon (Baigent), 229–30
fundamentalists, 314
funeral imagery
  at Çatal Höyük, 142–43
  at Yazilikaya, 152
  See also burials; pyramids; tombs
Furst, Peter, and Huichol shamans, 56–60

Galisteo Basin rock art, 252–55
Gampo, Songsten, 85, 91
  reforms of, 88–89
  tomb of, 87–88
Ganges River, sacred river, 95
Geertz, Clifford, and the hierarchy of power, 46–47
Genesis in Egypt (Allen), 226–27
Genghis Khan, 183–84
  astronomy, use of, 188–92
  Burkhan Khaldun, 186–88
  deification of, 185
  and Kököchu, 191
  rise to power, 186–87
  shamans of, 187–92
Geng xun, 233–34
genii, Mo li brothers, 24–25
geography, sacred, 95. *See also* center of the world; central lake; world axis; world-axis mountains
geometry, directional, 17–18. *See also* directions, cardinal
geranos (crane dance), 63
Giza funerary complex, 289–92
Goddess Movement, 130. *See also* Mother Earth; Mother Goddess
gods
  contacting through monumental architecture, 140
  interactions with, 283
  as personifications of power, 51
  system of, Hittite, 151
Goldi, 36
  shamans, costumes of, 249–50
Gómez, José Humberto, and the Cave of Balamkanché, 111–13
Gorden, Mary and Jim, and Yokuts rock art, 109–10
Gozo, tombs of, 132–33
Granary of the Master of Pure Earth, 163
Great Pyramid, 287–90
Great Wall of China, 20
Griaule, Marcel, and the Dogon, 163
Griffith Observatory, 11–13
Grove, David, study of Chalcatzingo, 5
Guo, Shoujing, and the Old Beijing Observatory, 7–11, 192

Haida, 77–80
  creation myth, 77–79
  Raven, 77–78
  and San, comparison of, 79
  shamanism, 79–80
hallucinogens, in shamanism, 56–60
Hallucinogens and Shamanism (Harner), 212
Hamayon, Roberte N., 125
Harner, Michael J., and shaman trickery, 212

Harrington, John Peabody, and the Chumash winter solstice ceremony, 158
Hawaiian chiefdoms, 168–73
astronomical knowledge, 170–73
chief, role of, 169–73
creation myth, 171–72
mana, 168–69
world view of, 168
Hawass, Zahi, and the Giza funerary complex, 289–90
Heart of the Goddess—Art, Myth and Meditations of the World's Sacred Feminine, The (Austen), 129
Heart of the Hunter, The (van der Post), 67–68
Heaven
Shang di as, 7–8
in shaman's world view, 35
heavens
of Dogon, 163
Four Celestial Kings, realm of, 24
of the Hawaiians, 169
of Tungus people, 36
Heider, Dave, and the white buffalo calf, 39
Heliopolis, creation myth of, 221–23
Hepat, Mother Goddess, 147–49
Hermopolis, creation myth of, 224–25
Hinduism
cosmographical ideas, 93–95
Milky Way, symbolism of, 94–95
Hittite rituals for cyclic renewal, 144–52
Hollywood, California, and its world-quarter shrine, 40–42
Hopi culture
cardinal directions, symbolism of, 28–32
kachinas, 4–5
traditionalists of, 314
horizon symbolism, 2
horoscopes, in ancient Rome, 240–41
horse, symbolism of, 188
How the Shaman Stole the Moon (Calvin), 209–10
Huichol, and peyote, use of, 56–60
Humphrey, Caroline, state development and shamanism, 192
hunter-gatherers
cyclic renewal, 34
world view of, 68, 124–25
See also San; shamans
Hurrians, and the sacred marriage at Yazilikaya, 146–49
Hypogeum of Hal Saflieni, 131–32

Ice Age cave paintings, 116–25
Ichneumon, 72–74
ideology, 153–54
functions of, 310–11
maintenance of, 312–13
and power centralization, 285–95
ritual art as a restatement of, 60
unification of, in present day, 315–17
Ife
the center of power, 180–82
and divine kingship, 179–82
Inanna, Queen of Heaven, 145–46
Indian Creek, sacred center at, 48–49
Ise
shrines of, 203–5
See also Japan, imperial
Izanagi and Izanami, 199, 203–4

Japan, identity with the sun, 196
Japan, imperial
creation myth of, 198–99
cyclic renewal, 200–203
Jimmu Tenno, 203
kami of, 200–201
mountains, sacred, 204
myth of first divine emperor, 202
shamanism of, 207
Sun Goddess, 196–202, 206–7
world axis of, 199
Yamato domination of, 197–98
Jayavarman II, 306–7
Jayavarman VII, 309–10
Jivaro Indians, and shaman trickery, 212
Julian calendar, 275
Juyong Pass, 20

kachinas, in Hopi traditional life, 29–32
Kakunupmawa (Chumash winter solstice ceremony), 157–61
Kalahari Desert. *See* San
Kawaiisu Indians, creation myth of, 101
Khafre, pyramid of, 290–92
Khitans, validation of royal power, 190
Khmer peoples, 303–10
  Angkor Wat, 304–9
  creation myth of, 307
Khubilai Khan
  astronomy, use of, 183–84, 192–95
  Dadu, founding of, 193
  and the Old Beijing Observatory, 7
Khufu, and the Great Pyramid, 287–90
kings
  advising, with astronomy, 227–36
  burial at center of the world, 26–28
  chiefdoms and shamans, similarities of, 178–79
  cosmic, 177–78
  and cyclic renewal, 179
  exploiting, with astrology, 236–40
  interaction with divine powers, 25, 144, 283
  and lions, associations of, 173
  motivating factors of, 176–79
  power, mechanisms of, 177
  power, and seasonal rites, 144–52, 181
  and state formation, 175–76
  supernatural power, manipulation of, 177–78
Kingship and Sacrifice—Ritual and Society in Ancient Hawaii (Valeri), 171
kivas, 28–32
Knifewing, 254–55
knowledge
  astronomical, 13, 154
  and dynamic equilibrium, 50–51
  skepticism of, 209–12
Knowth passage grave, 138–39
Kojiki, 198–99
Kököchu, Mongol shaman, 187, 191
Kraft, John, and European maze tradition, 62–63
Krascheninnikow, Stephan, and false shamanic magic, 211
Kroeber, Alfred L., and the Chumash, 154
Kukulcán, 257
  pyramid of, 267–69
!Kung San, 70. *See also* San
Kwammang-a, 72–74

labyrinths, 62–63
lake, central. *See* central lake
Lamaist Buddhism, 20–22. *See also* Four Celestial Kings
landscape, symbolism of, 6, 15, 95, 203. *See also* center of the world; central lake; world axis; world-axis mountains
landscape of the world, 15
Lascaux, cave paintings of, 119–20
Lehner, Mark, and solar alignment of pyramids, 290, 292–93
Leroi-Gourhan, André, and cave art, 120
Lewis-Williams, J. David, and San rock art, 71
Lhabab Ri, 83–84
Lhuillier, Alberto Ruz, and the Palenque pyramid, 301–2
Librado, Fernando, and the Chumash winter solstice ceremony, 158–61
lingam, symbolism of, 93
Lion King, The, 173–75
lions, symbolism of, 173–75
Lloyd, Lucy, and the /Xam Bushmen, 72
Lono, 169–73
Lost World of the Kalahari, The (van der Post), 67–68
Lounsbury, Floyd, and Bonampak mural, 259
Luiseño Indians, creation myth of, 97–99

Mahabharata, 94
Makahiki festival, 169–73
male aspect
  and the central mountain, 93
  role in cyclic renewal, 143–52
Malta, stone temples of, 128–32
Malta: An Archaeological Guide (Trump), 129–30

Manasarovar, 93
mandala, 85, 90–91
Mantis, 69
  the Creator, 72–75
Marshack, Alexander, and time-factored symbols, 122–23
Massaum ceremony, 213–18
Maya
  cardinal directions, symbolism of, 26–28
  caves, significance of, 111–16
  Earth God, 113
  hierarchy of power, 47–48
  pyramids of, 299–301
Maya Cosmos (Freidel and Schele), 26
mazes, power of, 62–63
McCluskey, Stephen C., and pagan calendrics, 280
McGowan, Charlotte, and yoni, 104–6
medicine wheels, 217–19
megalithic structures
  to contact gods, 140
  of Irish peoples, 133–41
  See also pyramids
Memphis, creation myth of, 221–22
Menes, 220–21
Mesoamerica
  monumental architecture of, 47–48, 298–303
  Venus-regulated warfare, 256–70
  world view of, 270
Mesopotamia
  astrology, skepticism of, 236–37
  astronomy specialists, 228–30
  cyclic renewal rites, 144–46
Mexico. *See* Chalcatzingo; Mesoamerica; Olmec
Meyerhoff, Barbara, and Huichol shamans, 56–60
Milbrath, Susan, and Kukulcán's pyramid, 268–69
Milky Way, in Hinduism, 94–95
Millon, René, and the Teotihuacán Mapping Project, 263–64
miracles, present-day expectations, 314–17
Mississippian culture
  astronomical knowledge of, 296–98
  burials of the elite, 297–98
  monumental architecture of, 295–98
  Woodhenge 3, 297
Mithraic religion, Nemrut Dag shrine, 281–84
Mo li brothers, 24–25
Mongolia
  modern-day symbolism, 189
  See also Genghis Khan; Khubilai Khan
Mongols, The (Morgan), 188
Mongols. *See also* Genghis Khan; Khubilai Khan
Monkey Cave, 85–86
Monks Mound, 295
monotheism, 33
monumental architecture
  to contact gods, 140
  of Irish peoples, 133–41
  See also pyramids
moon, significance for San, 74
Moose Mountain medicine wheel, 218–19
Morgan, David, and Mongol khans and shamans, 188
Morning Star. *See* Venus
Mother Earth
  and caves, ritual symbolism of, 110–16
  images of, 97–101, 142–43
  procreative powers, symbols of, 102–10
  as White Buffalo Cow Woman, 38–39
  See also cyclic renewal; Mother Goddess
Mother Goddess
  consorts of, 143–44
  Danu, 134
  Hepat, 147–48
  images of, 143
  role in centrally governed societies, 148–49
  See also Mother Earth
mounds, Mississippian, 295–98
mountains, sacred
  Burkhan Khaldun, 186–88
  Nemrut Dag, 280–85
  in Shinto Japan, 204

mountain, sacred *(Continued)*
Sumeru, 89
in Tibet, 85
See also pyramids; world-axis mountains
Mount Kailas (Bon Mountain), as world axis, 82–83, 91–93
Mount Meru, 93–95
Mount Nemrut (Nemrut Dag), 280–85
Mount Pinos, 103–4

Natchez chiefs, 154
natural phenomena, 50
Neandertals, burials of, 110–11
Needham, Joseph, and astronomical discretion, 238–39
Nemrut Dag (Mount Nemrut), 280–85
statues of, 281–84
New Age America, and traditional beliefs, 39
Newgrange, 133–41
age of, 135
fertility and renewal themes, 140–41
layout, 136–38
roof box, 137–38
winter solstice, alignment with, 136–40
Nganasan shamans, costumes of, 249
Nietzsche, and the quest for power, 209
Nigeria, center of power of, 181–82
Night of the Gods, The (O'Neill), 199–200
Nihongi, 198–99
nine, the number, 189
north celestial pole, 19
in Dogon cosmos, 163–64
significance, to the Tungus, 36–38
numbers, significance of, 39–40, 53, 180, 189. *See also* four, the number

O'Kelly, Michael J., and Newgrange, 135–38
Old Beijing Observatory, 7–11
Olmec civilization, 4
Olorun, Yoruba personification of sky, 180
omens
of ancient China, 227–36
of ancient Egypt, 228–30
See also astrology; divination
O'Neill, John, and Shinto world axis, 199–200
Orion Mystery, The (Bauval and Gilbert), 290
Orion's Belt, significance in Egypt, 288–89
Orochi, 36
Outer Shrine (Geku), 203–5

Pacing the Void (Schafer), 237
Painted Rock site, 102–3
Palenque, pyramid tombs of, 301–3
Panamint Valley rock alignments, 60–65
Papyrus Leiden I 350, 225–26
Parpola, Simo, and Assyrian omens, 229–30
passage graves, 133–41
petroglyphs, 1
at Chalcatzingo, 3–4
See also cave art; rock art
Pettazzoni, Raffaele, and sky gods, 34
peyote, 56–60
pharaohs
ascent of the soul, 293–94
and the sun, 289–93
See also Egypt
pictographs. *See* rock art
pilgrimages, accessing power, 85
Pivot of the Four Quarters, The (Wheatley), 310
planetaria, 12–13
Pleiades
Cheyenne, significance of, 214–15
Hawaiians, significance of, 170–71
Polaris, 19
in Dogon cosmos, 163–64
in Tungus cosmos, 36–37
political organization
and astronomy, 154
See also chiefdoms; kings
Polo, Marco
astrologers of Khubilai Khan, 194–95
and Chinese astronomy, 11
and "Day of the Red Disk" ceremony, 188
Polynesia
world view, 169–70
See also Hawaiian chiefdoms

Popol Vuh, 26
popular culture
  power, structure of, 313–17
  superhero revival, 245–47
power, 43–46
  and astronomy, 7–11, 13
  Bean, Lowell John, thoughts on, 50–51
  and calendars, 275–76
  and caves, 122
  centers of, 49
  centralization of, 149–52, 311
  of change, 17
  of chiefs, 153–54
  in Chumash cosmology, 157
  of costumes, uniforms, 246–52
  and Creation, 49–50
  definition of, 43
  demonstrated on earth, 34
  of directions, cardinal, 2
  distribution of, 67–96
  expressions of (*see* Diamond Kings of Heaven; Hopi culture; Tomb 12 at Río Azul)
  modern-day affirmations of, 313–14
  personification of, 51
  of place, 17, 75 (*see also* center of the world)
  and sacred geography, 5–7, 95
  and social complexity, 153–54
  and state formation, 176
  symbols of, 247–52 (*see also* kings)
  symbols of, accessing, 271
  of word, 224
Power of Images in the Age of Augustus (Zanker), 276, 278
Psychomental Complex of the Tungus, (Shirokogoroff), 36
Ptah, 221–23
Puranas, 93
pyramids, 285–95
  of Cambodia, 303–10
  Great Pyramid, 287–90
  Mesoamerican, 298–303
  New World, 295–303
  orientation of, 286
  solar connotations of, 289–92
  as temples, 310
Pyramid Texts, 223, 286–94

quadrant, 8–11
Quetzalcóatl, the god, 257, 265
Quetzalcóatl, the ruler, 257

Rafter, John, and Counsel Rocks, 107–8
rain, 4
  Dogon, significance of, 162
  Hawaiians, significance of, 169
  Hopi culture, significance of, 4–5, 30–31
  Maya symbolism, 113
  symbolism of, 99
rain king of Chalcatzingo, 4–7
Raven, the Creator, 77–80
Ray, T. P., and Newgrange, 139–40
Re, 222–23
rebirth. *See* cyclic renewal
Regourdou burial site, 111
Reilly, Kent, and Chalcatzingo, 5–6
religion
  in evolutionary terms, 33
  women in, 129 (*see also* Mother Earth; Mother Goddess)
renewal. *See* cyclic renewal
Renfrew, Colin, and the Maltese temples, 128–29
Reports of the Magicians and Astrologers of Nineveh and Babylon, The (Thompson), 228–29
Ricci, Matteo
  mathematics books, confiscation of, 240
  and the Old Beijing Observatory, 9–11
Rig Veda, 93
Rinpoche, Guru, 85
Río Azul, 26
ritual dance. *See* dance, ritual
rivers, sacred, in India, 94–95
rock art
  at Chalcatzingo, 3–4
  of the Chumash, 102–3
  at Counsel Rocks, 108
  of Dogon, 166–67
  fertility symbols, 107–10

rock art *(Continued)*
of Mother Earth, 105–6
in Panamint Valley, 60–65
and shamanism, 252–55
of the upper paleolithic age, 116–25
at Yazilikaya, 146–47
Rome
astrology, use of, 240–43
Rome *(Continued)*
the obelisk of Campus Martius, 274–77
See also Augustus Caesar
Roots of Civilization, The (Marshack), 122

Sacred Cenote, 267–69
Sacred Dance (Wosien), 63
Sacred Marriage ceremony, 144–52
mortuary component, 152
Sacred Pipe, The (Brown), 38–39
sacred places, and rock art, 252–55
sacrifices
at Chichén Itzá, 267
at Ife, 179
to Venus, 257–66
San
eland, importance of, 69, 71–75
eland dance, 71
and Haida, comparison of, 79
rock art, 71, 75–76
trance dance, 69–70, 73
sand dollars, in Chumash society, 160–61
San Joaquin Valley rock art, 109–10
scalping, 254
Schaafsma, Polly, and Comanche Gap rock art, 253–54
Schafer, Edward H., skepticism of astrology, 237
Schele, Linda, and Maya pyramids, 299–300
Schlesier, Karl, and Cheyenne ceremonialism, 212–20
Science & Civilisation in China (Needham), 238–39
scientific age, and belief systems, 314–17
Scipio's Dream (Cicero), 230
sculpture, of Yoruba, 179–82
Secret History of the Mongols, The, 185–86
serpent, feathered, 265–70
sex, ritual
origin of, 144–45
in Shinto Japan, 199
See also cyclic renewal
sextant, 8–11
shamanism
of the Cheyenne, 212–20
of Haida, 79–80
of Huichol people, 56–60
hunting and pastoral, 125
of the Mongols, 187–92
political dimension, 191–92 (*see also* astrology; astronomy)
ritual art of, 59–60
roots of, 61–62
of San, 69–70
of Siberian peoples, 248–52
in Turkic Altaic culture, 53
in Yupik Eskimo culture, 55–56
Shamanism, Archaic Techniques of Ecstasy (Eliade), 52–53
shamans
caves, significance of, 111, 119–20
and the center of the universe, 52
chiefs and sacred kings, similarities of, 178–79
community, reinforcement of, 211–20
cosmic power, access to, 51
drums of, 64–65, 250–52
hallucinogens, use of, 56–60
places of, 48–49
power of, 53
and rock art, 252–55
role of, 53–54
trickery of, 211–12
world view of, 35, 51–52
Shang di, 7–8
Shanidar Cave burial site, 110
Shintoism, 196–207
animistic nature of, 197
Shinto (The Way of the Gods) (Aston), 207
Shirokogoroff, S. L., and the Tungus, 36–38
Shotoku, Emperor of the Rising Sun, 196–97

shrines, 17, 203–5. *See also* cairns; temples
Shu ching, 231
Siberian shamans, costumes and drums of, 248–52
Sibylline Books, 241–43
Sibylline Oracle, The, 242–43
Sioux, and White Buffalo Cow Woman, 38–39, 215
sipapu, 30
Sirius
  Dogon, significance of, 164
  in Egypt, significance of, 291–92
  San, significance of, 68
Siva, 93
Six Sky, and the cardinal directions, 26
sky, the
  father god of, 144 (*see also* Mother Earth)
  horizon, symbolism of, 2
  Olorun, personification of, 180
  personifications of, 7, 33
  significance, in Hopi and Olmec culture, 4–5
  significance, in Mongol culture, 185–95
  See also world view
Sky People, of the Chumash, 157
Sleeping Lady statuettes, 131–32
Slippery Hills, Botswana, 75–77
social cohesion
  and cosmic order, 50
  king's responsibility to, 149
social organization
  and astronomy, 154–61
  and social geography, 67–96
Solarium Augusti, 277–78
solar year, 7
  Mongol measure of, 7, 10
solstice. *See* summer solstice; winter solstice
soul, the
  destination of, 279–80
  of Huichol shamans, 58–59
  journey and metamorphosis, symbols of, 63
  of pharaohs, 285
  of Siberians, 53
South African Bushmen. *See* San
Southern Cross, in Dogon cosmos, 164
Sphinx, 290–92
Spirit of Shamanism, The (Walsh), 211
spirits
  of Hopi ancestors, 4–5
  Newgrange, as abode of, 134
  realm of, 35–36
Sprajc, Ivan, and Mesoamerican world view, 270
stars
  and food, association to, 68
  See also astronomy
Stars and the Stones, The (Brennan), 138–39
states
  creation, place of, 220–27
  formation of, 175–76
  institutions of, 177
  See also kings
Stephen, Alexander M., and Hopi traditional life, 29–32
stone temples
  in Malta, 128–32
  See also temples
stupas, 20–21
Sumeru, world-axis mountain, 82, 89
summer solstice, and the Bighorn medicine wheel, 218–19
sun, the
  and Augustus Caesar, 278
  and Egyptian pharaohs, significance of, 289–93
  Japan, identification with, 196
  kings and lions, associations of, 173–75
  Mississippian culture, significance of, 296–98
  Hopi culture, significance of, 31–32
  gods of, 282
Sun Circle, 297
Sun Goddess of Japan, 196–202
  modern-day ties to, 205–6
  temple of, 203
sunstick, 158–61
superheroes, 245–47
supernatural phenomena, 50

Supernatural power
of mountains, 87
of place, 157, 162–67

Taakwic, 105–6
Taittiriya Aranyaka, 93
Tale of the Investiture of the Gods, 25
Talgua Cave, 113
Tarxien, temples of, 129–31
Tatewari (Grandfather Fire), 57–58
Tayaupá (Father Sun), 57–60
Tedlock, Barbara, and Maya directional symbolism, 28
temples, 49
at Cerros, 299–301
at Malta, 128–32
Mississippian mounds, 295–98
tombs, association with, 131–33
Temujin. *See* Genghis Khan
Tengri, Mongol god, 187–91
Teotihuacán
economic power, 266
and Venus-regulated warfare, 256–66
Terrace for Managing Heaven, 7
Terrace for Observing the Stars, 7
Teshub, weather god, 147, 149
Thebes, creation myth of, 225–26
Thompson, J. Eric S., interpretation of Maya glyphs, 262
Thompson, R. Campbell, and Assyrian omens, 228–29
Thoth, 224–25
Thuban, significance in Egypt, 288
Tibetans
agriculture, 85–86
Avalokitesvara, 86
Buddhism, introduction of, 81, 88–89
cardinal directions, symbolism of, 82
center of the world, 81–82
creation myths, 80–81, 85–86
kings, origin of, 83–85
sacred mountains of, 85
Tlaloc-Venus warfare cult, 257–66
tombs
of Antiochus I, 281–85
cosmic order, symbolism of, 26
of Gozo, 132–33
Newgrange, 133–41
temples, association with, 131–33
See also burials; pyramids
Tomb 12 at Río Azul, 26–28
Tong tian calendar, 7
traditionalists, 314
Tragliatella Oinochoe, 64
transitional zone between worlds
and mazes, 62–63
See also monumental architecture; pyramids; world-axis mountains
tribelet organization, 154
Trimble, Virginia, and the Great Pyramid, 288
Trump, David H., and the Maltese mother goddess, 129–30
Trupe, Beverly, and Counsel Rocks, 107–8
Tsangpo River (Brahmaputra), 95
Tsenpo, Nyatri, 83–85
Tsistsistas. *See* Cheyenne
Tsodilo Hills, Botswana, 75–77
Tungus peoples
cosmology of, 36–37
shamans, costumes of, 249
Turkey, Çatal Höyük, 142–43
Turkic Altaic people, shamanistic beliefs of, 53
Tylor, E. B.
animism, explanation of, 36
religion, evolution of, 33

UFOs, 314–15
underworld, the, 15, 35–36
and the central lake, 93
labyrinth, association with, 63
of the Maya, 26, 28
in mountains, 4
in shaman's world view, 35
See also world view

universe
center of, 15–17
See also center of the world
Upanishads, 24

Vaishravana, 21
Valeri, Valerio, and Hawaiian chiefdoms, 171
Van der Post, Laurens
and power of place, 75–76
and the San, 67–76
Vedic tradition, 93–95
Venus
and feathered serpent warfare, 265–70
Inanna, association with, 145–46
and ritual warfare, 256–70
significance of, to San, 68
Virudhaka, 21
Virupaksha, 21, 23
Vishnu, 307–10
Vishnu Purana, 93
Vogt, David, and Cheyenne medicine wheels, 218
von Heine-Geldern, Robert, and Angkor Wat, 308
Von Werlhof, Jay, and the Panamint rock alignments, 60–62

Walsh, Roger N., and shaman trickery, 211–12
Wang Chong, skepticism of astrology, 237
warfare
and fertility associations, 253–55
Venus-regulated, 256–70
water
sacred rivers, 94–95
symbolism of, 99, 113
See also rain
Wedded Rocks, sacred landmark, 203–4
Wheatley, Paul, and the Bayon, 310
White Buffalo Calf Woman, 38–39, 215
Wijiji ruin, 1
winter solstice
Chumash ceremony for, 157–61
Dogon determination of, 164–65
significance of, 140
Wittry, Warren, and Cahokia's solar observatory, 297
Wizard of Oz, The, and traditional cosmology, 39–40
Wolves of Heaven—Cheyenne Shamanism, Ceremonies, and Prehistor Origins, The (Schlesier), 212–13
Womb Rock, 107–8
women
in religion, 129–30
See also female aspect; fertility; Mother Earth; Moher Goddess
Woodhenge 3, 297
world axis
cairns as, 187
Cheyenne, 214–15
of the Dogon, 163–64
drums as, 251–52
and sacred mountains in Tibet, 85
in Shinto Japan, 199
Sumeru, 89
See also center of the world; world-axis mountains
world-axis mountains
Burkhan Khaldun, 186–88
and the central lake, 93
and Khmer pyramids, 303–10
Mount Kailas, 82–83, 91–93
Mount Meru, 93–95
Mount Pinos, 103–4
in *Puranas,* 93
world-quarter shrines, 17
in Hollywood, California, 40–42
of Kawaiisu Indians, 101
in Tibet, 81
See also directions, cardinal
world view
of Chumash, 157
of Dogon, 162–67
of Hawaiians, 168
of Mesoamericans, 270
of Polynesian cultures, 169–70
and power, 154
ritual art as reinforcement, 60
of shamans, 35, 51–52

world view *(Continued)*
of Tungus, 36–38
See also cosmovision
Wosien, Maria-Gabriele, and ritual dance, 63
Wuta si of Hohhot, 195
Wuwuchim, and the initiation of youth, 29–32

Xibalba, 26

Yakut shamans, costumes of, 249
Yamatai monarchy, 198
Yamato, domination of imperial Japan, 197
Yao, Emperor of China, 231
Yarlung Valley, 84–86
yarn painting, and Huichol sacred history, 59–60
Yazilikaya, and rituals for cyclic renewal, 146–52
yoni, 93
landmarks, 104–6
Yoruba culture, creation myth, 180
youth, Hopi initiation, 29–30
Yuan shih, 192
Yukaghir shamans, costumes of, 249
Yumbulagang, 84–85
Yupik Eskimos, shamanistic beliefs of, 55–56
yurt, symbolism of, 53

Zanker, Paul, and *The Power of Images in the Age of Augustus,* 276, 278
Zeiss Planetarium Projector, 12
Zeus, birthplace of, 128
Zeus-Oromasdes, 282
Zhou dynasty, 25
Zorthang (the Sickle), 85
Zuni myth of Knifewing, 254–55